HOOVER DAM

Hoover Dam. (Bureau of Reclamation)

HOOVER DAM

AN AMERICAN ADVENTURE

JOSEPH E. STEVENS

University of Oklahoma Press : Norman

Library of Congress Cataloging-in-Publication Data

Stevens, Joseph E. (Joseph Edward), 1956–
  Hoover Dam: an American adventure.

  Bibliography: p. 299.
  Includes index.
  1. Hoover Dam (Ariz. and Nev.) I. Title.
TC557.5.H6S74 1988   627'.82'0979159   87–40559
ISBN: 0–8061–2115–7 (cloth)
ISBN: 0–8061–2283–8 (pbk.)

6 7 8 9 10 11 12 13 14 15 16 17 18 19 20 21 22 23

Contents

Preface

Construction of Hoover Dam—the great pyramid of the American West, fount for a twentieth-century oasis civilization—began in 1931 and was completed in 1936. The first and most important link in a chain of dams, canals, and aqueducts built to harness the Colorado River, it was the supreme engineering feat of its day, a soul-stirring architectural and industrial achievement, the ultimate expression of machine-age America's ingenuity and technological prowess.

More than half a century later, the white concrete wedge on the Arizona-Nevada border still inspires awe. A new generation of bigger, more sophisticated engineering marvels has risen, yet Hoover Dam remains the benchmark, the most famous dam in the world, a monumental and distinctly American icon.

Words and pictures cannot convey the dam's majestic appearance or its overwhelming impact—something I discovered when I toured the structure in the fall of 1980. Like thousands of visitors before me, I was awestruck by the huge, graceful wall of concrete, the gargantuan towers and tunnels, the steady, full-throated hum of the turbines and transformers. I was stunned, too, by the enormity of the setting, by the stark, soaring grandeur of the river gorge and the forbidding red-and-black sweep of the surrounding desert.

How, I wondered, could this massive concrete barrier, its working parts functioning with the smooth precision of a finely crafted machine, have been built in so rugged and inaccessible a place? What manner of men had possessed the audacity to undertake such a daunting construction project in the depths of the Great Depression? How had the thousands of laborers who came to this lonely canyon coped with the terrible heat, the unforgiving rock, the dangers they faced at every turn? What had it been like to work and live in this inhospitable landscape during the 1930s?

To my surprise, I found that the answers were not readily available. Much had been written about Hoover Dam's political genesis and its economic and environmental effects on the Southwest and southern California; the structure had been analyzed as a work of architecture and as a modernist symbol; its size and otherworldly beauty had been celebrated in poems, in essays, and on film. But a book telling the full story of how the dam was built had not been written, and so my task became clear.

I turned to dusty treatises and yellowing journals, many of them untouched for more than fifty years; I read through stacks of moldering newspapers and magazines; I examined thousands of black-and-white construction photographs and plowed through reams of old records, correspondence, and personal papers. And I listened to the tales—humorous and heartrending, thrilling and terrifying—of the men who built Hoover Dam with their sweat and blood.

From the dim print, the ghostlike photographic negatives, the fading memories, the shape, color, and texture of a bygone era emerged, a time when the engineer and the builder were romantic figures, when taming a wild river was a heroic endeavor, when holding a job and feeding one's family was a feat of honor.

Against this backdrop, the story of Hoover Dam came to life. The facts and figures, the anecdotes and images, the sights, sounds, and emotions were revealed as interconnected elements in a sweeping drama, a technological and human epic, a great American adventure.

In writing this book, I have tried to concentrate on the human dimensions of Hoover Dam's construction while not ignoring the details of engineering, machinery, and construction techniques. It would be impossible to comprehend the dam's larger meaning, to understand what motivated its designers, to share the triumphs and tragedies of its builders, without some understanding of the formidable technical obstacles that were encountered and overcome at the dam site.

While researching and writing the Hoover Dam story, I, too, faced many obstacles, barriers that would have been insurmountable without the help of many individuals. For their knowledge, assistance, and patience, I would like to thank the librarians and support staff of the James R. Dickinson Library Special Collections Department at the University of Nevada–Las Vegas; the Boulder City Library; the Nevada State Historical Museum Library; the Library of the U.S. Bureau of Reclamation, Lower Colorado Region; the University of Nevada–Reno Oral History Program; the National Archives and Federal Records Centers at Denver, Colorado, and Laguna Niguel, California; and the New Mexico State Library.

I owe a special debt of gratitude to Bethe Visick and Gene Hertzog, of the Bureau of Reclamation Photo Laboratory, in Boulder City, for helping me collect hundreds of construction-era photographs, some of which are reproduced in this book; to Julian Rhinehart, regional public affairs officer, Bureau of Reclamation, Lower Colorado Region, for answering many questions about the dam, its day-to-day operations, and steps taken in recent years to upgrade it; to Frank Wright and David Millman, of the Nevada Historical Museum Library, for compiling the *Las Vegas Review-Journal* index, an invaluable research tool; to Professor Imre Sutton for reading and commenting on parts of the manuscript; to Dennis McBride, historian, writer, and video producer, for sharing his knowledge of Boulder City and photographs from his extensive collection; to John Meursinge, engineer and Hoover Dam construction worker, for allowing me to read and quote from his unpublished memoirs; and to Marion Allen, Steve Bechtel, Sr., Teddy Fenton, Elton Garrett, and Red Wixon for sharing their memories, which have added immeasurably to this book.

Finally, to my mother and father and to my wife, Anastasia, I would like to express my deep gratitude and love. Without their encouragement and unwavering support, I could never have started, much less finished, this book.

JOSEPH E. STEVENS
Santa Fe, New Mexico

HOOVER DAM

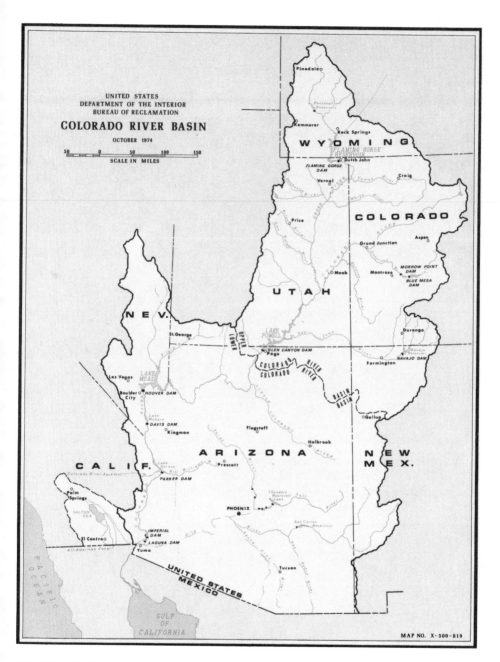

The Colorado River Basin. (Bureau of Reclamation)

CHAPTER ONE
A River and a Dream

William H. Wattis, seventy-two-year-old president of the Utah Construction Company, builder of railroads, highways, and dams, one of the wealthiest men in the West, was dying of cancer. Day after day he submitted to the painful injections prescribed by his physicians and watched for signs of improvement, hoping against hope that some of his old vigor would return. The ache in his tumor-ravaged hip, the withering of his arms and legs, the lassitude settling like a shroud over his mind and body only grew worse, however. To his wife, his nurses, and to the friends who came to visit him in his modest room in San Francisco's St. Francis Hospital, he continued to protest that he was growing stronger, but it was a transparent lie. Death was the prognosis, and he knew that his time had almost run out.

Swaddled in a cocoon of blankets, he sat before an open window that commanded a view of the San Francisco skyline; it was March 2, 1931, and warm spring sunshine bathed the city. On the bed beside him, a cigar smoldered in an ash tray. He reached for it, noting ruefully how the skin on the back of his hand, once taut and ruddy from long hours outdoors, was now crinkled and sallow. As he puffed on the cigar, his gaunt cheeks collapsed inward but his head remained fixed and motionless, as if he could not summon the energy to turn it. Only his eyes, bright blue and sparkling, seemed truly alive.

Just now they were focused on a group of men standing at the foot of his bed. Seeing those erect figures, heads bobbing in animated conversation, W. H. Wattis forced the joint in his cancerous hip to shift, putting him in a more upright sitting position. The groan that slipped from his chalky lips went unheard, drowned out by traffic noise drifting up from Hyde and Bush streets and by the sound of voices in the crowded hospital room.

The exchanges were crisp and rapid fire, carried on in a strange verbal shorthand that, if transcribed, would have consisted almost entirely of numbers punctuated by question marks and exclamation points. The numbers, affixed to units of measurement and containing six, seven, even eight digits, were offered up with casual assurance or flung down with defiant aggressiveness. Challenges and rebuttals couched in almost incomprehensible technical jargon, barks of laughter, and gusts of profanity formed a cacophony that seemed certain to grate on the ears of the pain-wracked old man. But rather than wincing, W. H. Wattis relaxed. As he settled back in his chair, wreathed by a cloud of cigar smoke, a faint smile curled the corners of his mouth.

Perhaps he was thinking that a cancer ward was a strange place to hold a business meeting, but then again it was no stranger than a tent or a railroad car or any number of other peculiar settings in which he had presided over similar get-togethers during his long career in heavy construction. He was of the old callus-on-the-hands, dirt-under-the-fingernails school of builders who scoffed at corporate folderol and felt more at home on a job site than in a boardroom. He could have called this meeting in a broom closet and the men now surrounding his bed all would have been there, just as eager and just as intent. Like him they were builders—aggressive, irreverent, incorrigibly ambitious—and they were preparing to go after the contract of a lifetime, the biggest job any of them had ever tackled: the construction of Hoover Dam on the Colorado River.[1]

Since late 1928, when Congress passed legislation authorizing it, Hoover Dam had been the dream and nightmare of every civil engineer and construction executive in the country. The largest dam on the face of the earth and the linchpin of a grandiose project to corral and harness the waters of the Colorado from the Grand Canyon to the Gulf of California, it would require organization, logistics, and industrial and engineering know-how on an unprecedented scale. It was destined to be one of the greatest monuments to man's ingenuity and conceit ever erected, but before the dream could become reality there were extraordinary risks to be taken and mighty obstacles to be overcome. The construction of Hoover Dam would pit men and machines against river and rock; it would be a battle between human resourcefulness and the raw, elemental forces of nature, fought in a desert canyon that seemed to have been torn open, carved, and scoured clean, just for this purpose.

For a man on his deathbed, it was a staggering undertaking. Even if the group he led won the right to build the dam, he would never see it started, much less finished. To stake fortune and fame on such a perilous venture without any chance of knowing the final outcome seemed a foolhardy gamble, but W. H. Wattis had committed himself to it in spite of his cancer. His partners were eager to proceed, and if they were ready to meet the challenge, so was he, terminal illness or no. Besides, what finer monument could a builder ask for than a dam that would stand for generations, inspiring wonder in all those who looked at it?

The old man's smile widened into a grin. A great adventure was about to start; yet the prologue had been so rich in drama and conflict that it could have stood alone as a complete, well-rounded story of surpassing color and excitement.

The protagonist was a river—*Río Colorado* the Spanish priest Francisco Garcés had named it in 1776—majestic, beautiful, and possessed of awesome power, running 1,400 miles through the heart of America's vast desert hinterland.[2] Born of snow and ice, it rose high in mountain ranges whose names evoked the wild splendor of the alpine West: Wind River and Medicine Bow; Uinta and Uncompahgre; Sawatch and San Juan. From peaks and plateaus where brooks carved blue tracks across pristine snowfields, from remote valleys where cataracts foamed down boulder-lined streambeds, it gathered strength. From north, south, east, and west, in an area covering seven states and 242,000 square miles, whitewater tumbled out of the high country in an explosion of sound and spray, pouring into the skein of tributaries that fed the Colorado.

The Green, river of the trappers' rendezvous and longest of the tributaries, flowed south out of Wyoming's Bridger Basin into northeastern Utah, swung east into Colorado to meet the Yampa, then turned back into Utah to be joined by the Duchesne and Uinta rivers. Twisting south, it plunged into Desolation Canyon, gathered up the waters of the Price and San Rafael, and rolled on toward its confluence, a few miles south of Grandview Point, with the broad, mud-laden Colorado flowing down from the northeast.

Immense and powerful after its mating with the Green, the Colorado sliced deep into a series of arid plateaus—the Kaiparowits, Kaibab, and Coconino—laying open a maze of spectacular canyons. From the northwest the Dirty Devil and Escalante rivers added their flood to the fast-swelling stream, and from the east came New Mexico's San Juan.

Through Glen Canyon the big river entered Arizona, passed Lee's Ferry, and slashed south into Marble Canyon, flanked on the east by the

Echo Cliffs and the Palisades of the Desert and on the west by the eroded spires and buttes of the Kaibab and Walhalla plateaus. The Little Colorado arrived from the sand hills of the Painted Desert after a hundred-mile journey through Navajo country, bearing the accumulated snowmelt of the White Mountains. Engorged by this fresh surge of water, the Colorado turned west and dropped into Granite Gorge, cutting through the heart of Grand Canyon in a series of violent rapids, before rolling on under the Grand Wash Cliffs to a junction with the Virgin River. Past the silent, shadowy walls of Boulder and Black canyons, it flowed west, then wheeled south in a broad, sweeping turn to begin its final run to the sea.

Leaving behind the labyrinth of canyons it had incised in the high plateaus, the Colorado spilled out across the low country under the unrelenting glare of the sun, sliding past the Mojave and Sonoran deserts. Its last major tributary appeared from the east: Arizona's Gila River, bearing the waters of the Salt, Santa Cruz, Agua Fría, and Hassayampa. Swollen by this final flood, the Colorado left the United States south of Yuma, flowed some fifty miles through Mexico, and spread out into a muddy, brackish delta at the head of the Gulf of California. In tidal sloughs fourteen thousand feet below the mountaintops where it was born, the river ended its long journey.

The first white men to see the Colorado were the Spanish conquistadors. In 1539 they sought to ascend the river in search of gold and new territories, but as an avenue of exploration and a pathway to conquest, it proved a bitter disappointment to them. Its currents battered and repulsed their boats; its canyons blocked their marches. As the centuries passed, the Spaniards' curiosity about the Colorado gave way to puzzlement and finally to superstition. For men of European birth and heritage, the muddy torrent was too wild to master and too alien to comprehend. It was a source of endless failure and frustration, an unfathomable mystery, and they made only feeble efforts to chart its course and ponder its possible uses. It would be left to a new generation of explorers—men less fearful, more analytical, keenly interested in the commercial potential of the river— to dispel the fog of rumor and legend the Spaniards had spread over the Colorado and its canyons and to map, measure, and manipulate its waters.

When the territories of California, Arizona, and New Mexico became part of the United States after the Mexican War in 1848, American officials and private entrepreneurs, their interest piqued by reports of such adventurers as William Ashley and James O. Pattie, began to think of the Colorado as a highway of commerce similar to the Missouri or the Mississippi. The War Department was especially eager to find a waterway that could link its far-flung southwestern outposts, and in 1857, Lieutenant Joseph C. Ives was ordered to sail up the Colorado and assess its suitability for steam navigation and its potential as a route for resupplying army forts.

Ives embarked on his voyage in January, 1858, leaving Yuma on the

58-foot steel-hulled steamboat *Explorer.* He and his crew of twenty-four struggled northward, encountering snags, sandbars, shifting currents, and rapidly rising and falling water levels. Four hundred miles upstream from Yuma, at the lower end of Black Canyon, the *Explorer* hit a submerged rock and was damaged badly, but Ives gamely pressed on in a wooden skiff. He reached Las Vegas Wash, three miles beyond the head of Black Canyon, before turning back.

When the *Explorer* had been repaired, half of the expedition members boarded it and returned to Yuma; the other half, led by Ives, marched northeast across the Colorado Plateau to the Grand Canyon, then southeast across the Painted Desert to Fort Defiance on the Arizona–New Mexico border. In spite of the obstacles encountered in piloting the *Explorer* upstream, Ives wrote in his report to the secretary of war that the Colorado could be used as an avenue to transport supplies as far as Black Canyon. But of the region beyond, he flatly stated: "Ours was the first, and will doubtless be the last, party of whites to visit this profitless locality. It seems intended by Nature that the Colorado River, along the greater portion of its lonely and majestic way, shall be forever unvisited and undisturbed." [3]

Eleven years after Ives wrote those pessimistic lines, John Wesley Powell, an intrepid, one-armed Civil War veteran, mounted the expedition that finally took the full measure of the Colorado River and punctured many of the myths and legends that had haunted its canyons since their discovery by the conquistadors three centuries earlier. Powell's odyssey of 1869 was one of the great adventures in the annals of world exploration; his second voyage down the Colorado in 1871, and subsequent geographical and geological surveys of the Colorado Basin in the mid-1870s, were landmark achievements in the study and interpretation of the West's natural history.

The records of Powell's voyages confirmed Ives's report: commercial navigation of the Colorado above Black Canyon was impossible. The one-armed major also agreed with Ives's assessment that most of the Colorado Basin was a "profitless locality" made virtually uninhabitable by rugged topography and unrelenting aridity. "Only a small portion of the country is irrigable," wrote Powell, "and practically all values for agriculture inhere, not in the lands but in the water." [4] He was convinced that development in the Colorado Basin, to the extent it was possible, would have to conform to the dictates of geography and climate and therefore would be very modest in scope. This was an eminently logical if not popular conclusion, but what Powell could not foresee was that in fifty years new technology would permit politicians and land promoters to alter geography, ignore climate, and pursue a dramatic scheme for agricultural and urban growth in the Colorado Desert of Southern California.

The key to this scheme was a geological anomaly: a valley beneath the level of the sea. In the very distant past, the Gulf of California had extended 150 miles farther north than its modern high-tide line. It lapped at the base of the Orocopia Mountains in the vicinity of present-day Indio and Coachella, California, and spread back into Mexico, where it filled an oval-shaped bay some fifty miles wide. The Colorado poured into the gulf's eastern side south of Yuma, where it dropped the cargo of silt it had carried hundreds of miles. Each year 140,000 acre-feet of silt, enough to cover 214 square miles with mud a foot deep, was swept past Yuma and dumped into a huge delta at the river's mouth.

Year after year, century after century, the deltaic deposits extended farther and rose higher until finally they created a natural dam, sealing off the upper end of the gulf and forming a huge, shallow, saltwater lake. The sun evaporated the lake and gradually exposed the sandy alluvia that would be known as the Colorado Desert. At the lowest spot, where the receding water concentrated before it finally was burned away, the Salton Sink, a shimmering, brine-encrusted depression appeared. The river continued to follow a winding course along the southeastern rim of the below-sea-level valley, steadily elevating its bed with fresh deposits of mud. From time to time it overflowed its banks and rushed downward, carving deep channels through the sand on its way to the Salton Sink, but in most years it meandered lazily to the sea, leaving the parched valley silent and empty.

In the nineteenth century, explorers began to probe this 2,000-square-mile wasteland, mapping its contours, sampling its soil, and speculating about its bizarre topography. They found that the desert was bounded on the east by a sixty-mile belt of wind-blown sand dunes, which they called the Walking Hills, and on the west by the rocky palisades of the Vallecito and Santa Rosa mountains. Within this outer ring the valley swept downward to the Salton Sink, three hundred feet below sea level, where midday temperatures surged past 120 degrees Fahrenheit and hot winds scorched away any hint of moisture.

This was some of the most desolate territory on earth, a place where even reptiles would not go, yet in 1849 and 1850 caravans of fortune hunters, on their way to California to dig for gold, dared to traverse it. One of the argonauts was Oliver Meredith Wozencraft, a 35-year-old New Orleans physician who had abandoned his family and his medical practice and set out for the gold fields.[5] He arrived at Yuma in May, 1849, weak from cholera and the rigors of his ride across Texas and New Mexico, but grimly determined to push on to the Golconda he was sure awaited him in California.

The four-day journey across the Colorado Desert was always hazardous, but in May, when the heat was near its zenith, the trip was virtually suicidal. The people of Yuma desperately wanted a doctor, and they urged Wozencraft to postpone his crossing until fall, when the temperatures moderated. They hoped that once he recovered from his illness and resumed practicing medicine he would forget the gold rush and stay in Arizona. But Wozencraft would not be deterred; joining a group of similarly gold-crazed men, he crossed the Colorado River on a ferry and rode off on his mule into the sand dunes. The wind and sun soon took their toll, and two members of the group collapsed. At last perceiving the folly of what they were attempting, the others turned back, all but the heat-befuddled doctor, who rode on until he reached the east bank of the Alamo Barranca, one of the dry channels cut by the Colorado in its flood stage. There he tumbled from his saddle, crawled to the brink, and stared in a delirium at the dusty, down-sloping riverbed that led north toward the Salton Sink.

"I felt no distress whatever," Wozencraft wrote of his ordeal. "I was perspiring freely and was as limber and helpless as a wet rag. It was an exhilirating experience. . . . It was then and there that I first conceived the idea of the reclamation of the desert."[6] The idea was a hallucination, a vivid dream of water rushing down the barranca and flowing across the desert to green fields and oases of palm trees. The dream might have died aborning if one of the doctor's companions had not arrived at that moment with a full water bag; the man revived Wozencraft, hoisted him onto his mule, and led him back to Yuma.

Wozencraft soon recovered from the effects of sunstroke, but he could not forget the mirage he had seen on the edge of the barranca. He questioned the people of Yuma and was told that when the Colorado was at flood stage, water did indeed run down the Alamo and New River channels and other smaller arroyos into the Salton Sink. He looked again at the heavily silted banks of the river and at the desert plain below and was struck by a simple yet wholly original idea: if he could dig a ditch through the riverbank to the Alamo Barranca, gravity would draw the waters of the Colorado down into the desert, causing the fertile soil to sprout the green paradise he had seen in his delirium.

Wozencraft eventually traveled to California, but by then he had forgotten the gold rush. In the cool climate of the Pacific coast he dreamed of the burning desert and began to work to make his idea of reclamation a reality. Large-scale irrigation was virtually unknown in the United States in the middle of the nineteenth century, but with common sense and the help of a surveyor named Ebenezer Hadley, the doctor devised a plan for dredging the Alamo Barranca, linking it to the Colorado, and diverting the water through side canals so that it would nourish fields throughout the desert valley.

The feasibility of his plan was soon confirmed by a government geologist, William P. Blake, who surveyed the Colorado Desert for a railroad route. "The Colorado River is like the Nile," Blake wrote. "If a supply of water could be obtained for irrigation, it is probable that the greater part of the desert could be made to yield crops of almost any kind. By deepening the channel of the New River . . . a constant supply could be furnished to the interior portions of the desert."[7]

Blake's report was not entirely optimistic, however; it cautioned that diversion of the Colorado might lead to uncontrolled flooding of the Salton Sink. Wozencraft, now completely enthralled by the commercial potential of his plan, chose to embrace Blake's comments about an American Nile and ignore the warning of potential disaster. He busied himself establishing a land company and getting himself elected to the California legislature, where he planned to drum up support for his irrigation scheme.

During the gold rush years, the doctor generated little interest among fellow lawmakers for his reclamation proposals, but he never stopped lobbying or accumulating political favors. Finally, in 1859, his persistence paid off. The legislature passed a bill giving him title to ten million acres of the Colorado Desert and its blessing for his irrigation scheme. The bill also provided that California's congressional delegation would work to secure federal rights to the land for Wozencraft.

With an empire almost in his grasp, the doctor hurried to Washington, where for three years he lobbied the House Public Lands Committee, marshaling every bit of evidence and bringing forward every cooperative witness he could find to support his project. On May 27, 1862, his land bill was reported favorably to the House by the committee chairman, but the majority was unconvinced; as Wozencraft watched in dismay from the gallery, the bill was killed.

Penniless and heartbroken, he returned to California to resume the practice of medicine, but the alluring vision of a garden in the desert continued to haunt him. As the years passed, he convinced himself that the legislation awarding him ten million acres had failed not because of its merits but because Congress was preoccupied with the Civil War. In 1887, at the age of seventy-three, he returned to Washington and again presented his land bill, only to see it derided in committee as the "fantastic folly of an old man." Several days later, Wozencraft, now broken in body as well as spirit, died of a heart attack in a Washington boardinghouse. The visionary was dead, but his dream of irrigating the Colorado Desert survived and was seized upon by a new generation of would-be empire builders. The triumphant realization of this dream, the fabulous results it produced, and the incredible natural disaster it unleashed ultimately would lead to the construction of Hoover Dam.

In 1892, Charles Robinson Rockwood, a young surveyor and irrigation engineer, came to Yuma in the company of a Denver land promoter named John Beatty.[8] The purpose of his visit was to assess the agricultural potential of a million-acre tract that Beatty held title to in the Mexican state of Sonora. Rockwood quickly determined that Beatty's scheme for irrigating his Mexican land with Colorado River water was impossible, but disappointment gave way to excitement when he crossed the stream into California, hiked through the Walking Hills to the edge of the Colorado Desert, and saw what Wozencraft had seen forty-three years earlier: the rich, alluvial soil could be watered by digging a canal through the natural levee of the Colorado's west bank and letting part of the river flow down into the gently sloping valley. Gravity would do all the work. A network of ditches and a few headgates to regulate the discharge were all that was needed to transform the desert into an agricultural paradise. He and Beatty would make a fortune by selling water to the thousands of farmers who would rush to stake a claim to this cheap and incredibly fertile land once its potential had been demonstrated.

While Beatty went east to raise money, Rockwood organized a survey crew and plunged off into the desert to map out a route for the main irrigation canal. From the standpoint of topography, his canal layout was masterful: it took full advantage of the Alamo Barranca to bypass the sandy barrier of the Walking Hills and to minimize digging. Politically, however, it was ill conceived. The Alamo wandered fifty miles into Mexico before recrossing the California border and running into the Salton Sink. This Mexican bottleneck was destined to cause forty years of international strife, but Rockwood did not foresee the problem. He was an engineer, not a politician, and all he cared about was finding the most efficient route for his canal.

When Rockwood returned from his three-month sojourn in the desert, he learned that Beatty had been using the irrigation and land development scheme to swindle gullible investors. Disgusted, he severed his relationship with the Denver con man, recruited new partners, founded the California Development Company (commonly referred to as the CDC), and went east himself to raise working capital by selling stock. It proved much more difficult than he had anticipated. The Panic of 1893, the collapse of the silver standard, and the outbreak of the Spanish-American War in 1898 combined to thwart his efforts; after seven years of frustration, he was about to give up when a well-heeled investor suddenly materialized to rescue his irrigation project.

The investor was George Chaffey, a financier and engineer who had made a fortune planting orange groves in the Los Angeles area. He knew what irrigation could do, and after determining that the alluvial soil of the Colorado Desert was as rich as had been claimed, he decided to underwrite construction of Rockwood's canal with a $150,000 check. Dirt began to fly and printing presses began to clatter, cranking out pie-in-the-sky promotional literature in which the sinister names *Colorado Desert* and *Salton Sink* were replaced with the grandiose title *Imperial Valley*. In the Imperial Valley, trumpeted the brochures and handbills, was the richest farmland in the United States, available for purchase under the provisions of the Desert Land Act in tracts of up to 320 acres at just $1.25 an acre. All that was needed to make the valley a Garden of Eden was water, which soon would be available courtesy of the California Development Company. Bewitched by the siren song of bountiful harvests and soaring land values, thousands of prospective settlers flocked to the desert aboard CDC-chartered trains. They were met at the railhead by CDC representatives and harangued about the Imperial Valley's fabulous future and the need to act quickly before the entire wonderland was snapped up.

It took great faith to move to the desert valley during the summer and fall of 1900, for there still was not a drop of water and, as a consequence, no visible sign that anything other than gnarled scrub would grow in the cracked, bone-dry soil. Nevertheless the homesteaders came, arriving in wooden wagons with their horses, chickens, and cows. They lived in tents and brush wickiups in shantytowns that sprang into existence overnight and sported such brave names as Imperial, Calexico, Blue Lake, and Brawley. Enduring heat and blinding sandstorms, they fanned out into the valley each day to clear the mesquite and greasewood from their plots and to await the promised water.

Rockwood and Chaffey finished their canal in the spring of 1901, and on May 14 a rickety wooden headgate was lifted and the muddy red waters of the Colorado gurgled into the unlined ditch and swirled toward the Imperial Valley some sixty miles away. "Water Is King: Here Is Its Kingdom!" blared the *Imperial Press,* the valley's first newspaper. The settlers, now some 1,500 strong, began preparing their land for planting. As quickly as the grid of irrigation ditches and laterals was extended northward into the valley, fields were brought into production and the miracle that Dr. Wozencraft and Charles Rockwood had dreamed of came to pass.

The transformation was astonishing. By 1904 the population of Imperial Valley had grown to 7,000 and its agricultural production was outstripping even the wildest forecasts of the CDC promoters. The value of the barley crop was estimated at $150,000; alfalfa, $125,000; dairy products, $100,000; cattle feed, $75,000; and various other crops, including fruits, vegetables, honey, and poultry, $700,000. Ten thousand head of cattle were grazing on Imperial Valley pastureland and almost as many hogs were being fattened in valley feedlots. Experimental plantings of melons and cotton had been very successful, and hopes were high that their value soon would equal or exceed that of established crops. It seemed that the cornucopia of Imperial Valley might prove bottomless, and the delight and optimism of the inhabitants knew no bounds. Exulted one valley farmer:

> I come from a real fairyland,
> Whose children are sure a happy band,
> Who when they reach our fertile shore
> Seldom leave it. They weep no more. . . .
> Three hundred feet beneath the sea,
> Where hearts beat fast and the blood runs free,
> Where you live like a live one, and when you die
> They plant you under the alkali.[9]

It was a time of prosperity and high spirits, but the boom was built on an unreliable foundation: a flimsy system of gates and canals and a wild river flowing within easy striking distance just over the horizon.

The first sign that the fairyland actually might be a fool's paradise appeared in the spring of 1904. Three years had passed since the opening of Rockwood's irrigation canal, and the vast burden of sediment the river dropped every day had clogged a four-mile stretch of the watercourse. When the river next slackened in fall and winter, the flow through the canal would be completely choked off and a fortune in crops would die of thirst. Charles Rockwood tried to solve the problem with brush dams to raise the river's level and a sluicing scheme to clear the canal; both efforts failed. Telltale swatches of yellow and brown began to appear in once uniformly green fields, and creditors made the rounds trying to collect overdue loans. A wave of anger and dismay, as strong as the previous year's flush of joy and optimism, swept across the Imperial Valley.

Rockwood and his CDC partners were desperate; not only did they face a blizzard of lawsuits for failing to deliver water as contracted for, but the United States government, in the form of the recently established Reclamation Service, was vigorously challenging their irrigation monopoly on the Colorado River. A bold expedient was hit upon: they would open a new canal to bypass the silted one, locating it in Mexico, where the U.S. government would not be able to interfere with its operation. The plan killed two birds with one stone, but it was fraught with danger. At the point where the intake was to be cut, the delta was flat and broad and the river snaked over it freely, running between crumbling banks that were only a few feet high. It was clear that the channel had shifted during the spring high-water season and would do so again when the next flood came downriver, perhaps choosing the freshly excavated canal as its new route. Normally such an occurrence could be prevented by closing a headgate at the canal's mouth, but because the CDC was on the brink of bankruptcy there was no money to build one.

Rockwood knew it was irresponsible to dig a canal without a headgate, especially at this site, where the grade of the canal as it ran north toward Imperial Valley would be much steeper than the grade of the riverbed leading south to the sea. To breach the bank was to invite the river to abandon its course and pour, in an uncontrollable and perhaps unstoppable rush, into its prehistoric repository, the Salton Sink. Such an event would have disastrous consequences, including the possibility that the stream might never find its way back to the Gulf of California, creating instead a vast inland sea. Rockwood had to do something to stave off financial ruin, however, so he proceeded with the excavation of the Mexican intake and the new canal.

At first the gamble seemed to have paid off. Water flowed freely through the fields around Brawley, Imperial, and Calexico, talk of lawsuits died away, and the valley returned to business as usual. But whether they knew it or not, the farmers, the CDC directors, and Charles Rockwood were living on borrowed time. During the valley's first four years the Colorado had been deceptively placid, feeding the canals and ditches when and where the engineers wanted it to. While troublesome, its transgressions had been manageable. Silting and occasional low water were predictable and could be dealt with if money and manpower were available. A major winter flood, on the other hand, could neither be foreseen nor be controlled, and that was precisely what the Colorado now unleashed.

It began in early March in the Arizona mountains with torrential rains that turned the Gila into a raging torrent and sent a powerful surge of water roaring down the Colorado. Carried along by the muddy wave was an uprooted oak tree, spinning and rolling like a matchstick. It rushed downstream on the breast of the flood until fate drove it hard into the riverbank at the mouth of the CDC's Mexican intake. In a matter of minutes it spawned a whirlpool that cut into the canal opening, widening it with lightning speed and permitting a solid wall of water to barrel through and race toward the Imperial Valley.

When the flood finally subsided, the intake had been turned into a gaping channel sixty feet wide. Fearing the worst, Rockwood and his colleagues began constructing a brush and sandbag dam to close the opening permanently, but on March 18, just as they were finishing, another flash flood struck, bursting through their barrier, ripping at the banks of the canal, and spreading the breach another ninety feet.

Again Rockwood's crews tried to close the gap, but it was a futile effort. The river was in charge now, and with the summer flood season just ahead and a yawning cut ready to draw the water into the Imperial Valley, disaster was all but assured. The flood crests came, striking with triphammer speed, slashing the crevasse deeper and deeper, and sending more and more water into the Salton Sink, where a dirty lake was beginning to form. The coup de grace came in November, 1905, in the form of a torrent that blasted through the Mexican intake at 150,000 cubic feet per second and turned the Salton Sink into the Salton Sea, 150 square miles of water sixty feet deep.

The Colorado had now abandoned its former course and was flowing unchecked into Imperial Valley. Thoroughly discredited, Rockwood and his partners transferred most of the CDC's stock to the Southern Pacific Railroad, which was earning large profits from Imperial Valley traffic. While Southern Pacific officials pondered how best to curb the wild river, the situation worsened. Floodwaters opened the Mexican intake a halfmile wide, drowned thousands of acres, and rendered thousands more forever barren.

Most terrifying of all was the natural phenomenon known as the cutback. It began as a small waterfall not far from the place where the Colorado was pouring into the Salton Sea. The sandy soil that formed the lip of the falls was too loose and powdery to withstand the force of the water churning over it, and it collapsed, widening and deepening the falls. Seeking to equalize its descent, the river kept eating its way backward from its lowest point, the Salton Sea, toward its highest point, the Mexican intake. Like a ripsaw tearing through rotten wood, the Colorado cut back through the desert at the rate of more than a mile a day. The waterfall's height increased rapidly and eventually reached a hundred feet. Its location was marked by a towering cloud of dust and spray, and its roar could be heard miles away. If the cutback reached the intake, the falls would be three hundred feet high and the river channel so deep that the Colorado would never be able to return to the sea. Some geologists were predicting that the cutback would not stop there but would continue upstream, destroying Yuma, carving out a canyon one mile deep and several hundred miles long, and eventually inundating southwestern Arizona and southeastern California. The river had to be turned before this catastrophe occurred.

Under intense pressure from President Theodore Roosevelt, E. H. Harriman, owner of the Southern Pacific, opened his checkbook and ordered his engineers to stop the Colorado at all costs. Every Southern Pacific freight train in the Southwest was made available to haul timber, rock, and other supplies to the scene, and four hundred Mexican Indians were hired to help the railroad crews. A bridge, set on 100-foot pilings, was thrown across the intake and rock dumping began. On February 10, 1907, after seven weeks of nonstop dumping, the breach was closed. It had taken two years and three million dollars to push the Colorado back into its former channel, running to the Gulf of California.

The Imperial Valley farmers tallied their losses: millions of dollars' worth of crops either drowned or dead of thirst, thousands of acres scalped by rampaging floodwaters, and an irrigation system reduced to a shambles by silt and stupidity. The scale of the devastation was breathtaking and the prospects for recovery discouraging, but there was no thought of letting the valley return to desert. The memory of its bounty and the vision of ever more plentiful harvests ripening in the years to come, was too strong for that. The valley dwellers had glimpsed the wealth that might be theirs and were determined to stick it out, battling nature to secure that golden future.

The Colorado would be a bullying, exasperating opponent. Less than two years after the Southern Pacific's herculean struggle to corral the river, it ran wild again, flooding into Volcano Lake on the Imperial Valley's southeastern edge and threatening a replay of the 1905–1907 deluge. Congress made an emergency flood control appropriation of $1 million in 1910, and a chain of earthen levees was thrown across the breach, affording the valley a temporary and rather shaky measure of security. As if to mock the Lilliputians who struggled to control its gargantuan outbursts, the Colorado then went from flood to drought, and the valley residents' fear of too much water became one of too little. When the farmers built a brush weir across the stream to raise its level and divert it into their irrigation canal, water from a flash flood backed up behind the weir and inundated Yuma.

Meanwhile, heavy silting raised the elevation of the Colorado's channel so rapidly that the protective levees were in danger of being swamped. There was nothing to do but keep raising them, for each year the river laid down another foot of mud. To make matters worse, the levees, the intake, and miles of the canal were in Mexican territory, subject to the whims of the local alcaldes, who made life miserable for the crews that had to cross and recross the border to maintain and repair the irrigation system. In time, much of the paranoia and hostility traceable to the Imperial Valley's precarious natural position came to be focused on Mexico; the real culprit was the capricious Colorado, but it was easier to direct anger at the Mexicans and to devise ways to stop them from sharing or exercising control over the irrigation flow than it was to face up to the awesome task of permanently harnessing the great river.

Out of preoccupation with the perceived Mexican threat came the idea of building an All-American Canal, which would run entirely through United States territory. In 1917 the Imperial Irrigation District, the corporate entity that had succeeded the ill-starred CDC as owner and operator of the valley's waterworks, sent a young lawyer named Phil Swing to Washington to present the All-American Canal idea to Congress and the Interior Department. Swing was a shrewd politician and an effective lobbier. He persuaded the Interior Department to join the Imperial Irrigation District in determining the valley's water needs and enlisted considerable congressional support for the proposed canal. The committee investigating the canal scheme reported favorably in 1919, and Swing, by then a congressman representing the district that included Imperial Valley, moved quickly to introduce legislation authorizing construction.[10]

Much to the surprise and discomfiture of Swing and the other sponsors of the All-American Canal bill, it was defeated, largely because of the opposition of one man: Arthur Powell Davis, director of the United States Reclamation Service. It seemed to the baffled canal supporters that the natural laws of politics and bureaucratic self-interest had been ignored, for the Reclamation Service was the last agency they would have expected to oppose a large water project. But Arthur Davis was no ordinary bureaucrat, as even a cursory review of his career revealed.[11] He had spent nearly four decades in government service as a topographer and hydrographer for the U.S. Geological Survey, as chief hydrographer for the Isthmian Canal Commission investigating the Panama Canal route, and as an engineer for the Reclamation Service since its birth in 1902. He had masterminded the design and construction of Shoshone, Arrowrock, and Elephant Butte dams, as well as many smaller dams, tunnels, and irrigation canals, and was without question the world's leading expert on reclamation. He was also a rare breed: a selfless visionary, able to look beyond the narrow concerns and petty politics of the moment and project for the Colorado River a grand design that would meet current needs and serve future generations as well.

The Colorado was in Davis' blood; he was the nephew of Major John Wesley Powell, the celebrated explorer of the river's canyons, and during his years with the Geological Survey and Reclamation Service he had come to know the Colorado Basin as well as any man alive. As a young surveyor and hydrographer he had mapped its mountains and deserts and had measured the flow of its streams; while he was a supervising engineer in the Reclamation Service, he had conceived a sweeping plan for developing the entire drainage, from the peaks of Utah and Wyoming in the north to the Gulf of California in the south. Over the years, as he rose to the directorship of the Reclamation Service, he compiled hydrographical and geological data on the river and its course and investigated potential dam sites. His studies had convinced him that harnessing the Colorado was too complex, costly, and important to be undertaken in piecemeal fashion at the behest of special-interest groups such as the Imperial Irrigation District. Development had to be comprehensive, he believed, and only the federal government, acting through the Reclamation Service, could muster the financial and engineering resources and wield the political clout necessary to ensure safe, coherent, equitable development of the Colorado and its tributaries. This judgment was the product of the professional engineer's careful sifting of data, but there was more to the plan than cool calculation. Like his famous uncle, Davis was inspired by the Colorado's challenge and potential; he saw, in the construction of the interlocking network of dams, hydroelectric plants, aqueducts, and canals that would blunt the river's destructiveness and distribute its benevolence, an undertaking to rival or even surpass in scale and importance the construction of the Panama Canal.

So it was that he torpedoed the All-American Canal bill, not because he opposed the concept, but because he believed that the project should be undertaken only as part of a much larger program incorporating flood control, water storage, and power generation. Congress agreed and ordered the Interior Department to make a comprehensive study of the Colorado Basin and the problems involved in developing it. The Reclamation Service, which was a branch of the Interior Department, conducted the study and issued its findings in 1922. Known as the Fall-Davis Report, the document contained an exhaustive hydrological and geological profile of the Colorado River and its canyons. A profusion of charts, tables, and graphs summarized discharge, silt content, evaporation figures, temperature and precipitation data, present and future irrigation needs, flood-control requirements, and the storage and power-producing capacity of various reservoir sites. But to the congressmen, growers, developers, real estate speculators, utility executives, and other interested parties who flipped through its hundreds of pages, the essence of the report was a brief recommendation on page 21: the United States should construct, with government funds, a giant dam "at or near Boulder Canyon" and recoup the construction costs by selling the electric power it generated to the growing cities of Southern California. Once the dam was in place to provide flood protection and water storage, the All-American Canal could be built in the Imperial Valley, but the dam was to be the centerpiece.[12]

Davis' bold master plan seemed to offer something to everyone in the Colorado Basin, but the question of exactly how the river's flow would be divided remained unanswered. The seven states affected—Wyoming, Colorado, New Mexico, Utah, Nevada, Arizona, and California—agreed that it was necessary to settle on an equitable split of the Colorado's waters before they joined in support of the proposed Boulder Canyon dam; this was easier said than done, however. A series of conferences on the question ended in disagreement, and it appeared that a stalemate had been reached. The negotiators could not decide how much water to allocate to each state, and without such a decision there could be no dam or All-American Canal.

The federal government's representative at these meetings was Herbert Hoover, secretary of commerce in the Harding administration. Faced with a seemingly intractable problem, he hit upon a compromise that neatly sidestepped the sticky issue of state-by-state allocation. He proposed that the Colorado Basin be divided into two parts: the upper division would include Wyoming, Colorado, Utah, and New Mexico, while the lower division would include Nevada, California, and Arizona. The two divisions would share the river's water; individual state apportionments could be hammered out later. Each division was allocated seven and a half million acre-feet of water from an assumed annual flow of at least eighteen million acre-feet. The lower basin was allowed to take an additional one million acre-feet as needed, and the remaining two million acre-feet was left unapportioned as a reserve from which water for Mexico could be drawn in the event a treaty were signed, but which otherwise would be set aside for forty years. The state commissioners signed the agreement, called the Colorado River Compact, on November 24, 1922, at Santa Fe, New Mexico, then took it to their state legislatures to be ratified. All of the basin states except Arizona adopted the compact, and Arizona's intransigence was overcome by the simple expedient of declaring the agreement binding if six of the seven states ratified it.[13]

A major impediment to passage of a bill authorizing construction of a high dam on the Colorado for flood control and water storage, and construction of the Imperial Valley's All-American Canal, had been removed. Congressman Phil Swing and California Senator Hiram Johnson led the legislative effort by introducing the first Boulder Canyon Project Act, popularly known as the Swing-Johnson bill, in 1923. Before Congress could appropriate the necessary funds, however, the location of the proposed dam and reservoir had to be pinpointed and the approximate size and cost of the structure determined. Toward this end, the Reclamation Service in 1920 had organized and launched a three-year expedition to explore the rugged canyon country along the Nevada-Arizona border and find the best site for the giant dam.

As far back as 1902, government hydrographers and geologists had been identifying potential sites on the Lower Colorado and speculating about the suitability of their underlying rock formations; now, with the dream of a Colorado River dam nudging ever closer to reality, speculation about which location would be best had to give way to hard data and precise calculations. There were three critical factors to be considered in making the choice: the geological and topographical nature of the site, the water and silt-storage capacity of the reservoir that would be created, and the location of the site in relation to a railhead that could serve as a base for the construction activities and in relation to markets for the hydroelectric power. By applying these criteria the Reclamation Service was able to eliminate six of the eight sites that had been prominently mentioned. The other two, located only twenty miles apart, were Boulder Canyon and Black Canyon.[14]

Both of these river gorges looked promising to Reclamation Service engineers. They were deep enough to accommodate a dam of great height and narrow enough so that the structure would not be prohibitively wide. They had ample reservoir areas upstream, were about thirty miles south of the Union Pacific tracks running between Moapa and Las Vegas, Nevada, and were within economical electrical transmission distance of the power-hungry cities of Southern California. The final decision, then, would be based on geological conditions: the strength and stability of the underlying formations, the hardness and permeability of the rock, and the depth of the silt and gravel deposits that would have to be dug up to prepare the dam's foundation. Precise measurement of these features was vital. Should the dam fail it would unleash a mammoth wave that would crash downriver, engulfing the towns of Needles, Topock, Parker, Blythe, and Yuma, obliterating the Imperial Valley levees, and opening a permanent channel to the Salton Sea. The death and destruction that would be left in the wake of such a flood was almost unimaginable, and with this in mind, Reclamation Service engineers and geologists approached their investigations of the two canyons with great care.

Preliminary reconnaissance by engineers Homer Hamlin and E. T. Wheeler in April, 1920, and by Reclamation Service Director Davis and Chief Engineer Frank Weymouth in November of that year suggested that Boulder Canyon was the better choice because its foundation rock was granite whereas Black Canyon's was volcanic tuff. Davis had enough confidence in this initial judgment to include the phrase "at or near Boulder Canyon" in his proposal for the dam in the 1922 Fall-Davis Report to Congress. This in turn led to the Swing-Johnson legislation's being titled the Boulder Canyon Project Act even though the dam site had not been selected.

A full-scale testing program was begun in Boulder Canyon in January, 1921, in the face of formidable physical barriers. The gorge, which the Colorado had sliced transversely through the northern flank of the Black Mountain range, was tight and steep; the river flowed wall to wall, leaving only a few slender, rocky crescents for men and equipment to stand on. Not a single tree or blade of grass softened the jagged contours of the canyon's cliffs. In summer the constricted space was an airless oven, in winter a frigid wind tunnel.

Leading the little band of explorers into this place was a slight, bespectacled engineer named Walker Young. Called Brig by his friends, even though he was not a Mormon and had been married to the same woman for ten years, the 36-year-old Young was a rising star in the Reclamation Service, which he had joined in 1911 after receiving his engineering degree from the University of Idaho.[15] After four years on the staff that built Arrowrock Dam east of Boise, he had moved to Denver to join the service's design team, where he remained until he was selected to head the testing program on the Colorado.

The assignment, for all the hardships it entailed, was really a plum. The Boulder Canyon Project would be the most ambitious government-sponsored civil engineering task ever undertaken in the United States, and the man who spent three years directing mapping, sampling, and testing at the potential dam sites was almost certain to be rewarded with an important post overseeing construction. From there the future was less certain, but what began with a handful of geologists in the depths of Boulder Canyon could eventually lead to the highest echelons of the Reclamation Service in Washington.

The banks of the Potomac were far away, however, when Young and his crew bounced along the dusty track that led from the Mormon farming hamlet of St. Thomas, Nevada, to the gates of Boulder Canyon in January, 1921. On a gravelly beach just upstream from the canyon cliffs, twenty-eight canvas tents were pitched for the fifty-eight men of the expedition. A small fleet of flat-bottomed boats, some equipped with outboard motors but most powered by long oars, had been floated down to the camp and would be used to carry the men and their equipment to and from work stations in the canyon.[16]

During the earlier surveys made by Hamlin and Weymouth, five possible Boulder Canyon dam sites had been identified and labeled A through E. Close inspection by Young's party soon revealed that Site C was the

A Bureau of Reclamation drilling barge extracting core samples in Black Canyon, February, 1922. The samples were used to determine the suitability of the under-lying bedrock for a dam site. (Bureau of Reclamation)

best, and probably the only, site where a dam more than five hundred feet high could be built. This decision made, the crews could begin studying the canyon's rock to determine its depth, texture, strength, and durability and to look for geological features that might affect the stability of the dam's foundation. Some of the information could be gathered on shore by exam-ining the canyon walls, but most would have to be collected by drilling deep into the riverbed and extracting core samples.

Attempting to rig stable drilling platforms on the Colorado was like trying to balance a tray of glasses on the rump of an angry saddle bronc. Outboard motors could not hold boats in position because of the river's fractious currents, which rose and fell, curved and swirled, surged and slackened with alarming unpredictability. There was only one way to set up a work station: drive ring bolts into the canyon walls, string steel cables

Preparing topographical maps of Black Canyon was an extraordinarily difficult task. Here, a rodman rappels down the Nevada cliff face to mark a triangulation point. (Bureau of Reclamation)

across the river, and anchor wooden barges to the cables. Four 15-foot-wide platforms, each mounted on two 36-foot-long pontoons, were built and sculled down to the first cable line, where they were lashed into place about 100 feet apart. Fastened to the deck of each barge was a timber derrick from which hung a drill tipped with a diamond bit. When everything was ready, a ten-horsepower gasoline engine at the base of the derrick sputtered to life, the drill was lowered, and the rotating bit chewed into the river bottom. Samples were withdrawn from holes as deep as 200 feet, the last 50 feet of which extended into solid bedrock.

The drillers had not expected their job to be easy, but the full extent of the difficulties they faced came as a depressing shock. Often, after boring through ten to twenty feet of soft sediment, they would strike a nearly impenetrable layer of jumbled cobbles. Painfully slow progress would be made piercing ten, twenty, or thirty feet into this stratum only to have the drill stick in or between large boulders. There was nothing to do then but relocate and start over. There was disappointment, too, when the drillers finally broke through the gravel layer and learned that the underlying granite was badly jointed.

As if the delays and frustration that attended the drilling were not bad enough, working on the barges was extremely dangerous. On February 22, 1921, less than two months after Walker Young's party arrived in the canyon, a storm-spawned surge of water caused several accidents. Two men trying to maneuver alongside barge No. 2 lost control of their small boat and were thrown overboard when it capsized. Fortunately they were on line with the barge and were rescued by the drill crew. The next day, a worker lost his balance on barge No. 1 and tumbled overboard. He grabbed a trailing rope just in time to avoid being swept to his death.

On the morning of February 28, as one of the night-shift crews was heading upriver toward camp, the motor on its scow suddenly quit. The current seized the flat-bottomed boat, smashed it into the front of barge No. 3, then sucked it under the heaving deck and launched it downstream like a torpedo. Three of the four men on board were able to cling to the barge when the scow struck it, but the fourth was caught in a whirlpool and dragged beneath the surface. Three hours later, a boat was dispatched to look for his body, but instead of recovering a waterlogged corpse the searchers found the man, badly bruised and thoroughly frightened, clinging to a rocky outcrop on the canyon's Arizona side. Fate was not so kind in December when a storm swelled the Colorado; one of the barges was torn from its cables and smashed to kindling and a driller was drowned.[17]

While the barge crews were struggling to obtain core samples for the geologists in the Reclamation Service's Denver laboratory, the topographical survey teams were having adventures of their own on the rugged cliffs above the river. Accurate topographical maps would be as important to the dam's designers as the geological data dredged from the riverbed, but establishing triangulation points so that the maps could be plotted turned out to be a surveyor's nightmare. The cliffs rose a thousand feet from the water at steep angles. They were pitted with shallow caves and scabbed by large pieces of flaking rock that gave way at the slightest vibration. Nevertheless, it was across these treacherous walls that the mapping crews had to go, fixing points and tying them into earlier surveys. Narrow trails were hacked across the rock, and ladders were lashed together and used to get from level to level. When all else failed, ropes and safety belts were broken out and rodmen rappelled to the most inaccessible spots, swinging under overhangs and into caves, marking control points so that the transit crews above could make readings.

The drilling and survey work in Boulder Canyon was finished in late December, 1921; a year's backbreaking labor—minus the summer months, when intense heat forced abandonment of the camp—had shown that although Site C could be used safely to build the high dam, it was far from ideal. Hoping for better results downstream, Walker Young moved his party to the head of Black Canyon, some thirty miles southeast of Las Vegas, and began investigating the two sites identified there.

The Black Canyon survey got off to an inauspicious start. One day soon after the new camp had been pitched on a wide riverside beach, the wind began to blow. It increased in velocity as the afternoon wore on, whistling menacingly as it funneled out of the narrow canyon, sweeping sand up off the beach and hurling it against the billowing canvas tents. By evening the wind's whistle had become a full-throated scream, and one by one, the tents tore loose from their moorings and wafted away before the gale, followed by pots, pans, cots, survey equipment, clothing, and even the duckboards of the floors. By the time the storm had blown itself out, the camp had been leveled; it had to be rebuilt and restocked from the ground up.[18]

The diamond drilling got under way in earnest during the winter of 1922. The upper Black Canyon site was eliminated, and attention was focused on the downstream site, designated Site D. Drilling was easier here than it had been in Boulder Canyon; the blanket of silt and gravel covering the bedrock was thinner and the bedrock less severely jointed. Work was suspended in May because of high water and blistering air temperatures, but it resumed in September and continued until the investigations were completed on April 24, 1923.

In Denver, Reclamation geologists and engineers pored over topographical maps, tested core samples, and sifted through pages of field notes

Black Canyon as it looked in 1922. The rocky point jutting into the river at right center marks the site chosen for Hoover Dam. (Bureau of Reclamation)

and numerical data. They soon confirmed what Walker Young and his crew had sensed when they started drilling in Black Canyon: contrary to the preliminary judgment that Boulder Canyon was the best place for the dam, Site D in Black Canyon was superior. The jointing and faulting that marred the bedrock of Boulder Canyon were not so prevalent in Black Canyon, there was less silt and debris to be excavated once the river was diverted, tunneling through the canyon walls would be easier, and less concrete would be needed to build the dam because the gorge was narrower. By way of a bonus, the reservoir area was larger, good sand and gravel beds to provide aggregate for concrete mixing had been found nearby, and access to Black Canyon via Las Vegas was much easier than access to Boulder Canyon via St. Thomas.

The data and observations from the Boulder and Black Canyon field investigations and the results of the comprehensive studies and laboratory tests were compiled into an eight-volume report submitted to the secretary of the interior on February 1, 1924. Tentative plans and cost estimates for a concrete arched-gravity dam were included along with conclusions about the relative merits of the sites and the recommendation that the dam be constructed at Site D in Black Canyon.

The report was received favorably by both the secretary and the congressional proponents of Colorado River development, but the shift of focus from Boulder to Black Canyon created confusion about the location of the dam that was destined to persist through the years. The Fall-Davis Report of 1922 had mentioned Boulder Canyon as the probable site, so Congressman Swing and Senator Johnson had followed its lead and labeled the bill they introduced in 1923 the Boulder Canyon Project Act. The press had started to refer to the Boulder Canyon Dam, or just plain Boulder Dam, in its accounts of debate on the bill, and by the time the report identifying Black Canyon as the best site had been made public, the name *Boulder* had become so closely associated with the project that it was impossible to change it. The official christening of the dam—for Herbert Hoover—was still six years in the future, but Boulder Canyon Project it was and would remain, regardless of where the dam was built.

＼ With rough plans, cost estimates, and two thousand pages of facts and figures about the Colorado River canyons in hand, Congress was at last ready to grapple with the political and financial complexities of the Boulder Canyon bill.[19] Johnson and Swing had redrafted it several times to circumvent the objections of various critics and now could muster considerable support; their opponents were powerful and well organized, however, and determined to prevent the bill from ever reaching the floor. Leading the attack was the congressional delegation from Arizona, the only basin state that had failed to ratify the Colorado River Compact. The Arizonans believed that California was trying to steal their water, a theft that would cripple Arizona's economy and choke off its growth. Their position had been strengthened in 1923 when the Los Angeles Water and Power District (later called the Metropolitan Water District) proposed construction of a 240-mile aqueduct to carry drinking water from the Colorado River at Parker, Arizona, across the desert and over the coastal mountains to the taps of two million people in Los Angeles County. This plan confirmed what the Arizonans long had suspected: California was a water-thirsty vampire intent on sinking its fangs into the Copper State's jugular and sucking it dry.

Joining Arizona in the fight against the Boulder Canyon Project was a bloc of eastern legislators who saw the undertaking as a white elephant that would deplete the federal treasury and in no way benefit their constituents. Harry Chandler, the wealthy and influential publisher of the *Los Angeles Times,* also opposed the Boulder Canyon Project because he feared the All-American Canal would siphon off water that was irrigating 830,000 acres he owned in Mexico just south of the Imperial Valley. The fourth head of the anti-dam hydra was the power lobby, championed by Utah Senator Reed Smoot, which saw the project as an attempt by the federal government to get into the electric power business in direct competition with private industry.

The pro-dam forces were led by the Boulder Dam Association, a coalition of southwestern business and farm groups; by Chandler's rival, newspaper magnate William Randolph Hearst; and by Congressman Swing, Senator Johnson, and other representatives of the basin states. At first this group made little headway; three versions of the Swing-Johnson bill were introduced between 1922 and 1926 and three times the bill was bottled up in committee. Finally, in February, 1927, it reached the Senate floor, where the full obstinacy of its foes was revealed. Senator Henry Ashurst of Arizona, determined that the legislation would not come up for a vote, began a filibuster, aided and abetted by his Copper State colleague Senator Ralph Cameron. After three days of nonstop fulminating by Ashurst and Cameron, the Swing-Johnson bill was withdrawn, but the Arizonans' victory had been costly: their sensational filibuster had focused national attention on the Boulder Canyon Project, and much of the publicity was favorable to the proposal and critical of the obstructionist tactics of the two senators.

In December, 1927, the fourth Swing-Johnson bill was introduced. This time the legislation made it to the floor of both houses of Congress; it passed easily in the House of Representatives on May 25, 1928, but an Arizona filibuster blocked it again in the Senate. House passage had given the bill momentum, however, and when the Senate opened its second session in December, cloture was invoked. The Senate passed the bill on December 14, and a week later President Coolidge signed it into law.

Obdurate to the end, Arizona moved its fight into the courts, but to no avail. The final details of the six-state ratification of the Colorado River Compact were worked out, and on June 25, 1929, the Boulder Canyon Project Act, authorizing expenditure of $165 million to build Boulder Dam and the All-American Canal, was declared effective in a proclamation signed by President Herbert Hoover.

While Congress was debating the merits of the Boulder Canyon Project, the design staff of the Bureau of Reclamation, as the Reclamation Service was now called, led by Chief Engineer Raymond Walter and Chief Design Engineer Jack Savage, was methodically researching various dam types. In all deliberations, height was the overriding factor. Rock fill, concrete arch, multiple arch and gravity, and straight concrete gravity dams were all considered, but only the massive concrete arched-gravity type was deemed workable for an undertaking of this size. Based on the principle that concrete stands up better in compression than in tension, this design called for a structure in the shape of an inverted wedge. It would be thin at the top and thick at the bottom, presenting a convex face to the immense body of water pressing against it. The curving arch would transmit the water's force into the abutments (in this case the rock walls of the canyon) while

Longitudinal Section of Hoover Dam and appurtenant works. (Bureau of Reclamation)

the weight of the dam's base pressed down on the canyon floor. Two other fundamental requirements added to the design problem: the dam was to be situated between two transverse faults, approximately nine hundred feet apart in the canyon bedrock, and the maximum full-load stress was not to exceed thirty tons per square foot.[20]

More than thirty plans incorporating different variations in horizontal curvature, cross section, and slope of the upstream and downstream faces were prepared at the Denver office and then subjected to exhaustive trial load analyses to determine what allowances should be made for various forces affecting structural integrity. The effects that differing concrete cooling and grouting techniques would have on the solidity and stability of the huge arch also were considered. Rubber and plaster models were crafted and put through stress tests at the bureau's laboratory and at the engineering laboratory of the University of Colorado in Boulder to check the results of mathematical calculations. Gradually the number of designs being studied and refined was whittled from thirty to eight and then to one: a gigantic yet graceful concrete monolith 726.4 feet high—as tall as a sixty-story skyscraper. The wedge, 660 feet thick at bedrock, swept dramatically up and out, tapering to a thickness of just 45 feet and leaving a delicate, smoothly curving crest 1,282 feet long to serve as the bed for a highway connecting Arizona and Nevada. The dam's base sat squarely between the two transverse fault lines in the canyon floor, but the requirement that full-load stress not exceed thirty tons per square foot proved impossible to meet because of the structure's height. This deviation from the preliminary specifications was accepted by the board of consulting engineers that had been appointed to monitor the design effort, and the final plan, which allowed for a maximum compressive stress of approximately forty tons per square foot, was accepted as safe.

＼   No less ambitious than the dam were the towers, tunnels, and outlet works accompanying it. The bureau's scheme called for the Colorado to be diverted around the dam site through four huge concrete-lined tunnels, two on each side of the riverbed, running approximately three-quarters of a mile through the cliffs. Temporary earthen cofferdams, one upstream and the other down, would seal off the construction area for the duration of the project, preventing water from seeping into the foundation while concrete was being poured. To prevent a flood from spilling over the dam's crest when the reservoir was full, spillways 650 feet long and 170 feet deep were to be excavated, flanking the lake behind the Arizona and Nevada dam abutments. Should the water level rise dangerously, steel drum gates could be opened, allowing the runoff to pour into the concrete-paved channels

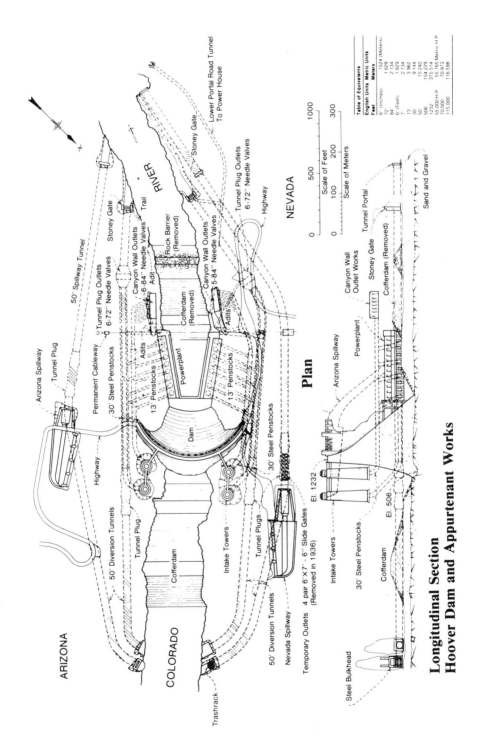

ARIZONA

COLORADO

Arizona Spillway

Tunnel Plug

Permanent Cableway

Highway

30' Steel Penstocks

50' Diversion Tunnels

13' Penstocks

Adits

Powerplant

13' Penstocks

Tunnel Plug

Dam

Highway

30' Steel Penstocks

Intake Towers

Tunnel Plugs

Temporary Outlets 4 pair 6'×7' · 6" Slide Gates
(Removed in 1936)

50' Diversion Tunnels

Nevada Spillway

Trashrack

Cofferdam

Intake Towers

30' Steel Penstocks

Cofferdam

Steel Bulkhead

50' Spillway Tunnel

Stoney Gate

Tunnel Plug Outlets
6-72" Needle Valves

Canyon Wall Outlets
6-84" Needle Valves

Rock Barrier
(Removed)

Cofferdam
(Removed)

Adit

Adits

Canyon Wall Outlets
5-84" Needle Valves

RIVER

Stoney Gate

Trail

Stoney Gate

Lower Portal Road Tunnel
To Power House

Tunnel Plug Outlets
6-72" Needle Valves

Highway

NEVADA

Plan

El 1232

El 506

0   500   1000
Scale of Feet
0   100   200   300
Scale of Meters

Arizona Spillway

Powerplant

Canyon Wall
Outlet Works

Stoney Gate

Cofferdam (Removed)

Tunnel Portal

Sand and Gravel

Table of Equivalents

English Units	Metric Units
6" (Inches)	1524 (Meters)
72"	1.829
84"	2.134
6' (Feet)	1.829
7'	2.134
13'	3.962
30'	9.144
50'	15.240
506'	154.229
1232'	375.514
55,000 H.P.	55,765 Metric H.P.
70,000	70.973
115,000	116.598

Longitudinal Section
Hoover Dam and Appurtenant Works

and down steep spillway tunnels that joined the two outer diversion tunnels, carrying it well past the dam. On rock shelves blasted out of the Arizona and Nevada cliff faces, twin intake towers 395 feet high would draw water out of the lake and deliver it, via a network of penstocks, to the power plant at the dam's downstream base and to the outlet works in the canyon walls.

Taken alone, any one of the diversion or appurtenant works would have been a noteworthy engineering achievement, but combined the mammoth parts formed an absolutely stunning whole. The Bureau of Reclamation designers knew that the dam, which now appeared only as crisp penstrokes on their meticulously drafted drawings, would be the largest in the world, and although it was first and foremost an engineering work, it also would be a monument to twentieth-century technology, a symbol of man's triumph over nature and his ability to shape and control his environment. Thus, while decisions about size, configuration, and layout were based on hard engineering data, the opportunity to make an architectural statement was not overlooked. Gordon B. Kaufmann, an English architect who had immigrated to Southern California in 1913, was hired as a consultant to advise the bureau engineers on shaping the various facades and adding the subtle detailing that would create a coherent and compelling style for the giant structure.[21]

In the engineers' early renderings, the powerhouse looked like an oversized mill, the parapet on the dam's crest was incised with a strange decorative motif, and a pair of enormous eagles, wings outstretched, perched atop the access towers. Kaufmann studied these plans and wisely wiped the slate clean. Seldom had the architectural mantra "form follows function" been more apt, and the California architect took it to heart as he worked out a visual scheme that would complement rather than clash with the engineers' design.

The dominant feature of the dam was its downstream face, a broad, curving, dizzyingly high expanse of concrete. Attempts at adornment would only detract from the aura of power that radiated from its smooth, concave bulk. Accordingly, Kaufmann stripped the towers and observation niches at the crest and had them rise directly and unobtrusively out of the body of the dam. The slim vertical extrusions would provide the only embellishment on the otherwise blank face: an orderly series of daggerlike shadows emphasizing the graceful, unbroken curve from canyon floor to sky.

The theme of power, expressed by monolithic concrete surfaces, was carried through in the powerhouse and canyon-wall outlet works by eliminating decorative details and emphasizing the dominolike ranks of evenly spaced vertical recesses formed by supporting columns and windows. The intake towers on the upstream side of the dam's crest were sited symmetrically, two on the Nevada side and two on the Arizona side, and echoed the stripped vertical lines of the powerhouse. Their tops were given an ornamental treatment, albeit a restrained one: roofs of layered, precast concrete slabs crowned by glass and cast-aluminum globes. These would be lighted at night, shedding a soft, candlelike glow on the lake's dark waters.

Last but by no means least, Kaufmann turned his attention to the spillways, again letting his aesthetic sense be guided by the engineering function of the twin structures. In times of flood, water would be let into the spillway channels by raising eight steel drum gates. The long, concrete lips of the channels were to have smoothly curving surfaces so that the runoff would flow easily down into the wide sluiceways and then into the gaping tunnels that would carry it through the cliffs and past the dam. The necessity of giving the channel crests an aerodynamic profile opened the way for experimentation with the design style known as streamlining. The reservoir sides of the drum gates were curved to complete the outline begun by the channel crests, and in profile the concrete piers dividing the gates looked like airplane wings standing on end, topped with zigzag setback buttresses and adorned with two finely incised lines heightening the impression of speed. This particular design effect was a great success, for when they were completed the streamlined spillway piers proved to be the most visually appealing of all the dam's architectural elements.

Even as the bureau designers were putting the finishing touches on their plans, complex and delicate negotiations were getting under way in Washington to put the Boulder Canyon Project Act's finances in order.[22] A key element of the legislation was the provision that the federal government be reimbursed over a fifty-year period for the huge sums it would expend on the dam and the All-American Canal. Repayment was to be guaranteed by executing contracts for the sale of hydroelectric power generated at the dam. Without these contracts, work on the project could not be started because the legislation specified that no money was to be appropriated for construction until the promise of revenue sufficient to meet the amortization payments had been secured. The ticklish task of setting hydroelectric rates that would pay construction costs, yet be competitive with rates for steam-generated power in Southern California, and then finding buyers willing to contract for power at those rates, fell to President Hoover's interior secretary, Ray Lyman Wilbur.

After considerable study and much complicated computing, a competitive rate was set, and by the time the filing deadline arrived on October 1, 1929, twenty-seven applications for Boulder Canyon Project power had been received. Apprehension that buyers could not be found was now replaced by concern over how to divide the power equitably among competing bidders, who were requesting more than three times the 3.6 billion kilowatt-hours the dam would generate. The city of Los Angeles and the privately owned Southern California Edison Company each asked for all the power produced; the Metropolitan Water District, which was going to build the aqueduct to bring Colorado River water to the coast, wanted approximately half the output; and the state of Nevada pegged its needs at one-third of the generating capacity. These bids, pitting public against private interests and state against state, formed a political minefield for Secretary Wilbur; it took him seven months to pick his way through it and award the contracts.

Wilbur placed the highest priority on getting more water to the Los Angeles area, and so the Metropolitan Water District was given the largest single share of the power, 36 percent, to pump part of the Colorado 240 miles over mountains and desert to the Pacific coast. The city of Los Angeles was to receive 13 percent of the power, other Southern California cities 6 percent, and the Southern California Edison Company 9 percent. In the interests of regional harmony, Nevada and Arizona were allotted 18 percent each, but because neither state was yet in a position to use that much electricity, the remainder was to be divided equally between Los Angeles and the Edison Company until such time as the states were ready to absorb it. Los Angeles and Edison would lease and operate the powerhouse at the dam and would, along with the Metropolitan Water District, finance, build, and operate the transmission lines linking the dam with Southern California.[23]

With signed contracts worth more than $327 million in hand, Wilbur asked Congress to appropriate $10,666,000 so that work on the Boulder Canyon Project could begin. On July 3, 1930, the funds were approved and the interior secretary ordered Elwood Mead, commissioner of the Bureau of Reclamation, to commence construction of the dam. Seemingly insurmountable political and financial obstacles had been overcome, and the final legal hurdle soon would be cleared when the Supreme Court rebuffed Arizona's last-ditch effort to have the Boulder Canyon Project Act declared unconstitutional. Ahead lay physical and logistical obstacles that were no less intimidating. It was one thing to build the dam with words and numbers in the effete environment of Washington's offices and hearing rooms, quite another to do it with concrete, steel, flesh, and blood in the crucible of the Nevada desert.

No one knew this better than Walker Young. Since his expedition into the Colorado River canyons, he had continued his ascent in the Bureau of Reclamation, supervising field studies for a proposed saltwater barrier dam at the mouth of California's Sacramento River, and construction of an irrigation system in Washington state's Kittitas Valley. He had waited patiently for his canyon labors to bear fruit, and in the summer of 1930 they did: he was appointed construction engineer for the Boulder Canyon Project. As the highest-ranking government engineer at the dam site, he would schedule and coordinate the activities of private contractors engaged in the various phases of construction, monitor the quality of their work, and serve as liaison between them and the bureau. He also would preside over the project reservation and the construction camp, which was expected to house three thousand people, and perform a host of other administrative duties. Like a general commanding an army in the field, Young was to execute the grand strategy handed down from above, but the disposition of the troops and the choice of tactics would be his for the most part, as would much of the credit or blame for the ultimate success or failure of the construction campaign.

After organizing his staff and opening an office in Las Vegas in July, Young turned his attention to the first step of the endeavor: linking Las Vegas and Black Canyon by railroad. Union Pacific was going to lay twenty-three miles of track from Bracken, a junction seven miles south of Las Vegas on its Salt Lake City–Los Angeles line, to the site Bureau of Reclamation surveyors had selected for Boulder City, the dam workers' yet-to-be-built town. A 400-car switchyard would be opened there, with sidings running to warehouses and repair shops. The government would be responsible for extending the tracks another ten and a half miles to the rim of Black Canyon, and from there it would be up to the contractor to take the line down to river level. Two and half million dollars from the first Boulder Canyon Project appropriation had been earmarked for railroad construction, and by September, Union Pacific's crews were ready to go.

The laying of the first ties and rails offered an ideal opportunity for celebrating the project's start-up, and Nevada Governor Fred Balzar and Interior Secretary Wilbur decided to capitalize on it. September 17 was selected as the date. Balzar declared a state holiday, and special trains were chartered to carry notables and spectators to Bracken for the festivities. For Wilbur the event was to prove especially memorable. First, his pocket was picked soon after he arrived in Las Vegas, an episode that strengthened his already low opinion of the town, formed during a stopover there the previous summer. Next he was hauled off to Bracken to stand in the dust and 103-degree heat while bands from the Mojave Indian School and the battleship *Nevada* blared and Governor Balzar, Senators Key Pittman and Tasker Oddie, Congressman Samuel Arentz, and other dignitaries preened before ten thousand onlookers. Then, immaculately dressed in three-piece suit, starched collar, and wingtips, Wilbur was obliged to drive a spike of Nevada silver into a railroad tie, a tricky and potentially embarrassing ceremonial duty, especially with a phalanx of eager photographers poised to record every misswing. At last it was his turn to speak, and now the sparks that had failed to fly while he hammered at the recalcitrant spike were struck, figuratively, by the conclusion of his short speech: "I have the honor to name this greatest project of all time—the Hoover Dam."

Instead of provoking the desired ovation, this announcement drew only scattered applause. As the crowd dispersed there was muttering about the propriety of naming the structure for an incumbent president and wry predictions that the dam workers' town would be called Wilbur City. The anti-Hoover press did more than mutter; Wilbur was blasted in print for his ignorance and "unapproachable gall," characterized as "a consummate ass," and compared to the Fascist Benito Mussolini for changing the dam's name by executive fiat. By choosing to honor Herbert Hoover, he had touched off a partisan political battle that would last seventeen years, during which the dam's name would be changed back to Boulder Dam before it was permanently made Hoover Dam in 1947.[24]

With the driving of the inaugural silver spike out of the way, the real work of tracklaying could start. It took just over four months to finish the Union Pacific branch line from Bracken to the Bureau of Reclamation camp at the Boulder City town site. Lewis Construction Company, low bidder for the contract to build the government's portion of the line, then began work on the extension to the edge of Black Canyon. Meanwhile, crews contracted for by the state of Nevada were grading a twenty-two-mile highway running southeast from Las Vegas to the Boulder City site. Other road gangs, working under a federal contract, were building an eight-mile highway from Boulder City to the dam site, and the Southern Nevada Telephone Company was stringing wires across the desert to connect Black Canyon with the outside world.

As the transportation and communication network rapidly took shape and the Bureau of Reclamation prepared to advertise the biggest construction contract in American history, representatives of the nation's major building firms began turning up in Las Vegas on informal reconnaissance trips. On January 10, 1931, the specifications and plans—a hundred pages of text and seventy-six drawings describing Hoover Dam, the power plant, and the appurtenant works in minute detail—were made available to interested parties at five dollars a copy. It was announced that bids would be opened at the bureau's Denver office at 10:00 A.M. on March 4. Each bid was to be accompanied by a $2 million bid bond and the winner would be required to post a $5 million performance bond, a huge sum in light of the depressed U.S. economy and banking system, but in line with the size and scope of the job.

Listed in the specifications were 119 bid items: 3.7 million cubic yards of rock were to be excavated, 4.4 million cubic yards of concrete poured, 45 million pounds of pipe and structural steel put in place, and so on. The government would provide all the materials entering into the completed work, such as cement and steel, but it was up to the contractor to furnish all the machinery, tools, vehicles, and supplies needed to carry out construction. The job was to be finished in seven years. Deadlines were set for various phases of the work and a schedule of penalties was fixed for any delays.[25]

Scores of firms hurried to purchase the specifications, but their enthusiasm sagged when they took a hard look at the drawings and figures and at the government's rigid timetable. The number and magnitude of the various tasks—drilling and lining the diversion tunnels, putting up the cofferdams, acquiring aggregates and erecting concrete-mixing plants, building the dam and powerhouse, providing living facilities and transportation for the labor force—presented problems so thorny that many experienced contractors said the job defied estimate. Those who were not intimidated by the construction program were discouraged by the comments of their field men who drove out from Las Vegas to look over the Black Canyon site and reported that the difficulties posed by the project's size and complexity would be made worse by terrain and climate. Typical of these gloomy assessments was the comment of B. F. ("Frank") Modglin, an engineer working for MacDonald-Kahn Construction Company. After boating through Black Canyon with a party of colleagues, gazing up at its walls, and thinking about getting men and machines into it, he said, "It hit us all at the same time, and candidly we were all scared stiff." Others reacted with traditional builders' bravura. "Hell, it was a deep canyon, but there's [been] lots of deep canyons, before and since," said Steve Bechtel of Bechtel Construction Company.[26] But even the most aggressive contractors were brought up short by the cost of the bid and performance bonds. Quite simply, it was too much money for any one company to foot alone, and so there began a round of telephone calls, board meetings, financial audits, and corporate and personal soul searching as construction executives throughout the nation pondered how they could get in on the job and worried about what might happen if they did.

Among the many builders anxiously weighing the risks and rewards of the Hoover Dam project and trying to devise a plan for raising the necessary capital that winter of 1931 were W. H. Wattis and his seventy-six-year-old brother, E. O., founders and chief executives of the Utah Construction Company, which had its headquarters in Ogden, Utah. Age and illness had made the once robust and aggressive Wattises ambivalent about the venture: there were no guarantees that it would be profitable, and they were well aware of other contractors' flat assertions that the job was too big, that it could not be done at all, that utter ruin was sure to be the fate of any organization foolish enough to tackle it. But instincts honed by fifty-five years in the construction business told them that for all its size and difficulty this job was still like any other: manageable and therefore profitable. It was a gamble to be sure, but they had gambled before and most of their ventures had paid off handsomely.

It is written in the Book of Mormon, "Inasmuch as ye keep the commandments of God ye shall prosper in the land," and for W. H. and E. O. Wattis this assertion had been amply borne out.[27] The sons of a forty-niner whose trek to California ended six hundred miles short in the Weber Valley of northern Utah, they were reared in the dynamic, enterprising environment of Brigham Young's Mormon commonwealth. In 1875, when

William was sixteen and Edmund twenty-one, they left their hometown of Uinta and began careers in heavy construction, working as teamsters on the Great Northern Railroad and later the Canadian Pacific and Colorado Midland. Soon they were contracting railroad jobs themselves, grading beds and laying track, and their business grew steadily until the Panic of 1893 all but wiped them out. While William continued to work in construction, Edmund devoted his energies to running a sheep ranch that the brothers had established in the Weber Valley, hoping that land and livestock would provide a solid financial foundation on which they could rebuild their contracting business.

By 1900 the ranch had generated enough cash for the Wattises to try again, and they established the Utah Construction Company and plunged back into the rough and tumble of contracting. This time there was no financial panic to strike them down, and their company thrived, capturing the lion's share of tunneling, grading, and tracklaying work in the Mountain West. They paid their tithes, plowed the rest of their profits back into the ranch, and looked for opportunities to diversify out of railroad construction. In 1917 they found one: building the seven-million-dollar O'Shaugnessy Dam, a 430-foot-high gravity arch on the Tuolumne River in California's Hetch Hetchy Valley. Success with this project encouraged them to bid on more dam jobs, and this part of their business boomed when they formed a partnership in 1922 with thirty-seven-year-old Harry W. Morrison, of the Morrison-Knudsen Company of Boise, Idaho.

Like the Wattises, Morrison was an up-by-the bootstraps contractor, having started his construction career at the tender age of fifteen working as a waterboy for the Bates & Rogers Company in Chicago.[28] In 1906 he joined the U.S. Reclamation Service as an axman and moved to Idaho, where he was promoted successively to chainman, rodman, levelman, inspector of timber and masonry, foreman, draftsman, and finally superintendent of construction, all within five years. His ambition unslaked by this rapid rise in the federal service, he resigned in 1912 to start his own company. He and a friend named Morris Knudsen founded the construction firm of Morrison-Knudsen and competed for contracts with Utah Construction so vigorously that, in his words, "they took me in on the theory that when the competition gets tough and you can't beat 'em, then join 'em."[29] For Morrison the alliance with Utah opened the door to participation in projects that his company was too small to bid on; for the aging Wattises, the arrangement with the youthful, hard-driving Idahoan pumped fresh blood into their company's management and gave them access to Morrison-Knudsen's most valuable asset: the services of America's foremost dam builder, engineer Frank T. Crowe.

Within the construction fraternity, a group not noted for paying deference to anyone or anything, Crowe commanded respect bordering on awe. He had gained his reputation through hard work, dedication to his craft, and consistent success bringing difficult projects in on schedule and under budget. His drive, ingenuity, organizational flair, and physical toughness epitomized the can-do spirit held in highest esteem by field engineers, and his efficiency, punctuality, and instinctive frugality endeared him to his employers, who profited handsomely from his efforts. Crowe's loyalty to the men he commanded on the job won him a large and dedicated following of construction workers who went from project to project with him and affectionately called him "the old man," even though he was only in his forties. "The best construction man I've ever known," Frank Weymouth, a brilliant engineer and builder in his own right, said of Crowe. "Nothing stumps him. He finds the way out of every difficulty. And he is not conceited." [30]

Crowe's father, a woolen mill operator who came to the United States from England in 1869, had hoped his son would enter the ministry, but mathematics and science appealed more to the young man than did religion, and he entered the University of Maine in 1901 to study civil engineering. During his junior year he attended a lecture about the work of the U.S. Reclamation Service and was so taken with the romantic vision of a magnificent but still wild West that, when the lecture was over, he asked the speaker for a job. The lecturer, none other than Frank Weymouth, obliged him, and that summer Crowe worked on a Reclamation Service crew surveying the drainage basin of the Yellowstone River near Glendive, Montana. Smitten by open western landscapes and the rugged outdoor lifestyle of the field engineer, Crowe finished his degree requirements in 1905, and without waiting to attend commencement exercises or receive his diploma, he returned to the Reclamation Service and the Mountain West.[31]

During the next twenty years the engineer from Maine worked on a variety of projects in Idaho, Wyoming, and Montana, earning rapid promotion and recognition as the government's best construction man. It was at the Reclamation Service's most glamorous and technically challenging work—the construction of high concrete dams—that he excelled, demonstrating both his engineering skills and his managerial and leadership abilities. These jobs appealed to Crowe not just because of their size but because they put him on the cutting edge of his profession; dam building was being revolutionized by dramatic advances in concrete technology and by introduction of motorized earth-moving equipment and pneumatic excavating tools. In 1911, while working as assistant superintendent of construction at Arrowrock Dam, he pioneered the development of two forward-looking mechanical methods that would help make possible a generation of superdams: a pipe grid for transporting cement pneumatically and an overhead cableway system for delivering large quantities of concrete rapidly to any point at a construction site. He was constantly devising new techniques for increasing speed and efficiency and always was on the lookout for equipment that might prove useful. His wife, Linnie, complained that the most interesting thing he could find on his honeymoon in New York City was a large automatic dump truck that unloaded eight tons of coal into a small chute in a matter of minutes without spilling a single lump on the sidewalk.[32]

In 1924, after completing Tieton Dam, near Yakima, Washington, he was appointed general superintendent of construction for the Bureau of Reclamation, making him responsible for construction projects in seventeen western states. He looked forward to new field challenges, but a year after he was promoted a policy change upset his plans: construction work previously handled by government forces now was to be performed by private contractors. Overnight, Crowe's status changed from hands-on builder to paper-shuffling administrator. For a man whose motto was "Never My Belly to a Desk" and who made it an unbreakable rule never to write a letter more than one page long, this was an intolerable state of affairs. His experience in the field and his dealings with other government engineers, such as Arthur Powell Davis, Raymond Walter, and Frank Weymouth, had convinced him that a new generation of dams, far bigger than any then standing, would rise during the next decade. He passionately believed that he was the man to build them, directing construction at the site rather than from an office hundreds of miles away. He was incapable of becoming a desk-bound bureaucrat; the odor of dynamite and concrete was in his nostrils, the music of rock drills sang in his ears, and in 1925 he resigned from the bureau and joined the Morrison-Knudsen Company, which was teaming up with Utah Construction to bid on government dam jobs.

As superintendent of construction for the Utah–Morrison-Knudsen combine, Crowe was completely in his element: in the river bottoms building bigger and better dams, working shoulder to shoulder with men whose drive and ambition matched his own. Guernsey, on the North Platte River in Wyoming, was finished in 1927; Coombe, on California's Bear River, was completed in 1929; and Deadwood, on the Deadwood River in Idaho, was brought in in 1930. With each project Crowe better integrated the mechanized equipment that was coming out of the factories, refined the construction techniques that allowed him to build bigger faster, and groomed the cadre of workmen that traveled with him from Wyoming to California to Idaho.

The profits earned and the experience gained on these jobs were satisfying, but Crowe knew that Guernsey, Coombe, and Deadwood were merely warmups for the biggest dam of all. He had followed the Boulder and Black Canyon investigations closely, had talked at length about the proposed dam with Arthur Davis, Frank Weymouth, and others, had even made a rough cost estimate for Davis in 1919 and worked with Raymond Walter and Jack Savage on the preliminary designs in the Bureau of Reclamation's Denver office in 1924. His interest in Hoover Dam was more than proprietary: it was a burning passion. "I was wild to build this dam," he said years later. "I had spent my life in the river bottoms, and [Hoover] meant a wonderful climax—the biggest dam ever built by anyone anywhere."[33]

Crowe's enthusiasm was shared by Harry Morrison, and after talking the matter over with W. H. and E. O. Wattis, they decided that Utah and Morrison-Knudsen should go after the job, even though they could not finance it entirely themselves. The idea of taking on additional partners did not sit well with William Wattis, who still fondly remembered the days when Utah had enjoyed a virtual monopoly on all major western contracts, but the rough bid estimates prepared by Crowe and his own senior construction men, Hank Lawler and J. Q. Barlow, and the $5 million price tag on the performance bond, left him no choice. An extraordinary project required an extraordinary contracting arrangement: Utah would contribute $1 million in cash toward the bond, Morrison-Knudsen $500,000. While Crowe, Lawler, and Barlow continued to work on the bid figures, Harry Morrison would search for partners who could ante up the remaining $3.5 million needed to underwrite the performance bond.

The first man Morrison called on was 47-year-old Charles A. Shea, president of the J. F. Shea Company of Portland, Oregon, the biggest builder of tunnels and sewers on the West Coast and the contractor responsible for laying out San Francisco's water supply system. The son of a Portland plumber, Charlie Shea was a stumpy, profane, cigar-chomping Irishman who had remained distinctly unaffected by his business success. He was an inveterate gambler who wore a battered fedora and rumpled clothing, ran his company from hotel rooms instead of an office, and delighted in boasting that "I wouldn't go near a bank unless I owed them at least half a million dollars—that way you get respect."[34] This was not the sort of attitude likely to inspire confidence in would-be partners, but among the men who had participated in joint ventures with him, Shea enjoyed a reputation for integrity and high-quality work that more than made up for personal quirks. Morrison, who was himself considered a maverick by some contractors because he did not smoke, drink, or gamble, had no qualms about approaching the flamboyant Irishman. He discovered, as he had expected, that Shea was eager to get in on the Hoover Dam bidding.

After agreeing to put up $500,000 toward the performance bond, Shea suggested that Morrison sound out Charles Swigert and Philip Hart, the top executives of Portland's Pacific Bridge Company, about investing in the partnership. Underwater work was Pacific Bridge's specialty; the company had put in the piers for the first bridge spanning the Willamette River at Portland and more recently had worked with Shea to build the aqueduct linking San Francisco with Hetch Hetchy Reservoir 130 miles to the east. Shea knew his former partners well, and in short order they had agreed to buy into the Hoover Dam combine for $500,000.

＼ With commitments for $2.5 million in hand, Harry Morrison shifted his attention from Portland to San Francisco, where he hoped to raise the rest of the money needed for the performance bond. It was not going to be an easy task. The Depression had cast a pall over the business community; cash was scarce, confidence low, and to complicate matters further, word was out that W. H. Wattis had cancer, raising doubts about Utah Construction's reliability as a participant in a joint venture. Morrison persevered, however, and at the urging of several banker friends he called on a well-heeled San Francisco contractor named Felix Kahn.

Pudgy, bespectacled, and soft-spoken, Kahn offered a stark contrast to the oath-spouting, bare-knuckled builders Morrison was accustomed to dealing with. Kahn was forty-eight years old, son of a Detroit rabbi, and holder of a bachelor of science degree from the University of Michigan. Unlike the typical western contractor who had begun his career as a mucker or mule skinner at some remote construction site, Kahn had got his start as a structural engineer in the offices of Truscon Steel Corporation in Pittsburgh. In 1908 he went to California as a sales agent and there met another Truscon representative, a mechanical and electrical engineer named Alan MacDonald. The latter was an impetuous, hot-tempered Scotsman who had been fired from no fewer than fifteen jobs since his graduation from Cornell University in 1904. The two men, so dissimilar in outlook and temperament, became fast friends, and in 1911 they left Truscon, incorporated their own firm, MacDonald & Kahn, and went into the construction business.

The pairing of the reserved, bookish Kahn and the abrasive, mercurial MacDonald was a strange one, but their different talents and personalities complemented rather than clashed. Kahn was at home in the boardroom, handling legal, financial, and organizational matters with a steady hand and a sure eye; MacDonald was at ease at a construction site, supervising work crews with a sharp tongue and a swift fist. Together they propelled their company to the forefront of the West Coast building industry, executing more than $75 million worth of contracts for sewers, storm drains, industrial plants, and skyscrapers, most notably the glittering Mark Hopkins Hotel high on San Francisco's Nob Hill. They had been thinking about bidding on the Hoover Dam job even before Harry Morrison came calling; after listening to the Idahoan's proposal, they decided to join forces with him and put up $1 million.

Morrison was now within striking distance of his $5 million goal, but time was running out, and other would-be bidders were putting together combinations of their own. Of these competitors the most energetic and formidable was Henry J. Kaiser, founder and president of the Kaiser Paving Company in Oakland, California.[35] Born in upstate New York in 1882, Kaiser had quit school at the age of eleven to earn his own living, and by the time he was nineteen he owned a chain of photo shops in New York and Florida. In 1906 he sold them and moved to Spokane, Washington, where he worked as a hardware salesman. From there he went into the sand and gravel business, working for companies in Washington and British Columbia until 1914, when he started his own construction organization.

The explosive growth of the automobile industry during the teens and twenties created demand for paved roads and opportunity for fledgling contractors such as Henry Kaiser. He had a hard time finding work, however, because of the cutthroat bidding practices of the better-established firms. The turning point for Kaiser came in 1921 while he was paving a thirty-mile stretch of highway between Redding and Red Bluff in Northern California. Warren A. Bechtel, one of the most powerful of the old-line San Francisco contractors, turned up one day, introduced himself, and asked for a tour of the construction site. Impressed by what he saw, he complimented Kaiser and invited the younger man to come in with him on a number of joint ventures.[36]

The Bechtel-Kaiser relationship resembled the Morrison-Wattis union in many ways. "Dad," as Kaiser affectionately called Bechtel, was an old-fashioned pick-and-shovel contractor who in his speech, appearance, and attitude still bore the stamp of the Oklahoma railroad camps where he had begun his building career as a common laborer. A "tall, beefy man with a bull-like roar," he had started his company in 1900 with a Fresno scraper and a string of mules and through hard work and dogged determination built it into a prosperous organization.[37] Kaiser, on the other hand, had not paid his dues with sledge and spike, and his business was hovering on the brink of failure. Bechtel saw beyond this, however, and recognized the latent entrepreneurial genius that was destined to build one of America's great industrial empires. But all that was in the future; for now, Kaiser offered energy, ambition, and a modern outlook, and that was enough to convince Dad Bechtel that the younger man deserved a break.

With the support and encouragement of his new partner and mentor, Kaiser's building career took off. His company expanded, became a leader in mechanization and technical innovation, and won larger and ever more lucrative contracts. In 1928 it subcontracted a whopping $19 million job to build highways in Cuba, where Henry Kaiser first heard about Hoover Dam. "Word of the project got down there," he wrote years later, "and I lay awake nights in a sweltering tent thinking about it over and over."[38] The more he thought about the huge contract the more anxious he became to secure it. When he got back to the United States, he went to see Warren Bechtel and proposed that they form a group to bid on the dam. Bechtel had misgivings. "Henry, it sounds a little ambitious," he purportedly said to his colleague, but Kaiser had made up his mind and continued to push until Bechtel agreed to join him in pursuing the contract. Next Kaiser went to see John Dearborn, chairman of the board of Warren Brothers, a big construction company with headquarters in Cambridge, Massachusetts. It was Dearborn who had subcontracted the Cuban highway job to Kaiser, and although Warren Brothers had been hurt badly by the stock market crash and was in precarious financial condition, Dearborn agreed to commit $500,000 to a Hoover Dam bid effort.

Kaiser now retired to his Oakland headquarters to plan strategy. The wealthiest contractors were on the East Coast, but they were likely to balk at becoming partners with an upstart like Kaiser and a roughneck like Bechtel. Still, they had the cash he needed. He was about to go east when he learned that Harry Morrison was putting together a group and already had lined up $3.5 million in capital. He talked it over with Dad Bechtel, and they agreed that, rather than chase a longshot in the East, they should seek an alliance with the Utah Construction–Morrison-Knudsen group. Such an alliance would offer two advantages. First, Kaiser and Bechtel knew and had worked with all the principals in the Morrison combine. They would be dealing with contractors of their own stripe, men W. H. Wattis had aptly described as "our kind." Second, by joining Utah-Morrison they would forge a group that was, with the exception of Warren Brothers, entirely western. Ever the visionary, Kaiser saw Hoover Dam as the key to large-scale industrial development in the West, which would create a bonanza of construction work and other business opportunities. He realized that if a coalition of western contractors could build the dam, then pick up the contracts that were sure to follow, it could end longstanding eastern domination of the construction industry and precipitate a major shift of economic power from the Atlantic coast to the Pacific.

In the fall of 1930, with Hoover Dam still on the drawing boards and the nation in a business tailspin, all this was a grandiose scenario, but it suited Henry Kaiser's vaulting ambition. Dad Bechtel, no small thinker himself once he had overcome his initial caution, was ready to share the dream. He went to see W. H. Wattis and a deal was struck: Bechtel, Kaiser, and Warren Brothers would come into the Utah–Morrison-Knudsen combination as a unit, staking $1.5 million to bring the group up to the $5 million needed to finance the performance bond.

Getting the bond written was going to be difficult, however. All the participants had made pledges, but there was no money in the bank and no partnership agreement on paper. This was customary among the western builders, who all operated under the principle that Henry Kaiser ascribed to Dad Bechtel: "He hated to sign papers, on the theory that if you couldn't trust a man's word, you couldn't trust his signature." [39] But that folksy aphorism was not likely to carry weight with the executives of the eastern insurance companies who were being asked to gamble on what would be by far the largest surety bond ever written on a single construction job. The actuaries nervously wondered whether such a large, loose group of contractors could hang together and successfully complete a project as difficult as Hoover Dam, and they questioned the reliability of individual members such as Charlie Shea, whose remarks about avoiding banks where he owed money were not considered funny. Then there was the problem of assets. The companies' balance sheets did not inspire confidence: Shea was giving a room number at San Francisco's Palace Hotel as his business address; Warren Brothers and to a lesser extent Pacific Bridge were having trouble meeting their cash obligations; and Utah Construction listed prominently among its financial resources twenty-five thousand cattle and thirty thousand sheep. What kind of businessmen operated out of hotel rooms and bankrolled their ventures with herds of livestock? wondered the conservative easterners as they fretted about whether to underwrite the bond. Finally they decided to do so, provided the contractors formalized their working arrangement in a satisfactory manner and produced the five million dollars, in cash.

Thus, in February, 1931, the builders recruited by Morrison and Kaiser, attended by a small army of engineers, lawyers, and bankers, met at the Engineers Club in downtown San Francisco to work out details of the partnership, discuss the size of the bid, and consider how difficult it would be to finance the job. The meeting was not without its awkward moments. Henry Kaiser, as was his wont, took charge of the agenda and acted as chairman, which caused no problems until he proposed that the group break for lunch. When he made this suggestion, the builders and their assistants were preparing forecasts that would be important to the afternoon's discussions and there was some balking at the interruption. Kaiser was insistent, however, and as his son Edgar remembered it, he picked out a quiet, youthful-looking man in a blue suit who had spent the whole morning scribbling notes and making calculations, then said loudly, "How about this young man over here? He seems to have all the figures down, let him work up the figures." The other executives capitulated and followed Kaiser out, leaving the man in the blue suit to work through lunch. When they had left the room, Harry Morrison gently asked Kaiser if he knew whom he had asked to stay and finish the forecasts. When Kaiser said no, Morrison told him it was Marriner Eccles, member of the Utah Construction Company's board of directors and president of First Security Corporation, operator of twenty-seven banks. A messenger was dispatched immediately to ask Eccles to join the others at the restaurant, but the man in the blue suit declined politely, saying that Mr. Kaiser was not to worry, that he really preferred to stay behind and work on the figures.[40]

Eccles' self-effacement and team spirit seemed to bode well for the new consortium, as did the projections that emerged from his lunch-hour calculations, particularly the estimate that at no time during the job would the group be out of pocket for more than $3.2 million, a figure likely to dispel the fears of the surety companies. The bid discussion was equally encouraging. Chad Calhoun of MacDonald & Kahn, J. Q. Barlow of Utah Construction, and Frank Crowe of Morrison-Knudsen had independently prepared estimates of what it would cost to build the dam; the gap between the high figure, $40.7 million, and the low, $40 million, was less than 2 percent, bolstering everyone's confidence that they had an accurate handle on the job others were claiming was too complicated to estimate. In an atmosphere of optimism and cordiality, they agreed to the terms of incorporation drawn up by their lawyers and financial advisers and chose officers. The last item on the agenda was selecting a name for the new organization. After rejecting Continental Construction and Western Construction, they adopted Felix Kahn's proposal, Six Companies, which alluded to the half-dozen units—Utah, Morrison-Knudsen, J. F. Shea, Pacific Bridge, Mac-Donald & Kahn, and Bechtel–Kaiser–Warren Brothers—that were banding together.

Six Companies was incorporated in Delaware on February 19, 1931, but its birth was not without labor pains.[41] Warren Brothers, drowning in red ink, was unable to come up with the cash for its $500,000 worth of capital stock and Dad Bechtel and Henry Kaiser had to dig deep into their coffers for an extra $250,000 apiece to keep the Warren defection from wrecking the entire arrangement. Harry Morrison also fell short at the last minute and had to give up 20 percent of his $500,000 stake to Alan Mac-Donald's brother Graeme. Likewise, Pacific Bridge could not scrape together enough cash and had to turn to Sidney Ehrman, partner in the new corporation's law firm, Heller, Ehrman, White & McAuliffe, to pick up $120,000 worth of its $500,000 share.

With these last minute additions to the roster of investors, Six Companies was fully capitalized, but it had been a near thing, straining the resources of the participating organizations to the limit. Nevertheless, five million dollars' worth of certified checks and certificates of deposit went into Crocker National Bank in San Francisco on February 25, and the insurance companies proceeded to write the performance and bid bonds. The executive lineup was duly installed: W. H. Wattis, the group's elder statesman, was made president, W. A. Bechtel vice president, Charlie Shea secretary, and Felix Kahn treasurer. Appointment of committees and assignment of specific administrative duties were put off pending the outcome of the bidding, but it was agreed unanimously that Frank Crowe was to be general superintendent of construction, the man responsible for actually building the dam, if Six Companies won the contract.

Barely a week remained before bid day, and the time was spent anxiously poring over the three cost estimates and harmonizing them to arrive at a single set of figures. At the February meeting it had been decided that the amount to be added to the bid price to cover profit would not be determined until the last minute—to prevent any leak to competitors. So it was that on March 2, 1931, the Six Companies directors and Frank Crowe squeezed into William Wattis' hospital room to settle on this crucial figure.

More than one hundred contracting firms, big and small, had purchased specifications from the Bureau of Reclamation, and rumors were flying in construction circles about who was going to turn up in Denver and what the bidding level would be. The reports were sketchy, contradictory, and unsubstantiated, but they were eagerly seized upon and nervously discussed by the Six Companies men. As the smoke of W. H. Wattis' cigar swirled over his bed, the numbers flew and the voices rose, fear that a figure was too high counterbalanced by the dread that it was too low, concern about losing the job to another outfit offset by a grim determination to leave no money "on the table."

Through it all Frank Crowe remained calm, confident that the minimum price he had worked out was the rock bottom, sure that he had foreseen all the contingencies and could bring the job in, not only at the figure stated, but well ahead of the bureau's schedule. As he ran through the 119 bid items listed in the specifications, outlined the steps involved in construction, and presented his plan for completing them, the contractors listened intently. They had known they were embarking on a high-stakes gamble, but the extent to which the outcome depended on the experience, instincts, and luck of this man now became apparent. The voice and hands of the tall, earnest engineer were steady, however, and as he made his points, referring from time to time to a wood-and-plaster scale model that had been wheeled into the room on a hospital cart, their doubts slowly dissipated. When Crowe had finished, they accepted his bid figure and added 25 percent for profit. With handshakes and words of encouragement for W. H. Wattis, they departed for Oakland to catch the train that would take them to Denver and the bidding.

On the morning of March 4, 1931, the Bureau of Reclamation office in Denver was swarming with contractors, insurance brokers, machinery and materials agents, newspaper reporters, and other interested onlookers.⁴² The room where the bids would be opened was so packed with bodies that movement was difficult. The smoke of cigars and cigarettes fouled the air; the babble of voices was deafening. At 10:00 A.M. the tumult subsided as portly Raymond Walter, his bald pate gleaming under the bright lights, elbowed his way to the front and spread the envelopes he was carrying on a table. Eager eyes quickly counted and an excited buzz ran through the crowd; many had considered competing for the job, but this, the moment of truth, revealed that only five actually had done so.

Walter, the bureau's chief engineer, opened the first envelope, quickly examined the contents, and began to laugh nervously. The bid, from Edwin A. Smith of Louisville, Kentucky, read "$80,000 less than the lowest bid you get" and was accompanied by a brief family history and several references of good character. Guffaws rang out as Walter read them aloud and then declared the bid invalid because it was not accompanied by the required $2 million bid bond. The second bid also provoked laughter. It was from the John Bernard Simon Company of New York and was for $200 million or "cost plus 10 percent." It, too, was ruled invalid because it contained no bond. The merriment abated as Walter announced that the third bid was from the Arundel Corporation, one of the eastern construction giants; the amount was $53.9 million and the bond was in order. The fourth bid, made by the Woods Brothers Corporation of Lincoln, Nebraska, was also valid and was for $58.6 million. Silence fell again as Walter opened the last envelope and glanced at the papers it contained. Clearing his voice, he read: "Six Companies, Incorporated, San Francisco, California. $48,890,955." [43] Pandemonium erupted as the crowd shouted and whooped, camera flashes popped, and reporters rushed for the door. Six Companies had won the right to build Hoover Dam; the bid worked out by Frank Crowe was only $24,000 more than the cost calculated by Bureau of Reclamation engineers. Secretary Wilbur had to examine the figures and approve the contract before it could be signed, but this was a mere formality.

At St. Francis Hospital, the press corps found William Wattis sitting up, wrapped in a silk bathrobe, puffing on a long, black stogie. His bushy white hair had been carefully brushed into place by his wife, and he was eager to talk. News of the winning bid had lifted his spirits. He grinned broadly for the photographers, then lectured the writers. "Now this dam is just a dam but it's a damn big dam," he said, eyes twinkling. "Otherwise it's no different than others we've thrown up in a dozen places. It involves a lot of money—more money than any one contractor has a right to have." Suddenly he grew pensive. "I don't know when I'll be out of here," he murmured, more to himself than to his audience, but then he brightened again. "I think I'm improving. . . . Don't worry, I'll be on this job." [44]

Even as the old man spoke, two trains were pulling out of Salt Lake City. One was westbound, carrying half a dozen excited, garrulous contractors toward the coast and a hero's welcome from San Francisco's city fathers. The other raced southwest into the rising desert heat, bearing a silent engineer toward Black Canyon, the Colorado River, and his destiny.

CHAPTER TWO
"A Deadly Desert Place"

A coal-black locomotive pulling a long string of dirt-streaked passenger cars chuffed into the Union Pacific depot in Las Vegas, Nevada, on the morning of March 11, 1931. With an earsplitting screech, the locomotive's steel driving wheels bit into the tracks, and the train rattled to a halt in the middle of the yard. On the station's platform a crowd of politicians, businessmen, and reporters pushed forward and peered anxiously through the cars' windows, jostling each other and shouting to be heard over the hissing of the airbrakes. The atmosphere crackled with tension. Like a hot wind off the desert, change had swept into Las Vegas, and the city would never be the same.[1]

Seven days earlier in the Denver office of the Bureau of Reclamation, sealed bids for the Hoover Dam construction contract had been opened and Six Companies had been declared the winner. Interior Secretary Ray Lyman Wilbur had formally awarded the contract, worth almost $50 million, to the consortium several days later. It was the largest single contract ever let by the United States government, and little Las Vegas was the funnel through which the millions of dollars would pour into Black Canyon, the rugged Colorado River gorge where the dam would rise.[2]

Many Las Vegans were inclined to think of their town's proximity to the dam site as a miracle. At a time when the rest of the nation was reeling from the stock market crash, a rising tide of bank failures, and skyrocketing unemployment, one of the biggest construction projects in American history was about to begin in their backyard. And on the train that now sat in the Union Pacific yard, shrouded in white steam, was the man who would make the miracle happen.

Inside one of the dimly lit cars, the unwitting savior of Las Vegas uncoiled his lanky six-foot frame from a too-small seat, picked up his valise, clapped a fedora on his balding head, and moved slowly down the aisle toward the exit. For Frank T. Crowe, superintendent of construction for Six Companies, the dream of a lifetime was about to become reality. Before him lay one of the great engineering challenges of this or any other age: damming the Colorado River. Directing the job was a glorious opportunity, one he had worked for his entire professional life, but it also was a crushing responsibility. Many knowledgeable engineers said that because of its huge dimensions Hoover Dam could not be built; others said the reservoir created by the dam would cause a catastrophic earthquake that would shatter the concrete arch and unleash a flood of biblical proportions.[3] Now it was up to Crowe to prove them all wrong.

During his twenty-seven years as a field engineer, Frank Crowe had supervised the construction of six dams, but never before had so many problems and hazards been combined in a single project. Hoover Dam would go up in the most rugged and inaccessible site that he had ever seen. Getting men, machines, and material into Black Canyon was going to be a logistical nightmare compounded by the thirty miles of open desert between it and the nearest town. Then there was a complication unprecedented in Crowe's construction experience: a national economic emergency so acute that President Hoover and Interior Secretary Wilbur had urged Elwood Mead, commissioner of the Bureau of Reclamation, to begin work on the dam as soon as possible to provide jobs for some of the nation's unemployed. Because of presidential pressure, the Boulder Canyon Project was to start six months earlier than originally planned, before the building of transportation and living facilities was finished. What effect this early start and the uncertain economic and political situation would have on his ability to finish the job Crowe was not certain, but he knew that millions of dollars and his own reputation were at stake, and that he would have to live up to the motto the newspapers were attributing to him: "To hell with excuses—get results."[4] He stood, hesitating for a moment, on the threshold of the railroad car door, then thrust out his chin and stepped from the dark interior into the dazzling glare of the desert sunlight.

Las Vegas, the small city that lay before him, was a raw western settlement of approximately five thousand souls, "the last frontier town in America," in the words of writer Frank Waters, "but really dying in its boots." It was a place where prospectors, cowboys, and tourists gathered to get blind drunk in the saloons that flourished in open defiance of Prohibition; to sing, shout, and gamble in the dingy clubs that lined Fremont Street; to stagger to the houses and cribs on Block 16 of North First Street, the seamy red-light district where gaudily dressed women sat on porches or in tipped-back chairs under cottonwood trees, laughing and calling out to passersby. On weekends, the frenetic revelers shared the downtown streets with dour Mormon farmers, come to town from St. Thomas, Logandale, or Overton to buy supplies and watch with undisguised disgust the antics of the bootleg-besotted Gentiles. Rip-roaring, no-holds-barred pursuit of pleasure was Las Vegas' stock in trade, and all that separated it from the frontier towns of the nineteenth century were the automobiles parked in front of the battered hitching posts and the flicker of neon tubes where weathered wooden signboards had once creaked in the wind.[5]

Las Vegas was not all liquor and lights, however. The town had its small elite of sober, civic-minded citizens who actually thought about the city's future beyond the next Saturday night and strove to give it a veneer of respectability. Foremost among them was newspaperman Charles P. Squires, who had moved to southern Nevada in 1905 with his wife, Delphine, and become the publisher and editor of the morning daily, the *Las Vegas Age*. Pop, as Squires was called by the locals, was a true believer in the future greatness of his adopted hometown, and with tireless zeal he promoted its positive features, many of which only he was able to discern

Fremont Street, downtown Las Vegas, in the early 1930s. The Union Pacific depot sits at the end of the street. (Ferron-Bracken Collection, University of Nevada–Las Vegas Library)

and appreciate. When the Bureau of Reclamation began making plans for damming the Colorado River, Squires was convinced that the project would be the key to Las Vegas' future, and he dedicated himself and his newspaper to championing the Black Canyon site. On the day that legislation authorizing construction of Hoover Dam in Black Canyon was finally passed, Squires jubilantly proclaimed victory in his campaign to "get the dam for Vegas." The engineers had picked Black Canyon more for its geological characteristics than for its closeness to Las Vegas, but Squires still deserved much credit for his years of determined effort on behalf of the town. "The present time will be looked back upon in future years as the turning point in the history of Las Vegas," he wrote in the *Age* of the government's decision to build at the Black Canyon site. "'The mills of the gods grind slowly, but they grind exceedingly fine.'"[6]

Pop Squires and the businessmen of Las Vegas were not the only ones eager for work on the dam to begin. Since announcement of the construction timetable in 1930, hundreds of jobless men had been streaming into southern Nevada in caravans of wheezing automobiles, in Union Pacific boxcars, on horseback, and even on foot, coming in a wave the likes of which the state had not seen since the heyday of the Comstock Lode. A few of these men were longtime construction stiffs, itinerant workers who drifted from building project to building project across the West. Some were cowboys from the Nevada and Utah range, men who had never seen a bulldozer or a dump truck in their lives; they had come to ride and wrangle the thousands of horses and mules they were certain would be needed to move earth and haul supplies. But most of the newcomers were greenhorns—unemployed factory workers, mechanics, salesmen, store clerks, lawyers, bankers, and students—who had never performed hard physical labor or lived outdoors. Many of them had brought their families and household belongings, gambling everything that at Hoover Dam they would find jobs and a new beginning.[7]

But the work had not yet begun, and instead of a new beginning the migrants encountered the same problems and hardships they were trying to escape. "A 'pitiful and pathetic sight' is describing in the mildest possible language conditions that exist here in Las Vegas due to the influx of the hundreds that have come seeking work," Pop Squires wrote in a front-page story. He reported that a hundred people a day were being fed in city soup kitchens and that hundreds more, including many small children, were going hungry. The situation was so explosive that government officials issued statements cautioning the public that work on the dam would not begin for months and that even when it did start the number of jobs available would be limited. "Under existing conditions none should go to Las Vegas unless he is assured of employment upon arrival, or is equipped with independent means to carry him over a period of several months," admonished the Bureau of Reclamation.[8]

In 1930–31 unemployed men camped on the outskirts of Las Vegas and along the banks of the Colorado River near Black Canyon waiting for construction of Hoover Dam to begin. (Union Pacific Railroad Collection, UNLV)

But nothing kept the desperate job seekers from flocking to the city, and as Frank Crowe made his way from the railroad station to the Clark Building, which would be Six Companies headquarters until offices were constructed near Black Canyon, he saw why President Hoover had ordered the dam started immediately. Almost in the shadow of the depot, scores of gaunt, dirty men were sprawled on a patch of open ground known as Union Pacific Park. Downtown, on the lawn surrounding the courthouse, another large contingent waited for jobs on the dam. Nearby, in the 800 block of Fremont Street, the U.S. Labor Department's employment service, in conjunction with the state of Nevada, had opened an office to process job applications and manage the hiring for Six Companies. A ragged mob swirled around this little one-room building, piling up in front of the windows where the harried official in charge, Leonard T. Blood, struggled to maintain order. Blood told a reporter that during the brief time the office had been open 2,400 men had filed applications and more than 12,000 letters inquiring about work had been received.[9]

On the outskirts of Las Vegas the situation was just as grim. North of town, in and around the city cemetery, a community of tents and shacks housing a thousand men, women, and children, had sprung up. To the southeast, on the slope of Las Vegas Wash, more shanty towns had sprouted like clumps of greasewood after a hard rain. Boulder Highway, a freshly graded dirt and gravel road leading to the dam site, passed through the ramshackle settlements of tents, lean-tos, and parked automobiles. Beyond them the highway angled upward toward Railroad Pass, a broad gap in the River Mountains. There the new Union Pacific branch line from Las Vegas reached the crest of its grade, then descended into the Boulder Canyon Project Reservation. The highway paralleled the railroad, running through the pass before dropping into the glaring desert valley that was the promised land to the thousands of desperate job seekers.

Journalists dispatched to southern Nevada by big-city newspapers and magazines were aghast at the squalor they witnessed. "Las Vegas [is] . . . thronged with wanderers looking for work . . . and at present the open space in front of the station is so full of sleeping men at night that it looks like a battlefield," wrote Edmund Wilson in the *New Republic*.[10]

The reporters also were stunned by the project reservation's stark, savage landscape. "Its sparse vegetation—greasewood, sage-brush, and spiny cactus—rises out of it like blisters baked black," wrote Duncan Aikman in the *New York Times Magazine*. "Furnace-like winds blowing alkali and gypsum dust through the sheer sunlight of the desert give the sand hills, the jagged little mountains a few miles off to the northeast, and to the very air a curious look of incandescence."[11]

"Our road wound through the desert; and let me here remark that my preconceived idea of a desert and the real thing as seen here in Nevada are as different as the wooden horse of Troy and a wild horse of the plains," observed Mrs. D. L. Carmody in an account of her trip to the dam site, published in the Bureau of Reclamation's monthly magazine *New Reclamation Era*. Her first glimpse of the river provided an even greater shock: "Who hasn't read of the mighty Colorado River with its treacherous rapids, its deadly quicksands, its everlasting tragedies? But this broad and apparently stagnant stream crawling between the sands on the Nevada side and a low bare 'bench' on the Arizona shore gave no hint of romance; I do not know what I expected to see, but surely not this." A boat trip through Black Canyon depressed her even more. "It really seemed to me as if things had not come up to any of my expectations and an intangible sense of disappointment vexed me; a sort of feeling of depression and danger. Those strange mountains, that oily river, those masses of black rock, those deep crevasses—what could mere man hope to accomplish against that overwhelming fortress built by nature—what dangers, what tragedies lay ahead of their pygmy efforts at mastery?"[12]

Bureau of Reclamation surveyors, including Mrs. Carmody's husband, had been living on the project reservation in a tent community known as Government Camp One since August, 1930. The camp was situated at the head of Hemenway Wash, on the northern rim of the Eldorado

Valley, described by journalist Theodore White as "a bowl formed by barren unsympathetic mountains. . . . a deadly desert place."[13] During the early months of its occupation the lonely outpost had been disturbed only by rattlesnakes, scorpions, and windstorms that came up suddenly and lashed the little line of khaki tents with stinging blasts of sand. However, the desert isolation was to be shattered permanently by the completion of Boulder Highway and the Union Pacific branch line early in 1931 and the arrival of crews from Lewis Construction Company, which was laying tracks to the edge of Black Canyon, and Le Tourneau Construction Company, which was extending and improving the road.

The first stirrings of construction activity galvanized the unemployed in Las Vegas, and by the hundreds they loaded their jalopies and made the jolting drive over Railroad Pass to search for a living place closer to the dam site. A cook named Michael McKeever, who worked at Government Camp One, had pitched his small tent on the north slope of Hemenway Wash; almost overnight it became the center of a squatters' town dubbed McKeeversville in his honor.[14]

McKeeversville soon became overcrowded and many would-be dam workers, their wives, and their children moved down to the foot of Hemenway Wash where the muddy Colorado swept in a wide bend around Cape Horn, a steep promontory at the mouth of Black Canyon. Boatman Murl Emory, who had been navigating the Colorado since 1921 and had served as outfitter and guide for many Bureau of Reclamation survey expeditions through Black Canyon, managed a dock and small grocery store in the shadow of Cape Horn. Like McKeever's tent, Emory's little store became a magnet to the settlers, and a new village materialized on the broad floodplain at the bottom of the wash. Within months of the completion of Boulder Highway, more than five hundred people were living on the Flat, as the sandy, brush-covered site along the Colorado was called, sleeping in tattered tents, crude wickiups, or under their cars, cooking over open fires, carrying drinking water from the river, reducing life to its most basic elements while they waited for work on the dam to begin.[15]

As more people arrived, the Flat was given a real name, Williamsville, in honor of U.S. Marshal Claude Williams. He and his wife, Dorothy, and their young daughter, Marjorie, lived in a tent among the squatters. Williams' primary responsibility was locating and destroying the stills that enterprising bootleggers were building in caves and remote side canyons, but in spite of this he was well liked by most of the citizens of the Flat, who respected his fairness and his generosity in aiding destitute families. Murl Emory also helped the hungry by extending credit at his store and by allowing shoppers to pay the same prices for groceries that they had paid in the towns they came from. For Emory this gesture meant operating the store at a loss because trucking food in from Las Vegas was expensive and grocery prices in other parts of the country had fallen very low, but he continued the practice, subsidizing the store with the money he made operating his boats for Six Companies.[16]

Williamsville was a respectable name for a village, but it did not really fit the job squatters' slipshod settlement on the river bottom; a more suitable if less complimentary title was soon in universal use: Ragtown. An even more accurate name would have been Cardboardtown, for that was the principal material used in building the hovels that most of the residents called home. Scrap lumber, flattened oil and gasoline cans, tarpaper, and burlap also were used to fashion rude structures that provided at least some shelter from sun and rain. The mesquite and tamarisk trees along the Colorado were festooned with blankets and ragged clothing, the ground was littered with refuse, and battered automobiles were parked everywhere, even in the river itself, so that the wooden spokes of their wheels would not dry out and shatter when they bounced along the washboard road to Las Vegas.[17]

Slumbering Ragtown was jolted out of its lethargy on the afternoon of March 12, 1931, when a caravan of government sedans rolled down the main street and stopped in front of Murl Emory's store. Like wildfire the word flashed from tent to tent and shack to shack, "The big boss is here!" and from out of the shelters, under the cars, and behind the bushes the men came running to catch a glimpse of Six Companies' superintendent of construction. Frank Crowe, accompanied by Walker Young, the Bureau of Reclamation's construction engineer, and several Six Companies and bureau officials, was striding toward the dock where one of Murl Emory's boats waited to carry the party down into Black Canyon. In an instant Crowe was surrounded by men shouting questions and entreaties. "When's the job gonna start boss? How many men you gonna need? I'll work hard chief! We ain't had much to eat lately. Please mister. . . ."[18]

Without answering, Crowe pushed through the throng and climbed into the boat. The plaintive chorus faded into the distance as the long wooden skiff nosed around the talus slope of Cape Horn and was swept into the mouth of Black Canyon. On both sides, rough walls of red and purple rock soared more than twelve hundred feet straight up, framing a narrow slash of desert sky and plunging the river into shadow. On these walls the first critical phase of construction would begin: the building of a railroad down the canyon to link the railhead at Government Camp One with the dam site. The skiff picked up speed as the powerful current sucked it deeper into the canyon; Walker Young pointed to a series of white blazes high on the Nevada cliff and shouted that they marked the place where the dam abutments would be. Murl Emory swung the rudder around and gunned the motor to propel the shuddering boat back upstream. It had taken only ten minutes to run the length of Black Canyon; bucking the heavy current, it would take more than half an hour to return to the landing.

A dam worker's family poses in front of a Ragtown home in the summer of 1931.
(Union Pacific Railroad Collection, UNLV)

That evening at the Six Companies headquarters in Las Vegas, Frank
Crowe reviewed the construction obstacles he had seen on the tour. Fixed
in his mind was the image of sheer cliffs with not so much as a toehold at
the bottom and of naked desert along the rim, a place without houses,
roads, or power lines. The human problem, as epitomized by Ragtown,
would have to be solved, too, just as surely as the engineering problems
involving materials and design. It was imperative that as many men as possible be put to work soon, but before that could happen Black Canyon had
to be opened, access roads and the railroad bed had to be blasted out of the
rock, water and power provided, transportation of work crews and equipment organized, and a new city built to house 80 percent of the men employed on the dam project as stipulated by Six Companies' contract with
the government. Experienced miners, powdermen, machinery operators,
electricians, pipe fitters, and carpenters were needed for the preliminary
work, and not until they had done their jobs could the unemployed horde
in Las Vegas, Ragtown, and points in between go on the payroll.

Word already had gone out to construction camps throughout the

West that Frank Crowe needed help, and by the hundreds his boys, the loosely knit fraternity of workers who had been with him on other jobs, were answering the call. From California and Idaho, from Montana and Wyoming, the dam builders were coming to southern Nevada ready to practice their trades. Some, like 22-year-old power shovel operator S. L. ("Red") Wixon, had only a few construction projects to their credit. Others, like master mechanic Sy Bouse, tunnelman Floyd Huntington, and engineers Woody Williams, Gus Ayres, and Frank Bryant, had followed Frank Crowe from state to state and job to job for many years. They were professionals—tough, resourceful, and indispensable—the nucleus around which the other, less-experienced workers would form the disciplined and efficient labor force essential to building a dam. "They never hit me for a job," Crowe said of this cadre of seasoned men, "they just ask when they go to work." [19]

The work was under way in earnest by April 1 as crews armed with pneumatic drills and dynamite floated into Black Canyon to begin blasting pathways out of the rock walls. Twenty-one miles of track had to be laid along the face of the Nevada cliff so that heavy equipment and supplies could be brought to the work areas at river level and so that aggregate from the gravel-screening plant to be built in Hemenway Wash could be carried to the two concrete-mixing plants that would be erected at the dam site. Roadbeds also were being excavated so that tractors, trucks, and cars could get in and out of the canyon; this would retire the expensive and cumbersome barges and launches that were carrying workers to and from their jobs.

From dawn until dark Black Canyon echoed with the sounds of an all-out assault on its rock walls. The chattering of jackhammers, the roar of air compressors, the clatter and clang of picks and shovels, the grumbling of bulldozers, the shouts and oaths of workmen, the shriek of whistles, and the boom of explosions swelled into a symphony that made earth and air throb with its resounding, insistent chords. Upstream, in the shadow of Cape Horn, the pace of activity was just as frantic. In a cluster of corrugated tin sheds, forty-eight tons of steel drill bits were sharpened by hand every twenty-four hours; mechanics, bare chests smeared with grease and hands black with oil, labored over balky compressors and generators; powder monkeys unloaded case after case of dynamite and blasting caps into temporary magazines. On the flank of Cape Horn, carpenters were hammering together the Six Companies River Camp, a half-dozen two-story wooden barracks, perched on trestles, that clung precariously to a steep, boulder-strewn slope. These crude, raw-lumber bunkhouses would provide quarters for approximately 480 single men; workers with families would have to fend for themselves until Boulder City, the permanent town that was being erected on the northern slope of Eldorado Valley near Gov-

Blasting out a railroad bed on the side of Cape Horn, October, 1931. Eight tons of dynamite were detonated in this shot, bringing down 160,000 cubic yards of rock. (Manis Collection, UNLV)

ernment Camp One, had been completed. Some of the married men commuted from auto camps in Las Vegas or along Boulder Highway, but most moved into Ragtown, swelling its population to 1,400.[20]

Excavation of the railroad bed and access roads in Black Canyon was proceeding rapidly as April turned to May, and like a steady roll of thunder the sound of explosions echoed in Ragtown and at the Boulder City town site. Caught between the anvil of rock and the hammer of dynamite, the men digging and blasting along the canyon wall paid a price for each foot they advanced downriver.

At 4:30 P.M. on May 8, a crew working at river level failed to hear the signal that a dynamite charge was about to be detonated two hundred feet directly overhead. Miner P. L. Lezie was blown fifty feet onto a boulder and suffered a severe back injury. Another miner, Herman Schmitto, was struck by a rock fragment that severed his ear and fractured his skull.[21]

Ten days later, miners Andrew Lane and Harry Lange, jackhammer operator Harry Ludwig, and chuck tender Floyd Hall were working on the Arizona cliff face when they heard a cracking noise and saw a stream of sand and pebbles cascading onto the 150-foot-wide rock shelf they were clearing. Before they could move or even shout, the shelf disintegrated beneath them and careened down to the canyon bottom. Hundreds of work-

Steamshovel excavating a roadbed into Black Canyon, spring, 1931. (Kaiser Collection, Bancroft Library)

men saw the avalanche, raced to the rubble heap, and began prying it apart with shovels and crowbars. Incredibly, Hall was pulled from the top of the pile, suffering only bruises and scratches. Ludwig was found moments later with a badly fractured leg. Heartened by the miraculous escape of Hall and Ludwig, the rescuers burrowed into the mass of debris searching for the remaining two men, but their effort was in vain. The bodies of Lane and Lange, crushed almost beyond recognition, were discovered eight hours later.[22]

The growing casualty list did not slow the advance downriver to the place where the portals of the four diversion tunnels had been marked, two on each canyon wall, with circles of white paint fifty feet in diameter. Ladders and rope-suspended floating stairways zigzagged down from the Nevada rim, allowing more laborers to get to the work sites. On May 12 the *Las Vegas Age* reported that the number of men on the Six Companies payroll had jumped to eleven hundred. Four days later the second critical

Six Companies workers drilling with jackhammers on a ledge high above the
Colorado. Operating power equipment on the unstable rock was extremely dangerous. (Bureau of Reclamation)

step of the dam construction, excavation of the diversion tunnels, began
when dynamite charges opened a construction adit on the Arizona side.[23]

The speed with which Frank Crowe and his crews opened the canyon
and began the tunnels was gratifying to the Six Companies directors and to
Bureau of Reclamation officials. "Those western dirt-moving fools are
building highways, starting tunnels, and laying railroads all at once, but
without any mix-ups," Reclamation Commissioner Elwood Mead exulted
to a newspaper reporter. He did not mention the principal reason for
Crowe's pushing the work ahead so hard: the contract clause specifying
that the diversion tunnels were to be holed out and the river turned by
October 1, 1933, or Six Companies would be fined three thousand dollars
for every day it ran past the deadline.[24]

＼ In May the summer heat began to settle over the Arizona-Nevada
border, and by June daytime temperatures were spiking above one hundred
degrees. Not even the faintest wisps of cloud softened the hard blue sky; by
midmorning the last traces of evening coolness were gone, and through the
long afternoon the sun beat down on the sandy hardpan. On the horizon
the black and red hills, wrapped in a veil of superheated air, seemed to
twitch and writhe as if they were alive, while in Black Canyon the dark

rocks absorbed the sun's rays and smoldered. Occasionally the wind spilled over the canyon rim in a skin-searing blast, but more often than not the air in the gorge was stirred only by the shock waves of the tremendous dynamite shots, which sent choking, yellow-white clouds of smoke and dust billowing out over the shimmering surface of the river.

Working in this hellish atmosphere, without shade and often without drinking water, was a harrowing ordeal. Some of the workers simply could not endure it and had to give up their jobs, but most kept returning, day after day, to struggle on. As temperatures continued to rise, heat prostration took its toll. One moment a man would be at work and the next he would convulse, vomit, then collapse into unconsciousness. Sometimes there was nothing his fellow workers could do but stare with mouths agape as the victim gasped out his life on the ground. More often, however, the limp, twitching figure would be put in one of the launches and taken upstream to the River Camp. "While I was [at the River Camp] two men were brought in who had passed out completely," wrote mess-hall waiter Victor Castle. "One, a commissary clerk, collapsed in his tent after he finished work and had convulsions. There was no doctor. Buckets of ice water were thrown on him. This was the only medical service he got, and we, the workers, gave it to him. . . . The other man went under on the job. We threw more ice water on him. It was the only thing we knew to do, and the only thing we could do without advice or aid." [25]

If a heatstricken man survived the hour-long trip to the hospital in Las Vegas, he would be packed in ice, given a heart stimulant, and slowly nursed back to health. But for many this treatment was too little too late. Dr. R. W. Martin told the *Las Vegas Evening Review-Journal* that some of the men brought in from the dam site had body temperatures as high as 112 degrees, and there was little he could do for them. Worker John Gieck described the situation more bluntly: "[They] arrived in Las Vegas dead, bloated, and looking like they had been parboiled." The swiftly mounting toll was frightening; from June 25 to July 26, fourteen workers—approximately one every two days—died of heat prostration. [26]

Observers attributed the heat-related deaths to the dam workers' physical weakness, pointing out that many had been living on little more than coffee, doughnuts, and relief-kitchen soup for months before they arrived in Black Canyon, and thus were unable to withstand the rigors of hard labor in harsh climatic conditions. Overeating was also thought to contribute to the problem. According to this theory, the new men could not resist gorging themselves in the mess hall, where there was no restriction on the amount of food an individual could pile on his plate. Then they went out into the sun where, "like horses newly turned out to pasture, [they] literally foundered." [27]

Not until a team of physiologists from the Harvard University Fatigue Laboratory came to Black Canyon in the summer of 1932 to study heat

Waterboys carried drinking water in canvas bags to crews working in remote areas of the dam site. (Kaiser Collection, Bancroft Library)

prostration was the actual cause of the problem, dehydration, identified. The men were not drinking enough water on the job because they were working in dangerous locations where water boys would not or could not go, and because many mistakenly thought that drinking more water would make them feel sicker and bring on heat prostration. After the physiologists' findings were announced, Six Companies made sure that water was available everywhere at the dam site and strongly encouraged the men to salt their food and drink more water with their meals. These steps drastically reduced the number and severity of heat-prostration cases, but during the summer of 1931 there were as yet no answers, and the wail of ambulance sirens echoing off the promontory of Cape Horn became a sound as familiar as the rumble of blasting to the men toiling in the canyon.[28]

On June 25, 1931, the *Las Vegas Age* announced that "with the arrival of an official temperature of 110 degrees yesterday, the *Age* is willing to admit that summer is here." That afternoon at the dam site, worker Raymond Hoptland crumpled to his knees, lost consciousness, and died of heat prostration. In Ragtown, women and children squatted in patches of shade cast by tents, bushes, and cars, or huddled up to their necks in the muddy Colorado, seeking relief from the heat. One of the women was Erma Godbey, who, with her husband, Thomas, their young children, and an old man named Scotty Grants, had come to Ragtown from Oatman, Arizona, several days earlier. The Godbeys had bought a tent from a woman whose husband had been disemboweled by a shovel handle when a dynamite charge exploded prematurely, and while Tom Godbey looked for work, Erma set up housekeeping as best she could. She cooked over a smoky campfire, bathed her children in the river, and learned ways to keep cool, such as hanging wet sheets between mesquite trees to catch the breeze.[29] A few days after arriving in Ragtown, Tom Godbey found a job with a Le Tourneau construction crew surfacing the highway from Boulder City to the dam site. As an employment fringe benefit he was allowed to take water from the company mess hall. Erma was grateful that Tom had a job, that her children had food, and that she did not have to draw silty water from the river, but she wondered when the suffocating heat would break and just how long she could stand it.

Living conditions at the Six Companies River Camp also worsened as June gave way to July. There was no shade on Cape Horn—only a few withered creosote bushes grew in the shattered rock that blanketed its slope—and the 480 men who called the camp home were confined to the barracks, which had no showers, no electric lights, and no cooling devices. Because Cape Horn was so hard to get to, the barracks did not receive drinking water from Las Vegas. Instead, water was pumped from the Colorado into large, outdoor storage tanks. These tanks were not covered, and the sunlight glaring on the silvery metal heated the water until it was virtually undrinkable. The heating also caused the bacteria in the water to multiply rapidly, and amoebic dysentery soon afflicted almost everyone in the camp. "There is [an] open toilet with three open seats," wrote one of the cliff dwellers, as the barracks residents were called, "and a long line of waiting men appears constantly before these inadequate facilities. . . . a physical disorder among the men makes this an almost unbearable condition. . . . flies swarm over the encampment."[30]

Even worse for the dam workers, exhausted by their long shifts in Black Canyon, were the stifling nights that made sleep, when possible, "a heavy, sweating coma." Edmund Wilson visited the River Camp in early August and wrote of the ovenlike bunkhouses "in which the men sleep close together and in which the disinfectant for the water-closets pervades the stupefying air. . . . The mice in the inhuman quiet make noises like people marauding. They are a kind of wild jumping mice that astonish me by hopping on the cots and running over the faces of the sleeping men, who are so used to them that they brush them away like flies."[31]

And still the temperature rose. By the end of July the average daily

River Camp bunkhouses in the shadow of Cape Horn, June, 1931. (Bureau of Reclamation)

high was 119 degrees and the average low was 95; in the dark, airless shafts of the diversion tunnels, readings of 140 degrees were recorded on several occasions.[32]

Day after day Frank Crowe stalked about the dam site, watching the crews lethargically perform their tasks, and fretting about the heat-induced slowdown. In May he had moved into temporary quarters at the Boulder City townsite. Work on the town was moving forward, but it would be several months before he could get the men out of the River Camp and Ragtown and into the cooler, more comfortable dormitories and cottages. In the meantime, the heat was an implacable foe, sapping vitality and blunting the momentum of the assault on Black Canyon. The workers were physically worn down (Crowe himself had lost twenty-seven pounds since March) and their senses and judgment had been dulled.[33] Accidents caused by carelessness or exhaustion were occuring with alarming frequency, and each day brought a fresh spate of heat-prostration cases. Some men were even losing their sanity in the inferno of Black Canyon and on the baking sands of the river flat. Marshal Claude Williams was summoned to Ragtown one especially hot afternoon to investigate a report that several young boys were being abused by their father. Williams found the boys staked out, naked, in the burning sunlight, their deranged father sitting nearby. A mob threatened to lynch the man, but Williams maintained order. He gave the man gasoline to leave Ragtown at once and sent the boys to an orphanage in Las Vegas.[34] Unfortunately, this was not an isolated incident but part of a growing epidemic of bizarre behavior and domestic violence attributable to the unrelenting heat.

Like many other women, Erma Godbey had been working valiantly to make the Ragtown encampment as decent a home as possible for her husband and children, but she was approaching the limits of her endurance. She had been sunburned very badly, and the dry desert air and hours spent stooping over a smoldering cook fire had blistered and cracked her face. She was sure she had caught a skin disease by bathing in the river and was treating the condition by putting undiluted listerine on the affected areas, further irritating the inflamed skin and causing her great pain. On July 26 she finally reached her breaking point. That day, three Ragtown women died of heat prostration. Two of the corpses were taken to a mortuary in Las Vegas, but no vehicle was available to transport the third, and it lay for several hours in the afternoon sun, bloating horribly, before an ambulance finally arrived to take it away. When Tom Godbey got home from work, he found his wife in a state of panic. Fearing for her children's health and in desperate need of a doctor to treat her ravaged skin, she insisted that they leave Ragtown; three days later the family moved to an auto camp on Bonanza Road in Las Vegas.[35]

Since March, the press had found the Boulder Canyon Project an excellent source of upbeat news stories, but now more negative accounts began to appear, describing the danger the dam workers faced and the wretched settlements they lived in. "Out on the desert," wrote a columnist for the San Francisco Examiner, "men are killing stray burros for food. A child, blackened by the sun, gnaws a bone in the doorway of a rusty tin shack. A mother drags her way about the rude encampment; a tired-eyed man tinkers with a flivver in the head-high clump of mesquite that fringes the graveyard camp. . . . And just over the hills is a thundering pit of industry." Another article quoted Bureau of Reclamation engineers as calling Black Canyon a "hell-hole" where men got "goofy with the heat." The weekly newspaper Industrial Worker condemned living conditions at the

River Camp, charged that safety precautions were being ignored, and said that Six Companies was using the Las Vegas unemployment situation to force workers into a dangerous job speedup.[36]

In spite of these accusations and the heat's dramatic effect on the pace of work in Black Canyon, the Six Companies directors were determined to forge ahead. To scale back operations until cooler weather returned was economically unfeasible; large orders for equipment and building materials had been placed, and suspending work would make it harder to meet the government's deadline for finishing excavation of the diversion tunnels by October 1, 1933. Furthermore, it was not clear where the thirteen hundred laborers and their families would go or how they would support themselves if the job were shut down. As for the safety record, the company directors were convinced that it was as good as could be expected given the inherently hazardous working conditions. Ragtown was deplorable but not their responsibility; it was President Hoover who had ordered the project started in March instead of October, when Boulder City was to be finished. "Next year," said Elwood Mead, echoing the Six Companies executives, "there will be an entirely different story."[37]

And so the work went on, but like thunderheads piling up over the distant mountains, rumblings of discontent were rising out of Black Canyon; in the stifling bunkhouses of the River Camp and in the tents and tarpaper shanties of Ragtown, a violent storm of protest was brewing.

At the River Camp in early May a short, stocky laborer with a hard-set mouth and bristling gaze hired on with Six Companies as a truck tender. His name was Frank Desmond Anderson, and he was a professional organizer for the radical labor union the Industrial Workers of the World. He and a small group of followers had been dispatched to Nevada by the union leadership to establish a local in Las Vegas, recruit members among the Hoover Dam workers, and agitate for higher pay and better working conditions in Black Canyon.[38]

The IWW, which had been founded at the turn of the century, was notorious for its revolutionary goals, including the overthrow of the capitalist system by the working class and the creation of a "new society within the shell of the old." At its peak in 1917, the union had boasted nearly a hundred thousand members, but government repression and internal strife during and after World War I had curtailed its influence. By 1930 the number of card-carrying Wobblies, as IWW members were called, had shrunk drastically, recruiting was at a standstill, and the union was on the brink of extinction.[39] Nevertheless, IWW leaders remained optimistic that their vision of a workers' commonwealth would gain new appeal as the Great Depression deepened, and they selected 1931 as the year in which they would save their dying organization. "There was never a better time in the history of the IWW to get out our propaganda than now," proclaimed *Industrial Solidarity,* one of the union's two weekly newspapers, in February, 1931. "The slaves that would not listen to the message of the IWW a few years ago are ready to agree with us now that something must be done. . . . The time has come in the IWW for action." [40]

Because of its national prominence, the Boulder Canyon Project was chosen as the target for IWW action. Twenty-eight-year-old Frank Anderson was ordered to infiltrate the dam job and instigate a confrontation with Six Companies that would generate national publicity and restore the union to a position of influence in the American labor movement. "We hope to have this dam job one hundred percent organized before it is half completed and be in a position to dictate under what conditions and wages the construction workers will build this gigantic monument of human accomplishment," proclaimed *Industrial Solidarity.* [41]

When Anderson and the other Wobblies arrived at Black Canyon in May, the situation appeared ripe for agitation. The squalor of the River Camp and Ragtown and the increasing number of injuries and deaths on the job seemed powerful inducements for workers to join a union, and after setting up headquarters in Las Vegas the Wobblies began recruiting. They accused the contractors of ignoring Nevada and Arizona mining safety laws and engineering a dangerous job speedup, denounced them for indifference to the suffering of the people living on the riverbanks, and urged the workers to fight back by joining the IWW. The allegations against Six Companies, as well as general information about life and work at the dam site, were wired to the IWW offices in Chicago and Seattle, where they were written up and published in *Industrial Solidarity* and the other IWW weekly, *Industrial Worker.* Papers were sent back to Las Vegas, where the Wobbly organizers sold them on street corners and in front of casinos and dance halls where the dam workers congregated. Copies also were smuggled into the project reservation and distributed surreptitiously in Ragtown, the River Camp, and the Boulder City Camp. [42]

As the summer dragged on in Black Canyon and living and working conditions became worse, Frank Anderson felt confident that the workers would respond to the IWW's demands and join the union en masse. And yet the criticism of Six Companies and the appeals for worker solidarity failed to galvanize the men. Most of the laborers supported the Wobblies' call for increased job safety and better housing, but they did not trust the IWW and were afraid to join it. Many had had previous experience with Wobbly-led strikes, and the violently confrontational tactics and rhetoric of the Idaho Wild Women, as hard-core IWW agitators were derisively called, had created a legacy of bitterness and ill will. [43] The workers also knew that affiliation with any labor organization could cost them their jobs when there were thousands of unemployed men in Las Vegas desper-

ate to work for any wage and under any conditions. Memories of the lean
months before they hired on with Six Companies were still fresh; a regular
paycheck and three square meals a day were not matters to be taken
lightly.

\   The workers' reluctance to embrace the union was not the only problem facing the IWW organizers: Six Companies, Bureau of Reclamation,
and Las Vegas city officials were taking steps to drive the Wobblies out of
southern Nevada. The first salvo was fired on July 10 when Clark County
Deputy Sheriff Eddie Johnson, moonlighting as a bouncer at the Boulder
Club, a saloon and gambling hall on Fremont Street in Las Vegas, arrested
Frank Anderson on a vagrancy charge while Anderson was selling IWW
newspapers outside the club. Next morning a bold headline on the front
page of the Las Vegas Age screamed: "I.W.W. Group at Dam Revealed!"
The accompanying article quoted Anderson as claiming that three hundred
dam workers (nearly a quarter of the Black Canyon labor force) were IWW
members, that twenty-one new members had signed on during the last
month, and that the Wobblies would control Boulder City when it was
completed. The IWW's ultimate objective, the paper reported, was to call a
crippling strike that would shut down the job.[44]

Later on the morning of July 11, two of Anderson's lieutenants, C. E.
Setzer and Louis Gracey, were arrested as they left the Western Union
office, where they had wired news of Anderson's detention to IWW headquarters in Chicago and asked that legal defense funds be sent to them immediately. A Western Union operator had tipped off the Clark County
sheriff that the Wobblies were in the office, and several deputies waited outside while the text of the message to Chicago was copied down. Like
Anderson, Setzer and Gracey were charged with vagrancy and tossed into
the city jail.[45]

Word of the crackdown soon reached other members of the Las Vegas
IWW local. They retained a lawyer to defend their colleagues and began
planning a demonstration to protest the arrests. The demonstration took
the form of a gathering that night in front of the Boulder Club. Wobbly
speakers denounced the incarceration of Anderson, Setzer, and Gracey, renewed their accusations that Six Companies was exploiting the dam workers, and announced that their goal was *not* to bring on a job-ending strike
but to organize for better hours, more pay, safer working conditions, and
living improvements, such as water coolers, flush toilets, and more wholesome food. The rally was broken up when Deputy Sheriff Johnson came
out of the club and arrested five IWW men who were selling copies of *Industrial Worker* and *Industrial Solidarity*.[46]

There now were eight Wobblies crowded into a single cell in the tiny
Las Vegas jail, and Clark County officials had to decide how to hold them
on the trumped-up vagrancy charges. The U.S. attorney's office in Carson

City was asked whether the federal government would be interested in prosecuting Anderson and the others for violation of federal criminal syndicalism statutes, but Assistant U.S. Attorney George Montrose replied that the government did not want to get involved. No crime had been committed on the federally controlled Boulder Canyon Project Reservation, he pointed out, and said that prosecution was "wholly and solely in the hands of county and city officials."[47]

Meanwhile, Las Vegas Police Chief Clay Williams was doing his best to make the Wobblies' stay in the city jail as uncomfortable as possible. On the morning of July 13 he tried to force two of the IWW prisoners to work on a city road gang. When they refused on grounds that they had not been convicted of a crime, he withheld food from all eight for the rest of the day and told a *Las Vegas Evening Review-Journal* reporter that "he was giving I.W.W. members no assurance as to how much they would be fed when housed in the Las Vegas jail."[48]

The Wobblies were arraigned before Municipal Judge W. G. Morse on July 13, at which time they pleaded not guilty to the vagrancy charges and demanded jury trials. This demand was denied, and hearing of the first vagrancy case, against Frank Anderson, was scheduled for July 15. At four o'clock on that day Anderson and his lawyer, T. Alonzo Wells, entered Judge Morse's courtroom. City Attorney Frank A. Stevens, acting as prosecutor, called Deputy Sheriff Eddie Johnson as a witness and questioned him about the details of Anderson's arrest at the Boulder Club. Johnson claimed that he had seen Anderson in the casino several times during July, had heard him make critical remarks about working conditions at the dam site, and had observed him selling radical newspapers outside the club on the night he was arrested. When Stevens had finished his questioning, Wells cross-examined Johnson, asking him whether he had knowledge of Anderson's financial situation that led him to believe there were grounds for a vagrancy charge. Johnson replied that he did not. Did Johnson know that Anderson was paid twenty-eight dollars a month by the IWW? He did not. Had Johnson ever read one of the papers Anderson was selling? He had not. Why, then, Wells asked, had the deputy sheriff made the arrest? Because, replied Johnson, Clark County Chief Deputy Sheriff Glenn E. ("Bud") Bodell, who owned and operated the Boulder Club, had told him to. The following morning Judge Morse dismissed the vagrancy charge against Frank Anderson and, on City Attorney Stevens' motion, dropped the charges against the other Wobblies as well.[49]

The IWW men were free to go back to selling newspapers and recruiting dam workers, but in the wake of their court victory they knew they could expect more harassment from the Las Vegas police. Still, they were heartened by Morse's evenhandedness and by the U.S. attorney's decision to stay out of their prosecution. "Much fine sentiment among the local people has come to light as a result of the affair," wrote W. F. Burroughs, one of the jailed Wobblies, in an article for *Industrial Solidarity*. "Nevadans have not forgotten the workers' condition in the unparalleled Goldfield which was controlled by the IWW."[50]

As July melted into August, the Las Vegas Wobblies continued to criticize the bad food and foul drinking water at the work camps, the lack of cooling systems and flush toilets in the bunkhouses, and the inadequate safety precautions at the dam site. Six Companies responded by making several changes. The practice of pumping the River Camp's drinking water from the Colorado was discontinued, and water from Las Vegas' artesian wells was hauled in as it was at the other camps. A water cooler was installed in the River Camp mess hall so that at least some cold water was available to workers between shifts. The magazine containing the blasting supplies was relocated to the Arizona side of the river, away from the electrical lines and welding equipment in the machine shops. Use of gasoline as a cleaning fluid on the job was banned after a series of explosions and fires burned several workers.[51]

But neither Wobbly agitation nor Six Companies concessions could do anything about the major cause of danger and discomfort at the dam site: the searing summer heat, which, according to official weather records, was running almost 12 degrees above normal.[52] On July 20 the mercury hit 117 degrees in the shade, five men were stricken with heat prostration in Black Canyon, and Edna Mitchell, the fifteen-year-old daughter of worker Elmer Mitchell, died in Ragtown. Six days later the temperature again reached 117 degrees; three more Ragtown women died, and four laborers were incapacitated. The heat had burned away the workers' patience and sense of humor, leaving them edgy and disgruntled; all that was needed to ignite the volatile mixture of anger and frustration was a single spark.[53]

On Friday, August 7, Six Companies struck the spark. The swing-shift crews reporting to work at 3:30 P.M. were told there would be a wage cut for men working in the diversion tunnels. The details were hotly disputed later, the Wobblies claiming that the daily pay of various classes of tunnel workers was to be reduced 10 to 29 percent, and that 180 men were affected. Six Companies said the cut affected only 30 muckers, unskilled laborers who shoveled broken rock into trucks after explosions. They were to be transferred out of the tunnels with a reduction in pay, the contractors said, because they had been replaced by new power equipment.[54] In any case the unexpected announcement infuriated the swing-shift crews, and they greeted the departing day-shift crews with the news. An impromptu meeting was called at the boat landing, where a fleet of barges was waiting to take the day-shift workers back to the River Camp. Two Wobblies seized control of the meeting and began haranguing the sweating throng, reiterating the standard IWW grievances and asserting that the wage cut confirmed their longstanding contention that Six Companies was greedy and cared nothing about the welfare of its workers.

The men milled about, trying to decide what to do. Some swing-shift laborers wanted to go to work and discuss the matter later, while others were in favor of striking on the spot. Finally, at the urging of the Wobblies, it was decided that both the day- and swing-shift crews would return to the River Camp, where a mass meeting could be held to consider a response to Six Companies' action.

The meeting was convened in the River Camp mess hall at five o'clock. Four hundred men, almost the entire population of the camp, jammed into the building to hear swing-shift workers recount what had happened at the dam site earlier that afternoon, tell what was known about the wage cut, and propose retaliatory steps. The air was thick with profanity and the odor of unwashed bodies as speaker after speaker denounced Six Companies. When the motion to go on strike was made, a roar of assent shook the long, narrow room. A committee was chosen to prepare a list of demands, and a delegation was dispatched to the Boulder City Camp to enlist the support of workers there. At seven o'clock six hundred men squeezed into the Boulder City Camp's mess hall to hear what the River Camp workers had to say. When the report ended, they voted overwhelmingly to join the strike and elected a committee to help the River Camp committee draw up a list of demands and plan strategy. In the meantime work would halt, beginning with the 11:30 P.M. graveyard shift.

Mess hall lights burned late into the night as the joint strike committee debated whether its action should be limited to trying to reverse the wage cut or whether it should address a broader range of issues. The majority was in favor of expanding the scope of the strike so that men who worked outside the tunnels would have some reason other than sympathy for the tunnelers to stay off the job. Various grievances, large and small, were argued, and various demands were proposed, discussed, and voted on. The list that was finally adopted stipulated that Six Companies

— Cancel the pay cut and raise the minimum daily wage for tunnel workers from $5 to $5.50; for laborers outside the tunnels from $4 to $5; and for skilled workers, such as miners and carpenters, from $5.60 to $6;

— Rehire all striking workers without harassment or discrimination;

— Include the time spent traveling from camp to dam site and dam site to camp as part of an eight-hour shift;

— Provide pure water and flush toilets at the River Camp;

— Supply ice water on and off the job until fountains with electric-coil coolers were installed;

— Fix the daily charge for three meals in the messhalls at $1.50;

— Build changing rooms at the portals of the diversion tunnels so the workers could get out of their wet, dirty clothes before making the long trip back to the camps;

— Station a safety miner at each tunnel heading to provide first aid to injured workers; and

— Obey all Nevada and Arizona mining safety laws.[55]

With the demands agreed upon, the committee turned its attention to the question of how the strike should be conducted and what the IWW's role should be. The Wobbly agitators had been in the forefront of the walk-out that afternoon, rallying the men to leave the dam site. They had been talking for months about the problems enumerated in the list of demands, and had predicted correctly that Six Companies would try to use the almost unlimited supply of cheap labor in Las Vegas to force workers on the job to accept whatever wage scale, shift schedule, and safety standards it saw fit to impose. Furthermore, the Wobblies were experienced in labor-management confrontations and had a national organization to give them financial, legal, and public-relations support. And yet the committee members agreed that the IWW should not be allowed to control or in any way put its imprimatur on the strike. Even though Frank Anderson and the other Wobblies had assiduously avoided talk of class warfare and revolution, any association with the IWW would taint the strike with extremism and make negotiations with Six Companies much more difficult. It was decided that the Wobblies would be excluded from formal participation in the direction of the strike, and the meeting concluded with the selection of L. L. ("Red") Williams as chief spokesman and negotiator.

Word of Friday's actions reached Las Vegas quickly, and a reporter from the *Age* succeeded in reaching Frank Crowe by telephone late that night. The protest "was largely a result of I.W.W. agitation," Crowe said, and added that Six Companies "would be glad to get rid of such." He did not think the workers' action would slow the project. The reporter then asked a strike-committee member about Crowe's contention that the IWW was responsible for the strike. "We wish to make it plain that the strike has nothing to do with the I.W.W.s or the United Mine Workers," the committee member told the reporter. "It is a matter distinctly among the workmen on the project. We're not Wobblies and don't want to be classed as such."[56]

At 10:00 A.M. on Saturday, August 8, Red Williams gave Frank Crowe the list of demands. Crowe told Williams he would need a day to consider it, and ordered the project shut down. In spite of this, the fourteen hundred striking workers were in high spirits; the grind of the seven-day work week had been broken. There was considerable optimism, too, about the outcome of the strike. Crowe had the reputation of being tough but fair in labor disputes, and because he knew so many of the men, and valued their skills and experience so highly, they thought he might intercede on their behalf with the Six Companies directors.

It was not to be. During an interview with a *Review-Journal* reporter shortly after his meeting with Red Williams, Crowe criticized the workers for walking off the job and denied that any of the problems outlined in their list of demands existed. The so-called wage cut was not really a wage cut at all, he said, but a job transfer. A small number of muckers had been shifted to jobs outside the tunnels because their work was now being done by power drills and mechanical shovels. The wage scale for jobs outside the tunnels was lower than that for jobs in them, but the scale was not being changed as the strikers charged.[57] Crowe went on to say that safety in the tunnels was satisfactory and that his records showed no job-related deaths or accidents at the dam site during July. Bureau of Reclamation records revealed that at least fifteen men had died there that month—eleven of heat prostration, two of drowning, one of a ruptured appendix, and another in a car crash—but because none of the deaths had been caused directly by work activity, Crowe's claim was technically correct. His contention that July was accident-free also may have been technically correct, but it was certainly disingenuous, given the scores of heat-prostration cases recorded that month.[58]

Sunday at 10:00 A.M., exactly twenty-four hours after Red Williams had presented him with the list of demands, Crowe called the entire strike committee into his office and gave its members Six Companies' official reply: their grievances had been rejected. All Six Companies employees, with the exception of a skeleton office staff and a few carpenters, were to report to the paymaster to pick up the three days' pay they were owed and a notice of dismissal. Then they were to pack their belongings and get off the project reservation. Hiring of new crews would begin as soon as all fired workers had left or been removed.[59]

After the meeting Crowe talked to reporters and put more pressure on the strikers. "We [Six Companies] are six months ahead of schedule on the work now and we can afford to refuse concessions which would cost $2,000 daily," he said. His repudiation of the strikers' demands was echoed by W. H. Wattis, who spoke for Six Companies from his hospital bed in San Francisco. "They're not going to tell us what to do," he said. "They will have to work under our conditions or not at all."[60]

The harsh rebuke stunned the strike committee, as did the news that Walker Young had alerted U.S. Army troops at Fort Douglas, Utah, to be prepared to come to Black Canyon on short notice if government property were threatened.[61] When the strike leaders relayed Crowe's ultimatum to the workers, the festive mood that had prevailed since Friday night evaporated and solidarity began to crumble. The Wobbly agitators who had been banned from playing a formal role in leading the strike now addressed the men, exhorting them to remain on the reservation and resist Six Companies' strong-arm tactics. The strike committee took the same position and tried to organize a contingent to occupy the camps and picket the dam site, arguing that the project reservation was under federal control and Six Companies had no right to order the workers out. If the government remained impartial in the strike and prevented the contractors from forcing the workers off the reservation, the strike might succeed, they said, and announced that they had sent a telegram to Secretary of Labor William Doak: "As American citizens, we ask protection on Hoover Dam in case of deportation. Strike called in protest against wage-cuts."[62]

The strike committee's plea fell on deaf ears, however; twelve hundred of the fourteen hundred men had had enough and decided to pick up their paychecks and go to Las Vegas. They wanted no part of the confrontation that was brewing, and were more concerned about keeping a low profile and recovering the jobs they had just lost than continuing the strike.

A pall of white dust hung over Boulder Highway all Sunday afternoon as a long procession of automobiles passed a police checkpoint at the edge of the project reservation, crawled over Railroad Pass, and rolled down the long desert slope past the shanty town of Midway into Las Vegas. At the River Camp the two-hundred-odd strikers who had chosen to force a showdown with Six Companies gathered on the barracks porches and watched the evening shadows stretch across Cape Horn, softening the rough edges of its gullies and rock outcrops. The sun sank below the Nevada rim of the canyon, and to the east, above the Arizona precipice, the first stars of evening began to twinkle faintly. Stripped to the waist and sweating in the 100-degree heat, the strikers debated the day's pressing questions: What would the Six Companies bosses do? Were troops on the way from Utah? How long could they hold out without food or money? Eventually conversation ended, and a deep, uneasy silence settled over the canyon, broken only by the soft gurgle of river eddies at the foot of the hill.

The next morning sentries posted along the narrow road leading into the River Camp shouted a warning as a line of trucks lurched along the rutted track, gears grinding, and came to a stop in front of the mess-hall steps. Armed men leaped from the first few trucks and quickly fanned out among the barracks, rounding up strikers at gunpoint and herding them back toward the mess hall. Resistance would have been suicidal, so the strikers climbed into the waiting vehicles. The convoy was about to pull out when a car driven by a deputy U.S. marshal rolled into camp. To the amazement of the strikers, the deputy marshal jumped from his vehicle and arrested a strikebreaker. He then ordered the other strikebreakers to hand over their weapons, told the strikers to get out of the trucks, and announced that Six Companies had no authority to remove anyone from a federal reservation. The strike was still on.[63]

Late that night a carload of Wobblies, including Frank Anderson and several other organizers from Las Vegas, accompanied by writer Edmund Wilson, entered the project reservation. After being searched for weapons and liquor by marshals at the reservation boundary, they drove past the Boulder City Camp, down Hemenway Wash, through sleeping Ragtown, and on to the River Camp, where the two hundred striking workers were holding out. The Wobblies had brought eggs, coffee, and bread donated by sympathetic Las Vegas citizens and also bad news: the U.S. government was abandoning its position of neutrality and sending officials next morning to force the strikers off the reservation. With this announcement, spirits that had soared earlier in the day when the deputy marshal made his bold stand plummeted, except among the Wobblies, who continued to talk of resisting. "Some people wouldn't believe that a fellow can get a kick out of this kind of work," one of them told Edmund Wilson, "but you get a kick out of pulling off a strike—get sent to San Quentin and get a kick out of it!" To demonstrate their bravado and continuing defiance, the Wobbly leaders entered the bosses' bunkhouse and lolled on the cots, arguing how best to continue the strike. In one corner they found a small water cooler, a discovery that unleashed a flood of bitter humor among the strikers. The black night became darker as a thick layer of clouds obscured the stars; some of the men tried to sleep, but most crouched on the rickety wooden porches and smoked cigarettes in silence.[64]

The sky was still cloudy in the morning, and the air was damp and smelling of rain. At eight o'clock the guards posted on the road signaled that vehicles were approaching, and moments later several dark government sedans, two trucks, and a bus rolled into the camp. Walker Young was at the wheel of the lead car, and Jacob H. Fulmer, the chief U.S. marshal for the state of Nevada, was at his side. Assistant U.S. Attorney George Montrose and several deputy U.S. marshals rode in the trailing cars. Members of the strike committee greeted Young, Fulmer, and Montrose and invited them into the mess hall where the strikers were congregating. The patter of raindrops falling in the dust, a strange, muffled, musical sound after weeks of unbroken sunshine, accompanied the heavy tread of the government men's boots as they clambered up the steps leading from the road to the camp buildings and entered the mess hall.

Walker Young stood on a bench and told the assembled throng that he and Marshal Fulmer appreciated the peaceful, orderly way in which the strike had been conducted and that he hoped their behavior would continue to be good when they left the reservation, as he was now asking them to do. The two trucks and the bus would take them to Las Vegas, he announced, then added, with an unmistakable edge in his voice, that everybody was to clear the reservation. "We ask you to go, and we depend upon you to go," he concluded. "There's been no question at all of force." [65]

At that moment, as if to punctuate Young's blunt pronouncement, several boulders, loosened by the rain, bounded down the hillside just outside the mess hall and crashed into the river. When the rumbling and quivering had ceased, several strikers got to their feet and made short, angry speeches accusing the government officials of conspiring with Six Companies to empty the reservation so that scabs could be brought in. Young replied that he hated to call anyone a scab but that the best thing for the men to do was leave the reservation peacefully and "make a clean start."

Red Williams, head of the strike committee, now stood up to say his piece. They all had to leave peacefully, as Young had asked, he told the crestfallen strikers. To do otherwise would lead to violence and the loss of whatever public support they had gained. Vacating the reservation did not mean an end to the strike because they could still picket at the reservation border, but any resistance at the River Camp would end with the reopening of the job at bayonet point. [66]

It was raining harder now, and a sheet of water sluiced off the mess-hall roof, drummed noisily on the planks of the porch, and separated into sandy rivulets that ran blood red down the flanks of Cape Horn. On the sodden road the bus and the two trucks sat, engines idling, while the strikers filed out of the mess hall and dispersed into the barracks to collect their belongings. Soon the vehicles were loaded with the despondent men and pulled away, leaving the buildings of the River Camp empty and silent except for the steady drip of water leaking through the roofs.

In Las Vegas the twelve hundred men who had abandoned the reservation on Sunday were running out of money. They anxiously awaited news of the showdown at the River Camp, hoping the job would be reopened soon so that they could eat again. The wage cut and other grievances had dwindled to insignificance compared with empty wallets and empty stomachs, and any terms that Six Companies chose to offer would be accepted with alacrity by most of the fired dam workers. A few still wanted to push for improved conditions, however, and decided to appeal to state and federal authorities. In conjunction with the Las Vegas Central Labor Council, which represented the city's craft unions, they drafted a letter to Nevada Governor Fred Balzar, American Federation of Labor President William Green, and the editors of the Las Vegas newspapers, calling for action to force Six Companies to address the workers' grievances. Copies also were sent to Reclamation Commissioner Mead, Labor Secretary Doak, and President Hoover. Six Companies, they charged, was forcing the dam laborers to work at "a wage rate that does not guarantee a decent living and under working conditions which are unsatisfactory and indefensible." They then challenged the government officials to intervene:

Big headlines in the daily press tell from day to day how anxious the government heads are to relieve unemployment and maintain the standard wage of those who are already employed. If these high-sounding phrases have any meaning at all, then this Boulder Dam situation calls for action. We believe the Six Companies . . . are simply taking advantage of the depression throughout the country and of a mob of broke and hungry men to establish wage scales that are entirely unreasonable. . . .

We feel that it is a crime against humanity to ask men to work in this hell-hole of heat at Boulder Dam for a mere pittance, just enough to keep their bellies full and clothes on their backs. If labor conditions are permitted to go on as they are, we believe it is going to effect [sic] labor like a cancer. These conditions will spread if the Six Companies get away with it at Boulder Dam.[67]

The government leaders were unmoved. Elwood Mead had already vented his feelings: "The present wage rate on Hoover Dam is considerably above that of the surrounding region," he said and characterized the strikers' demands as "impossible." Doak echoed Mead, announcing that in his view the Boulder Canyon Project "does not come within the provisions of the prevailing-rate law."[68]

On August 13, the day after the last strikers were evicted from the River Camp, Walker Young announced that Six Companies would begin hiring workers and construction of the dam would resume immediately. He also announced that a gate and a guardhouse were to be installed where Boulder Highway entered the project reservation and that henceforth no one would be allowed on the reservation without a government pass. Labor Commissioner Leonard Blood was to move from his employment office in Las Vegas to a new office at the reservation boundary and supervise the hiring, giving preference, at Frank Crowe's request, to "old worthy workers" while screening out all agitators. Marshal Fulmer had deputized twenty Bureau of Reclamation engineers to prevent picketers from entering the reservation or harassing returning workers.[69]

The men removed from the River Camp on August 12 had established a tent city, dubbed Camp Stand, on Boulder Highway between Railroad Pass and Las Vegas. "We are not beaten," they told reporters. "We went out on strike in protest of a wage scale based on a $4 day for muckers and as that scale is still in force we consider men returning to work as strikebreakers. The strike is still on and we intend to stick with it until it is settled."[70] They formed a picket line along the highway and waited to see whether the other workers would honor it and stay off the job.

The answer came quickly. Between noon and 8:30 P.M. on August 13, 350 men entered the project reservation and occupied the Boulder City Camp. The following morning, this crew was trucked to the dam site, and blasting and digging were resumed in the diversion tunnels. While the day

Rehired dam workers entering the Boulder Canyon Project Reservation through the new gate, August 13, 1931. The strike against Six Companies ended two days later. (Union Pacific Railroad Collection, UNLV)

shift was at work, another 380 laborers signed up with Six Companies, crossed the picket line, and went into the tunnels at 3:30 P.M. to work the swing shift. Whatever solidarity these men felt with the little band of strikers who stood beside Boulder Highway, choking on the dust that boiled up in the wake of the cars and trucks streaming through the police gate into the reservation, it was not enough to keep them away from their old jobs and the groaning tables in the company mess halls. Hunger and fear had broken the strike.[71]

The Wobblies and the strike committee members were not ready to admit defeat, however. They announced that a public rally would be held at the Airdome Theater in Las Vegas on the afternoon of August 14. At the appointed hour, the remaining strikers and six hundred curious onlookers packed the auditorium to listen to a round of fiery speeches. "Are we going out now to where these fellows are going back to work as strikebreakers and take a shillelagh and beat the ears off the guys, or are we going to say 'please, Mister'?" bellowed one of the militant orators. But the committee was determined not to resort to violence and proposed instead that pickets be posted near the reservation gate, at the Las Vegas bus station, on the road from Searchlight to Boulder City, and on the highway from Las Vegas to Los Angeles. Pickets at the latter three sites were to recruit new strikers among the unemployed men converging on southern Nevada. In the meantime efforts would be made to keep the strike in the public eye and to raise funds to buy food and other supplies. This proposal was approved by voice vote, and the strikers headed back to Camp Stand while the committee went to the Western Union office to send a last-ditch plea for support to Governor Balzar: "Boulder Dam strike still on. One dollar wage cut still in effect for forty per cent tunnel crews. Six Companies employing strike breakers. Is the State of Nevada going to uphold Six Companies cutting wages in this time of depression? Strikers wish investigation of the project."[72] The committee knew that without government recognition the strike would collapse completely. More than a thousand men had been rehired already, and work in Black Canyon was quickly returning to normal.

On August 15, Balzar's reply was received: "State of Nevada absolutely neutral in all labor disputes. Matter should properly be referred to department of labor, Washington." With no support forthcoming from Nevada or the federal government, the committee proposed that a vote be taken on August 16 to end the strike. Only 118 of the strikers, many of them Wobblies, were still on hand in Camp Stand, now called Camp Despair by the papers, to cast their ballots; the final tally was 68 to 50 in favor of returning to the job. The strike was officially over, eight days after it had begun.[73]

For the workers, the walkout had not been entirely futile: the reduced wage scale remained in effect, but Six Companies announced that it was guaranteeing that scale for the duration of the job. It also moved quickly to install electric lights and additional water coolers at the River Camp, to build changing rooms at the diversion-tunnel portals, and to hasten completion of Boulder City. Best of all from both Six Companies' and the workers' perspective, the heat began to abate, and by the end of September temperatures had dropped to bearable levels. In the fall, families began to move out of the tents and shacks of Ragtown and into the new cottages of Boulder City; single men left the River Camp barracks to take up residence in the Boulder City dormitories, each one equipped with toilets, showers, and heating-cooling systems.[74]

In the end, the only real losers were Frank Anderson and his Wobbly organizers. Instead of breathing new life into the IWW, the strike had begun with worker repudiation of the union and had ended with victory for the bosses. The most potent organizing issue, the dreadful living conditions at the River Camp and Ragtown, was lost with the opening of Boulder City. To make matters worse, Six Companies and the federal government were now working together to keep Wobblies out of the project reservation. The U.S. marshal's office had been beefed up, and a corps of Reservation Rangers was organized to police Boulder City and the dam site and prevent further labor trouble. The man in charge of the new force was none other than Bud Bodell, the Clark County chief deputy sheriff who had ordered Frank Anderson's arrest at the Boulder Club in July.[75]

The Wobblies remained defiant, proclaiming in *Industrial Solidarity* that "'Red cards' are on the Boulder Dam today and will continue to be there until the project is finished." Frank Anderson stayed on in Las Vegas to continue agitating for better working conditions in Black Canyon, but his vow that "the second strike on the Boulder Dam will write a different finis," had a distinctly hollow ring.[76] The hiring leverage given Six Companies by the thousands of unemployed men in Las Vegas, the workers' loyalty to Frank Crowe, and the IWW's reputation as an organization committed to violence and revolution, were too much for the Wobblies to overcome. In September, 1931, worker Claude Rader wrote a poem that summed up how many of the men felt about their jobs on the dam and served as a fitting epitaph for the IWW in Black Canyon:

Abe Lincoln freed the negroes
And old Nero he burned Rome,
But the Big Six helped depression
When they gave the stiff a home.
In a nice bunk house their sleepin'
Their workin' every day,
The hungry look has vanished
For they got three squares a day.
You'll find tall Lou from Kal-a-ma-zoo,
And Slim from Alabam,
Mixed in with all the rest of us
Old boys on Boulder Dam.

And the fallin' rocks can't scare us
Nor the scorchin' rays of the sun,
We've rode the rods and brakebeams
Ragged and on the bum.
And they gave us jobs and fed us
When we needed it you bet,
And we all are truly thankful
With no feelin' of regret.
So we're stickin' till the finish
There's me and Ike and Sam,
And we're gettin' fat and stakie
Us old boys on Boulder Dam.

There are thousands we know that knock it
And holler that they are cheap,
But to us it brings no worry
Not a moments loss of sleep.
For we've been here since it started,
We're used to all the slam,
And we're stickin' to the finish
Us old boys on Boulder Dam.[77]

CHAPTER THREE
To Turn a River

Although the day shift had just begun and the sun had not yet risen high enough to rout the shadows from the depths of Black Canyon, rivulets of sweat were coursing down the face and arms of the jackhammer man. He pulled a long bandanna from the hip pocket of his dusty overalls, mopped his glistening brow, and then adjusted the heavy leather gloves that encased his broad, powerful hands. With a sound that was half sigh and half grunt, he tugged his broad-brim felt hat down snugly on his head and reached for the pneumatic drill that was propped against a nearby boulder. The long, ropy muscles of his forearms flexed as he hefted the chunky drill, positioned the steel bit on a smooth expanse of rock, and hit the trigger. With a harsh cough, the oily cylinder jerked, spat out a puff of white exhaust, and jumped to chattering life. In a moment a dozen other drills had started up, adding their staccato notes to a pounding chorus that reverberated off the beetling face of the Arizona cliff, bounced across the chocolate-brown ribbon of the Colorado, struck the Nevada wall, and echoed back, filling the steep, narrow corridor of Black Canyon with an ominous roar.

Almost as one, the gleaming lengths of sharpened steel chewed into the side of the cliff and wisps of smoke, redolent of pulverized rock and white-hot metal, curled from the drill holes and drifted through the morn-

Jackhammer operators at work, August, 1932. (Kaiser Collection, Bancroft Library)

ing air. The jackhammer man spread his legs a little wider for support, hunched his back, and leaned into the shuddering drill, bearing down hard with his arms and shoulders. The bit pounded and twisted with relentless percussive force, and the octagonal steel shaft disappeared, inch by inch, into the dark rock. A few feet away, powder monkeys carefully opened boxes of dynamite and blasting caps. When the pattern of drill holes was finished, the cylindrical cartridges would be inserted gingerly and tamped into place with long wooden powder sticks. Then a shrill whistle would warn the workmen to clear the area, and with a dull, cracking boom the round would be detonated, a single shot in the seemingly endless fusillade against the canyon walls.

The tableau of sweating, straining figures wrestling with big drills in the midst of a spaghetti-like tangle of air hoses was one of the most common on the Hoover Dam job, but on May 16, 1931, the scene had special significance. On that day the crews working just above river level on the Arizona side of the Colorado were drilling holes for an explosion different

Powder monkeys inserting primers in dynamite sticks preceding a blasting round. (Kaiser Collection, Bancroft Library)

from those that had extended the roads and railroad bed down the Nevada side of the canyon during the preceding month. When this shot was fired, like the "broadside of a dreadnought," into a small quadrant of the cliff, it would gouge out the portal of an adit, a horizontal access shaft, that would penetrate some eight hundred feet into the canyon's flank and intersect the course of the two diversion tunnels running parallel to the river through the Arizona wall. A similar adit would be started soon on the Nevada side to intersect the two tunnels that were going to be driven through that wall. The adits would make it possible to drill, blast, and muck each tunnel on four separate headings—from an upstream portal, a downstream portal, and from two directions in the middle—thereby hastening their completion.[1]

Of all the operations that would be part of building the largest dam in the world, diverting the Colorado was the most important and most difficult. The riverbed had to be exposed, pumped dry, and excavated down to bedrock so that the dam's foundation could be securely anchored and the U-shaped powerhouse at its base constructed. Room also had to be made for men, materials, vehicles, machine shops, and a large concrete mixing plant on the canyon floor close to the place where the dam would rise. In all, nearly a mile of riverbed would have to be securely sealed off from the Colorado's waters for up to four years while the concrete was being poured. It was imperative that the diversion system be absolutely reliable, even in the face of one of the Colorado's legendary floods, for a breach could drown the dam site, wipe out millions of dollars' worth of work, and threaten the lives of hundreds of laborers.[2]

It was, of course, impossible to divert the river around Black Canyon; the only choice was to force it out of its bed and through the walls. To do this, four gigantic tunnels, each fifty-six feet in diameter and averaging more than four thousand feet in length, were to be driven through the solid rock. Then the dam site would be bracketed by two massive earthen cofferdams, one just below the tunnel inlets and one just above the tunnel outlets. Six Companies was required by its contract to complete the four tunnels by October 1, 1933, or pay a $3,000 fine for every day it ran over the deadline.[3]

Frank Crowe and his engineering staff knew that meeting this deadline would be difficult even if the tunnels were started and finished without consideration for the time of year and the level of the river. But here the powerful, unpredictable Colorado forced the dam builders to plot their campaign in accordance with its timetable rather than one set by desk-bound officials in Washington and Denver. During the spring and early summer, when the river was gorged with snowmelt and carrying a heavy burden of silt and debris, diversion was out of the question. Rather than abandoning its deeply scoured channel and plunging into the tunnels, the red torrent would simply smash aside any barrier put in its path and continue on its ancient course. The only time the Colorado could be herded out of its bed was during the late fall or early winter, when the water level was lowest and the likelihood of flash floods minimal. Thus the contract deadline actually meant that Six Companies had to divert the river during the winter of 1932–33. If this goal were not met, the contractors would, in all likelihood, be forced to suspend operations until the winter of 1933–34, when the water was again at its lowest ebb, and pay the $3,000 fine for every day beyond October 1 they had to wait for the river to drop. To the Six Companies directors, and therefore to Frank Crowe, the prospect of losing several months of construction time and paying hundreds of thousands of dollars in fines was a powerful incentive to get the tunnels started at once and to make every effort possible to finish them and divert the river during the winter of 1932–33.

Before a full-blown assault could be mounted, however, many logistical problems had to be solved, not the least of which was gaining access to the remote and rugged sites of the tunnel portals. Because the roads into the canyon bottom were not finished, the crew that started work on the Arizona adit in May had to approach the job like a Marine Corps amphibious unit attacking an enemy-held coastline. Large barges were constructed at the river landing above Cape Horn and loaded with portable diesel-driven air compressors, gasoline-powered generators, jackhammers, drifter drills, drill steel, a portable blacksmith shop complete with furnace and drill-steel sharpener, lumber, cables, and other supplies. The barges were then floated down into Black Canyon, where the drilling crew hit the beach at the only spot where a foothold could be obtained: the base of a jumbled rockslide on the Arizona side near the blaze of white paint marking the spot where the dam abutment would be. Ring bolts were driven into the rock face to moor the barges with cables; the compressor, generators, and blacksmith shop were unloaded and set up; and work began on enlarging the staging area, which was so small at first that it could not even accommodate a Caterpillar tractor, forcing the miners to hand-shovel the muck after each blast. On the Nevada side of the canyon there was no foothold at all, which made it necessary for the miners to drive ring bolts, string a cable suspension footbridge, and hack a shelf out of the cliff before tunneling could begin.[4]

In addition to the delays caused by the as-yet-uncompleted access roads and the remoteness of the Arizona and Nevada adit portals, there was the problem of getting enough electricity into the canyon to operate heavy equipment and light the tunnel interiors. Gas and diesel-driven generators had been pressed into service, but they would not provide enough power once work began in all four diversion tunnels. Southern Sierras Power Company had been awarded a $1.5 million U.S. government contract to furnish electricity to the dam site and was constructing a 132,000-volt transmission line 222 miles across the Mojave Desert from San Bernardino and Victorville, California, to a substation on a rocky promontory near the canyon rim.[5] When the line was finished and the electricity turned on, sometime in late June, 1931, there would be ample power to drive all the machines and illuminate not just the tunnels but the entire dam site, permitting employment of full shifts twenty-four hours a day. Until the voltage arrived from California, however, the work force would have to make do with low-capacity portable generators and compressors, which slowed progress on the adit shafts and delayed the start of the main tunnels.

Still another problem—one more ticklish and less easily solved than the others—was adding to Frank Crowe's difficulties: the Six Companies directors were meddling in the day-to-day management of the project. On previous jobs Crowe had answered only to Harry Morrison, but now he had to contend with a whole slew of headstrong executives. On a given day, one of the directors would arrive from California, tour the dam site, and then issue explicit instructions on how to handle some aspect of the work. A week later a different director would show up and insist that the task be done another way. The result was confusion, wasted effort, and growing frustration on the part of the superintendent of construction, who felt that his authority was being undermined and the organization of the project sabotaged by conflicting orders. The situation was becoming intolerable, and Crowe finally informed the board of directors, through Charlie Shea, that their interference was hindering the work and that a clear chain of command had to be established.[6]

＼   While Shea was delivering Crowe's ultimatum to the board in San Francisco and the blasting of access roads and the stringing of power lines were being completed, the superintendent, his engineering staff, and the construction foremen were putting the finishing touches on a plan of attack they hoped would allow them to hole out, trim, and line the diversion shafts in time to turn the river during the winter of 1932–33. The plan was based on the time-honored strategy of divide and conquer. The fifty-six-foot bore of each tunnel, as wide as a four-lane highway and as tall as a five-story building, was too big to drill and blast in a single operation. The enormous tunnel faces had to be separated into sections and each section excavated separately in a fixed sequence.

The first step was to drive a 12-by-12-foot pioneer heading, also known as an attic tunnel, along the top of each 56-foot cross section to provide ventilation and access for the enlargement that would follow. Crowe's Black Canyon scheme provided for the pioneer headings to be opened off both the Arizona and Nevada adits and at the tunnel portals themselves to increase the number of rock faces available for drilling and blasting. Once these square shafts were open, the wedge-shaped wing sections on either side could be removed. The next step was excavation of the 30-by-50-foot bench sections, which would leave a horseshoe-shaped tunnel 42 feet high, allowing working headroom for the big electric shovels that would muck out the blasted rock. The crescent-shaped invert section would be removed last, completing the full 56-foot-diameter bore.[7]

The tunneling strategy was sound, but without the development of new tactics for rapid and highly efficient drilling, blasting, and mucking, the job could not be completed by the winter of 1932–33. Particularly troublesome were the massive bench sections, each thirty feet high, fifty-six feet across at its widest point, and measuring fifty feet in average width. A flat-hole drilling pattern (approximately seven parallel rows of fourteen holes up to twenty-three feet long driven horizontally into the bench) promised the best blasting results, but the Six Companies engineers were not sure how to get the drills and the men to operate them up to the portion of the face that was high off the tunnel floor.[8] To erect scaffolding after every successive blast would be too time consuming and would create a clutter of timbers and steel pipes that would slow the movement of mucking vehicles in and out of the tunnels.

The task of getting the mining crews into the tunnels, up to the bench face with their drills, and back out so that blasting rounds could be fired as rapidly as possible fell to 29-year-old Assistant Superintendent Bernard F.

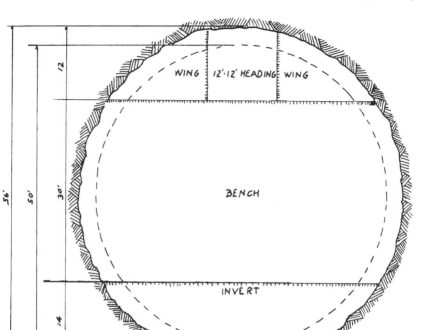

The twelve-by-twelve-foot heading was excavated first, followed by the wing sections and the bench. The invert was excavated last. (Adapted from Compressed Air Magazine)

("Woody") Williams, a thick-necked, barrel-chested construction veteran who had been selected personally by Frank Crowe to be second in command on the dam job. Woody Williams, known to his men as "the Bull of the River," was tough, determined, and resourceful. He brought all these attributes to bear on devising a method for massing large numbers of men and huge quantities of equipment for an attack on the tunnel breast without sacrificing speed and mobility.[9]

In March, Frank Crowe had answered a reporter's question about the grand scale of operations in Black Canyon by remarking that "this will be a job for machines."[10] Trucks, Caterpillar tractors, electric shovels, draglines, and other motorized vehicles of all shapes, sizes, and descriptions would be used on the job in unprecedented numbers, making Hoover Dam one of the first major construction projects in which horses and mules would play virtually no part. Williams was determined to mechanize the tunnel work as well, and using parts and material he had on hand, he de-

The drilling jumbo was a massive, motor-driven rig equipped with thirty 144-pound rock drills. The miners are at the controls of the drills while the chuck tenders stand ready to replace the drill steel. The tunnel superintendent standing in the center wearing a pith helmet is C. T. Hargroves. On January 22, 1932, the Hargroves crew drilled, blasted, and mucked a record forty-six feet of tunnel in eight hours. (Bureau of Reclamation)

signed and built a mammoth motor-driven rig that would allow his miners to drill half the bench face at once.

The heart of this strange-looking contraption, which was referred to as a jumbo, was the ten-ton chassis of an International chain-drive truck built for the army during World War I. A steel skeleton supporting four large wooden platforms was welded to the chassis. Two of the platforms ran the length of the truck and were the stations for the miners who operated the drills and for the chuck tenders who maintained them; the other two platforms were shorter and carried the racks of drill steel. Five horizontal bars were attached to the frame uprights and six 144-pound drifter drills, also known as Leyners, were mounted on each bar. Sheet-steel aprons protected the exposed miners from falling rock, and a web of pipes and hoses, running from the tunnel mouth to the platforms, provided compressed air to power the drills and water to cool the hollow, 1 1/4-inch-diameter drill steel.[11]

The squat drilling jumbo, bristling with lengths of sharp steel and dragging its tangle of lifelines behind it, resembled a disemboweled porcupine, but its remarkable utility more than made up for its ungainly appearance. It could be backed into the tunnel and up to the bench face, blocked solidly into position, connected to the air and water lines, and have its drills pointed and ready to go in twenty minutes or less. After the left side of the bench was drilled, the jumbo simply pulled away and backed up to the right side and the drilling operation was repeated. Despite its bulk, the machine's contours were such that it could drive down one side of the tunnel and leave room for a dump truck or an electric shovel to pass on the other side. It was an ingenious, highly effective piece of equipment that proved to be one of the keys to beating the deadline for tunnel completion. Several of these rigs were built, and Woody Williams was justifiably proud when the men on the dam took to calling the prototype the Williams Jumbo.

On June 25, 1931, the finishing touches were put on the transmission line from San Bernardino, a switch was hit at the substation on the edge of Black Canyon, and eighty thousand volts of power surged to the dam site.[12] The long-awaited arrival of the electricity and rapid progress on the roads into the canyon paved the way for installation of three large-capacity compressor plants that would provide the breath of life to the hundreds of drills and jackhammers poised to attack the diversion tunnels. Good news also arrived from San Francisco, where the Six Companies directors finally had realized that, as Felix Kahn put it, "a board of directors can establish policy, but it can't build a dam." The chaos of conflicting orders that had snarled operations for three months was to be eliminated by creation of a four-man executive committee, which would oversee the various aspects of the project and would alone deal directly with the superintendent. Kahn was to manage financial and legal affairs and the feeding and housing of the work force, W. A. Bechtel's son Steve would be in charge of purchasing, and Charlie Shea would be responsible for field construction and would serve as liaison between Frank Crowe and the others. The chairman of the executive committee, and Six Companies' chief representative in Washington, D.C., was to be Henry Kaiser.[13]

Buoyed by the promise of no more executive interference and by the availability of ample power to drive machinery and illuminate the dam site, Crowe let it be known that he was ready to start "hi-balling" the job. This prediction proved accurate: in spite of the summer heat, which was especially brutal in the tunnel shafts where there was no nighttime cooling and where the temperature consistently hovered in the 120- to 130-degree range, the pace of tunnel excavation quickened. During June the pioneer headings were advanced 410 feet; in July the advance was 1,045 feet; and in September the figure jumped to 3,235 feet.[14]

With the pioneer headings well under way, the rail and road system for hauling excavated muck out of the canyon in place, and the compressor plants and drill-steel sharpening furnaces fired up and ready to operate at full capacity, everything was set for the kickoff of the jumbo-led assault on the hulking bench sections of the tunnels. Three days before the drills were to roar to life, however, there was a brief pause in the noisy whirl of preparations: W. H. Wattis had finally succumbed to cancer in San Francisco, and Frank Crowe ordered a minute of silence in his honor. The seventy-two-year-old builder had never set foot on the dam site, the scene of his last and greatest undertaking, but his determination to see the project through had remained firm to the end; his final message to his associates was "full speed ahead." On September 21 the directors met and chose W. A. Bechtel to be Wattis' successor as president of the consortium, while six hundred miles away in Black Canyon the jumbos moved into the tunnels and went to work.[15]

Startup of the diversion-tunnel enlargment signaled the beginning of a new phase in Hoover Dam's construction and led to a significant expansion in the size of the work force. Each tunnel crew was composed of ninety men, including forty miners, forty chuck tenders, eight nippers, a safety miner, and a foreman (commonly known as a shifter). Each mucking crew was made up of a shovel operator, an oiler, a pitman who worked on the electric shovel, and a cat skinner who operated a Caterpillar tractor with a bulldozer blade attached to the front and a cowdozer scraper on the rear. Hundreds of drivers had to be hired to operate the truck fleet that would haul the muck from the tunnels, and scores of electricians, pumpmen, powdermen, and other support workers were also needed. By early 1932, when excavation was moving forward on all four tunnels, twelve hundred men on average were simultaneously engaged in tunneling; on occasion the figure rose as high as fifteen hundred.[16]

One of the new men who signed on to work in the tunnels in the fall of 1931 was Marion Allen, of Jackson, Wyoming.[17] He had come to southern Nevada at the urging of his father, who was an acquaintance of Frank Crowe and was already working on the dam. The elder Allen interceded with the superintendent on his son's behalf, and Crowe, who treated reliable men who had worked with him before as part of his extended family, was happy to give the son a job in the tunnels. Marion Allen was surprised and relieved by the swiftness with which he was hired—he had arrived in Las Vegas with only $1.50 in his pocket and found that thousands of men were registered at the employment office—but as a former carpenter and concrete finisher he had to wonder what hazards awaited him inside the gaping maws of the diversion tunnels.

Miners drilling in one of the twelve-by-twelve-foot construction adits, September, 1931. The summer heat was suffocating in these cramped, stifling shafts. (Kaiser Collection, Bancroft Library)

Allen would learn, as did all the other men who spent half a year toiling inside the walls of Black Canyon, that tunneling was dirty, dangerous, back-breaking work. The hard-rock miners who bore the brunt of this labor were a special breed; many were Irish, with kelly-green surnames like Regan, Ryan, Malan, and McCabe, and rough-and-ready nicknames like Mickey, Boxcar, Hardway, and the ubiquitous Red. They came from the copper mines of Butte and Bisbee, the silver mines of Leadville and Coeur d'Alene, the zinc mines of Missouri's Joplin district, and the gold mines of Nevada and California. They were rowdy, profane, and absolutely relentless in their assault on the rock faces, determined to outdrill and, in their off hours, outdrink the man next to them. There was single-minded, sometimes violent competition, not only between individual miners but between crews working in different tunnels and among the three shifts as a whole, to see who could drill the fastest, fire the most blasting rounds, and advance the farthest in an eight-hour period. This combative, competitive spirit was promoted vigorously by the tunnel superintendents, or walkers, as they were called, men like Floyd Huntington, Leigh Cairns, Tom Regan, Al Wentz, Jack Lamey, C. T. Hargroves, Paul Guinn, and Red McCabe, who, Marion Allen was told, was "so tough he could bite a nail in two and would fire a man for even looking like he was going to slow down."

Tunnel Superintendent Red McCabe (left), who would "fire a man for even look-ing like he was going to slow down"; Assistant Construction Superintendent Woody Williams (center), known as the Bull of the River, and Tunnel Superinten-dent Pete Hansen (right) pose in front of a diversion tunnel portal in October, 1931. (Kaiser Collection, Bancroft Library)

A typical shift in the tunnels began with the crew arriving at the por-tals in transport trucks from Boulder City. The men changed out of their street clothes in the drying rooms and, at the walker's command, spilled into the tunnel and ran to the heading, passing the men of the preceding shift who were on their way out. Above head level, along the jagged sides of the excavated bench section, were strings of electric lights, their bluish-white glow reflecting off puddles of water on the tunnel floor and casting eerie shadows on the vaulted ceiling high above. Despite the discharge from blowers and fans at the portals and the adit mouths and the natural drafts caused by the differences in temperature between the warm rock and the outside air, the atmosphere was thick with the odor of wet slag, the exhaust fumes of muck-hauling trucks, and the acrid stench of exploded dynamite.[18]

The steady rumble that was barely audible in the drying rooms be-came a full-throated, tooth-rattling roar as the workers approached the tunnel heading and the face of the bench. There, in the blinding white glare cast by its banks of 1,500-watt floodlights, stood a hulking jumbo, shiver-ing like a thing alive. The miners clambered onto the rig and manned the drills. They were followed by the chuck tenders, who checked the supply of steel on the racks and then took up position, standing or kneeling in front of the pipe mountings beside the drill chucks. The din increased to almost unbearable levels as the steel shafts bored into the bench, and the whole tunnel quaked with the intense, grinding vibration that radiated from the jumbo. Talk was impossible in the tumult of drilling, even between two people standing next to each other, so the nippers, whose job it was to bring up fresh steel, powder, drinking water, or whatever else might be needed on the jumbo, stayed back behind the big rig and watched for hand signals from the men on board.

Because the blasting produced a rough, sloping rock face, the miners at different levels on the jumbo used different lengths of drill steel to start their holes. A miner at the lowest level, near the tunnel floor, was likely to use a short length while a miner on the highest jumbo platform, near the top of the bench, might be working with a ten-foot or sometimes even a twelve-foot piece of steel. Accurately manipulating such long lengths of drill steel was very difficult—something akin to turning a small screw with a three-foot-long screwdriver blade while keeping both hands on the butt—but the miners quickly became adept at it. As a miner finished with a particular length of drill steel, the chuck tender would insert the next length into the drive rod and clamp it into the chuck. Occasionally a miner or a chuck tender would turn to the waiting nippers and hold up several fingers, indicating he wanted a certain length of steel. Four fingers, for example, meant an eight-foot length, five fingers a ten-foot length, and so on. One of the nippers would then run to the jumbo, scramble up to the drill-steel rack, and deliver the desired piece of steel. A nipper who did not jump quickly when given a hand signal, thereby slowing down a miner and costing him precious minutes in the race to see who could finish his quota of holes fastest, could expect to have his ears blistered when the shift was over.[19]

When drilling on the first half of the bench had been completed, a truck loaded with blasting materials backed into the tunnel. The nippers unloaded the truck and passed boxes full of 40 percent gelatin dynamite up onto the jumbo, where the miners began loading the drill holes with dynamite cartridges, tamping them into place with powder sticks. The primers, which would fire the charge, were tamped in last. Before the primers were inserted, however, the lights on the jumbo had to be turned off and a squad of electricians called in to set up portable floods; if one of the electrically ignited primers came into contact with a live wire on the jumbo, the resulting explosion would wipe out most of the tunnel crew. The leg wires from the primers were attached very carefully to the bus bar, the conductor through which the electric detonating charge would flow; a loose connection could lead to a hangfire, forcing a miner to reenter the tunnel and tinker with the unexploded charge, a nightmarish task that was to be avoided if at all possible.[20]

➤ Once the blasting round on the first half of the bench was set, the

A drilling jumbo backed up to the sloping bench face of a tunnel heading. The two miners in the foreground are manipulating long lengths of drill steel as their chuck tenders look on. In the upper right-hand corner, a nipper is removing a fresh piece of drill steel from a rack, while in the lower left-hand corner the shifter keeps an eye on the work of the miners drilling at floor level. (Kaiser Collection, Bancroft Library)

Smoke belches from the tunnel outlets as a blasting round is fired, July, 1931.
(Bureau of Reclamation)

jumbo was driven forward and then backed up flush against the second
portion of the face. The nippers hastily blocked it into position with
12-by-12-inch timbers and wedges, the air and water lines were recon-
nected, and drilling resumed. Records kept at the beginning of the tunnel
enlargement in September, 1931, showed that the average drilling time for
each blasting round, including jumbo setup and dynamite loading, was
approximately four and a half hours.[21] This time dropped, however, as
the crews became more accustomed to their tasks and more facile with the
equipment. The speedy drilling was also made possible by the rock of the
canyon walls, andesite tuff breccia, which turned out to be almost ideal for
tunneling. The more experienced miners said it was "dead," meaning that
it was not under pressure, could be drilled easily, and broke true when
blasted. No major faults or open seams were discovered; heavy ground—
dangerous loose or hollow spots in the tunnel roof—was not encountered;
and spalling did not develop, meaning that expensive and time-consuming
timbering was unnecessary.[22] The crews could concentrate on repeating the

cycle of drill, blast, and muck as rapidly and efficiently as possible without fear that the tunnels they were driving would come crashing down on them.

When a bench face had been drilled and stuffed with dynamite, the jumbo withdrew, the crew left the tunnel, the safety switches for the 444-volt blasting circuit were unlocked, and a blaster fired the shot. At the moment of detonation the ground shivered; seconds later, the muffled thunder reverberated through the tunnel and a thick cloud of smoke and dust belched out of the portal. Safety miners, veterans whose senses were so well honed that the other crew members swore they could smell loose rock, were the first men back into the tunnel to inspect the blast site, usually within a few minutes of the explosion. If they gave the all-clear signal, the Caterpillar tractor and electric shovel moved forward to begin mucking; if not, a group of scalers was sent in with ladders and pry bars to knock down the loose rock. These men were picked for their ability to think fast and move even faster because, as an old miner told Marion Allen, there were only two kinds of scalers: "the quick and the dead."

After the safety miners had determined that the tunnel was safe to re-enter, a one-hundred-ton Marion Type 490 electric shovel clanked down the shaft to the dust-shrouded muck pile. The eight electric shovels used for mucking had been modified, according to Six Companies specifications, for the tunnel work; their booms had been shortened so they could swing without striking the tunnel sides, and their standard 2 1/4-cubic-yard dippers had been replaced with 3 1/2-cubic-yard dippers, reducing the time it took to remove the muck piles, which averaged 1,000 cubic yards each, from 14 hours to an average of 9. During the course of tunneling, this seemingly innocuous equipment alteration saved approximately 974 hours of mucking time per shovel, or 40 days.[23]

The loose rock was concentrated by the Caterpillar tractor, scooped up by the electric shovel, and deposited in the dump trucks, which then roared out of the tunnel and up steep access roads to disposal areas in side canyons several miles from the tunnel portals. To visitors touring the construction site, one of the most memorable—and terrifying—sights was the procession of empty muck-hauling trucks racing backward down the winding, precipitous grades into Black Canyon, the drivers standing up in the open cabs with one foot planted firmly on the accelerator and the other on the running board, craning over their shoulders to see where they were going. The time-saving expedient of backing the trucks into the canyon, which eliminated the need for turnaround areas at the disposal sites and tunnel portals, was the idea of subcontractor Carl Bryant, who had a deal with Six Companies to haul muck from the canyon at fifteen cents a cubic yard. Negotiating canyon roads in reverse did save time, but it required drivers with unusual skill and nerves of steel. Even more skilled were the electric-shovel operators; these men were kings of the road in terms of their wages, which at ten dollars a day were the highest on the dam.[24]

At the heading of one of the diversion tunnels a one-hundred-ton Marion Type 490 electric shovel deposits a load of muck into a dump truck. (Kaiser Collection, Bancroft Library)

Work in the diversion tunnels was fraught with hazards, not the least of which was getting lost, as Marion Allen discovered not long after he started work at the dam. The four tunnels were numbered consecutively, with the Nevada tunnel farther from the river labeled No. 1, the Nevada tunnel next to the river No. 2, the Arizona tunnel next to the river No. 3, and the Arizona tunnel farther from the river No. 4. There was only one problem with this straight-forward identification system: the portals, which were almost identical in appearance, were not well marked and easily could be mistaken for one another, especially in the dark. Allen's misadventure occurred after he had been on the job about a week, just long enough, he wrote in his memoir *Hoover Dam*, to feel that he was an "old, seasoned hand." It was an inky-black night; a stiff northerly wind had swooped into the canyon and was whistling across the tunnel portals, and the murky surface of the Colorado and the looming face of the Arizona cliff exuded a damp, icy chill. Another swing shift was almost at an end and nipper Allen was sitting in the relative warmth of one of the changing rooms, completing the time slips for the rest of the crew and thinking

Part of the truck fleet assembled at Hoover Dam. (Bureau of Reclamation)

about the soft bed that was waiting for him in Boulder City. His reverie was interrupted when the swing-shift walker, Red McCabe, appeared and asked him to carry a message to the shifter in charge of the drilling jumbo in tunnel No. 3.

It would be quitting time in a few minutes, so Allen took off on the run, heading into what he thought was the right tunnel. He was well inside when he realized that something was very wrong: he did not hear the roar of the jumbo or the clatter of the drills, just the sound of his own steps echoing off the tunnel walls. In a flash he comprehended what the deathly silence meant, turned, and sprinted back toward the portal, but it was too late. Like a pair of giant hands clapping over his ears, the thunder of an explosion enveloped him. He was hurled to the floor by the shock wave and dust and rock fragments peppered his back. Stunned, he lay face down for several minutes, then staggered to his feet and limped out of the tunnel, ignoring the incredulous stares of miners coming in to check the results of the blast. A man was supposed to be posted at the tunnel portal when a shot was about to be fired, but he had either missed the running nipper in the dark or had never taken up his assigned position.

Allen finally delivered his message to the shifter, who was crossing the catwalk over the Colorado on his way to the waiting transports. He then rode home and collapsed in his bed, hoping that sleep would quiet the ringing in his ears and ease the pounding in his skull. The next day his head still felt "big as a barrel," but he reported to work anyway because he knew that missing a shift for any injury less serious than loss of a limb or for an illness that was not life threatening was tantamount to signing his own firing slip. Fortunately for Allen, the fearsome Red McCabe proved to have a soft spot. He took a close look at the bruised and obviously suffering nipper, rejected his explanation that it was "just a cold," and sent him back to Boulder City with orders not to return until he was better. The swelling went down and the ringing subsided after two days of bed rest and Allen went back to work, thinking himself no worse for wear but determined to be more cautious when he entered a tunnel in the dark. Not until ten years later, during a physical examination for another job, did he learn that both of his eardrums had been ruptured.

Danger was ever present in the diversion tunnels; it hung overhead in jagged rock outcrops, it ran underfoot in snake-like air and electric lines, and it hurtled to and fro in exhaust-spewing trucks loaded with shattered muck and crates of dynamite. Speed, not safety, was the overriding concern of the tunnelers. For the Six Companies directors it meant greater profits; for the Six Companies engineers it meant beating the contract deadline and diverting the river during its low-water cycle in 1932; for the crews sweating underground it meant meeting the demands of the walkers and outdrilling the miners in the other tunnels. But speed also meant taking risks, ignoring state mining safety laws, and, tragically, sacrificing the health, and in some cases the lives, of tunnel workers.

The miners had no choice but to accept the contractors' quest for speed and the unsafe conditions it created. Following the aborted strike of August, 1931, Six Companies, in concert with the federal officials running the project reservation, cracked down hard on labor organizers. Men who protested openly about dangerous job practices or the hazardous work environment were summarily fired and replaced by laborers drawn from the huge pool of unemployed men loitering in Las Vegas.[25] Most of the tunnel workers were not disposed to complain anyway. Hard-rock mining had always been dangerous work, and safety laws were a relatively recent innovation. If conditions in the diversion tunnels were not good, at least they were no worse, the veteran miners could rationalize, and perhaps were even marginally better, than those that had prevailed in the dank drifts and dust-choked stopes where they had learned their trade.

Miner stoicism and company intimidation could not stifle all complaints, however. The cavalier way in which dynamite was handled came in for special criticism. The practice of stacking powder boxes in open areas where flying debris could hit them and where glaring sunlight caused the explosive nitroglycerin to melt and ooze was condemned, as was the standard operating procedure of loading drill holes with powder at the same time jackhammers were operating a few feet away.[26] "I've been a miner all my life and I've never seen conditions so bad," an anonymous worker told a newspaper reporter in 1932. "Chances of a big explosion are taken daily. Blasting has been carried out within 100 feet of where powder and fuses were stored together. On orders of my foremen, I've carried armloads of mixed powder and fuses into tunnels. If I had stumbled—goodbye tunnel and about 80 men."[27]

Another peril of the tunnels was the deadly combination of high-voltage electric lines and water. Although seepage from the rock was almost nonexistent, water from other sources did collect in puddles on the floor, particularly near the portals. Electric cables for the lights in the tunnel and for the blasting circuit were strung along the tunnel sides and floor. The cables were insulated with rubber, but the constant dragging across rough rock cracked and wore away this protective jacket; during the course of the tunneling operations several workers were electrocuted when they touched one of the exposed cables or when a charge was conducted into a puddle they happened to be standing in.[28]

Not all tunnel accidents ended in injury or death. In fact, some of them, although terrifying when they occurred, proved in retrospect to be quite humorous. One such episode involved miner Lee Ryan and a rock fall in tunnel No. 2, on the Nevada side of the river. A shot had just been fired and tunnel superintendent Floyd Huntington, living up to his nickname, Haywire, was acting as the crew's safety miner. Accompanied by Ryan and a cat skinner on a small Caterpillar tractor, he went into the tunnel to inspect the blast results and to look for loose rock. The trio was admiring the muck pile when a little stream of dust and pebbles sifted down from the ceiling and a grating, splitting sound reverberated ominously through the tunnel. All three men ran for their lives, and Huntington yelled for Ryan to warn the operator of the electric shovel that was coming down the bore toward the heading to turn back. The words were barely out of his mouth when a portion of the ceiling overhanging the blast face gave way and came crashing down, filling the tunnel with a blinding cloud of dust. When Huntington and the cat skinner picked themselves up, the parked Caterpillar tractor had been buried and there was no sign of Ryan. The two men assumed that he, too, had been buried and started digging frantically into the rubble with their bare hands. Other crew members soon joined them, using shovels and pry bars in the desperate search for the buried man.

Meanwhile, Ryan, who had kept on running when Huntington and the cat skinner hit the floor, had warned the electric shovel operator as ordered, helped himself to a drink of water, and reentered the tunnel with the other men to attack the rock heap. After a few minutes of strenuous digging, he found himself working next to Huntington. "Who's buried?"

he asked the sweating, grim-faced superintendent. Huntington stopped, stared, threw down his shovel, put his hands on his hips, and bellowed in a voice that echoed through the tunnel almost as loudly as the booming cave-in had, "You, you son of a bitch!"[29]

The risks associated with dynamite, high-voltage lines, and cave-ins were obvious, but the most insidious safety hazard in the diversion tunnels, and the one with the greatest potential for long-term harm to the tunnel crews, was carbon monoxide. The noxious gas from the exhausts of the muck-hauling trucks and Caterpillar tractors mixed with the haze of dust and powder smoke that hovered in the shafts and slowly poisoned the men who breathed it day in and day out. Nevada mining safety laws specifically forbade the use of gasoline engines underground, and on November 7, 1931, the state inspector of mines, A. J. Stinson, ordered Six Companies to stop using gasoline-powered trucks to haul muck out of the tunnels. Six Companies responded by filing suit against Stinson to enjoin him from interfering with company tunneling operations, and continued to use its fleet of gasoline-engine trucks in the tunnels under the protection of a temporary restraining order.[30]

The case was postponed for six months while tunneling went on full speed in Black Canyon. On April 28, 1932, Six Companies' motion for an injunction barring the Nevada mine inspector from interrupting its operations was finally heard by a panel of federal judges sitting in San Francisco. The Six Companies lawyers, aided by government lawyers from the Bureau of Reclamation and the U.S. attorney's office, argued that Nevada did not have jurisdiction to enforce its laws on a federal reservation and that if the law prohibiting the use of gasoline-powered trucks was enforced, Six Companies would be deprived of its property without due process. The contractors claimed that their investment in gasoline-engine trucks totaled $300,000, and that the cost of converting them to electric power, and the layoffs and work delays that would be caused by conversion, would cost $1.5 million. They also maintained that the health of the tunnel workers was not endangered by carbon monoxide because natural air currents from the adits and tunnel portals, aided by blowers and pressure fans, provided adequate ventilation.[31] The Nevada attorneys responded that the cost of converting the muck trucks' power source to electric motors would be only $150,000, that carbon monoxide exhaust from gasoline combustion did indeed pose a health hazard in the tunnels, and that use of gasoline increased the risk of fire and explosions.

The judges, arguing that the safety issue was moot because the diversion tunnels were nearly finished, ruled for Six Companies and issued a temporary injunction against the state mine inspector. Ten months later the same panel again found for the contractor and made the injunction permanent on grounds that Nevada's prohibition against the burning of gasoline in underground excavations was limited to mining operations and therefore was inapplicable to work of the type being done in Black Canyon.[32] The issue of state versus federal jurisdiction in the project reservation was, for the time being, left unresolved.

Six Companies had been given a free hand to dictate whatever working conditions it saw fit in the Hoover Dam diversion tunnels. "Our directors would rather take a loss of $100,000 than to hurt one man," Frank Crowe said in a speech to workers assembled for the 1931 Thanksgiving dinner,[33] and Norman Gallison, Six Companies' public-relations man at the dam site, claimed in print that testing in the diversion tunnels showed carbon monoxide pollution there to be no worse than that found in traffic tunnels, such as New York's Holland Tunnel.[34] But notwithstanding these reassurances, the men who worked in the Hoover Dam tunnels were certain that carbon monoxide affected their health. "[T]he fleet of gasoline-propelled muck trucks . . . managed to blow out enough carbon monoxide gas to make life in the tunnels rather uncomfortable," wrote engineer John Meursinge.[35] Another man described symptoms of carbon monoxide poisoning and concluded: "Having worked underground before, I knew that the powder smoke could not be responsible for the beads of cold sweat which broke out on my forehead or the unexplained weakness in my knees. . . . In the six shifts I lasted there were dozens of men carried out or staggered to the portals deathly sick from the gas."[36]

The presence in the Six Companies hospital of many tunnel workers suffering from severe, debilitating respiratory problems appeared to lend support to the contention that carbon monoxide poisoning was occurring. Pneumonia was the catchall diagnosis for these cases,[37] and the apparent coverup by company doctors of the symptoms' actual cause fostered considerable bitterness. "The hospital had always been a topic of the workmen's conversation," John Meursinge wrote. "'Up there a fellow always dies of pneumonia, never of anything else.' That was a standard joke all over the job."[38]

With the arrival of the new year, tunneling activity reached its peak; January, 1932, was a record month, both for the number of men employed and the footage driven. As much as 16,000 cubic yards of rock was hauled away every day by the truck fleet, and on January 26 a record combined advance of 256 linear feet in twenty-four hours was recorded. The winner of the hotly contested individual record was the C. T. Hargroves crew, which advanced its heading at the upper portal of tunnel No. 2 by 46 feet.[39] Just before midnight on January 29, the crew working the graveyard shift in tunnel No. 3, which at 3,560 feet was the shortest of the four tunnels, triggered a shot. When the muckers moved in to begin cleaning up, they saw light streaming over the heap of broken rock from the downstream direction and felt a strong draft blowing in their faces. Whoops and whistles resounded up and down the length of the dusty shaft and an impromptu celebration began: the first tunnel had been holed out.[40] Just four days later, tunnel No. 2 was holed out, and in both No. 3 and No. 2 the final enlargement—excavation of the invert section and trimming and scaling of projecting rock on the walls and ceiling—began immediately.

The driving of the diversion tunnels had been the focus of attention since the fall of 1931, but during the long months of drilling, blasting, and mucking, other important activities had been proceeding. The thirty-mile railroad system linking Boulder City, the gravel pits in Hemenway Wash that would provide aggregate for mixing concrete, the gravel-screening plant, the sites of the high-mix concrete plant on the canyon rim and the low-mix plant at river level, and the dam site itself was finished. Laying track from the foot of Hemenway Wash, around the side of Cape Horn, and along the Nevada side of the canyon down to the dam site at river level was an especially noteworthy accomplishment involving the drilling and blasting of more than one thousand yards of tunnel and the dumping of thousands of tons of muck to build up a right-of-way along the vertical canyon wall.[41] Construction of the two huge mixing plants had also begun, and blasting was started at the site of the Nevada spillway.

The most spectacular job, however, belonged to the men known as high scalers, whose task it was to strip the canyon walls of all debris. Wind, water, and extreme temperatures had split and crumbled the volcanic rock of Black Canyon. The cliffs were scored with crevices and pocked with pits and blisters where slabs of rock had broken loose and were about to fall. Above the cliffs, the steeply angled slopes rising to the canyon rims were studded with jutting outcrops of fractured rock and littered with boulders and unstable scree slides. All this loose rock would pose a serious threat to the thousands of men excavating the dam's foundation in the exposed riverbed after the Colorado had been diverted. A pebble dropping a thousand feet could split a man's skull like a cleaver going through a ripe melon, and one of the bigger, automobile-size boulders falling on a work crew would have the same deadly effect as an exploding fragmentation bomb. Besides threatening worker safety, fissures and scales in the rock where the concrete of the dam met the sides of the canyon would undermine the stability of the entire structure and permit potentially disastrous water seepage around the arch. The tons of cracked and broken rock had to be removed before any concrete could be poured.

High scaling was not work for the weak, the clumsy, or the faint of heart. Heavy ropes were anchored to the canyon rim and cast out over the precipice. Bosun's chairs—rectangular boards two feet long and a foot wide—were rigged to these main, or lead, ropes with a chair rope about three feet long. The rigger's knot, or stopper hitch, connecting the chair rope to the main rope could be opened or closed by the man in the bosun's chair, permitting him to rappel down the cliff face to his work station. The return trip to the rim was more strenuous: a grinding hand-over-hand climb up several hundred feet of cliff. It was reported that many of the high scalers employed at Hoover Dam were former sailors or circus performers;[42] sailor, acrobat, or otherwise, they had to be agile and unafraid of dangling at the end of a slender line with nothing but air between them and the canyon floor far below. They also had to be strong, for they went over the side weighted down with wrenches, crowbars, water bags, and other paraphernalia. Once they were in position, the 44-pound jackhammers and bundles of drill steel were lowered to them; they tightened the stopper hitches, attached their safety belts, planted their feet against the wall, leaned backward out over the abyss with only the narrow plank of the bosun's chair for support, and began drilling.

When Red Wixon, a Frank Crowe protégé from Boise, Idaho, arrived at the dam site in September, 1931, the electric shovel he was to run had not been delivered, and Crowe told him to join the high scalers until his machine was available. Wixon had his doubts about this kind of work but knew better than to argue about job assignments with the superintendent of construction. He wrestled with one of the heavy jackhammers on the cliff face for a week or so until his shovel showed up. The day he transferred from the bosun's chair to the shovel cab was one of the happiest in his construction career. He later recalled that the drill's relentless pounding left his neck and shoulders feeling as if they had been beaten with a baseball bat and that his hands were swollen and covered with red, raw blisters.[43]

Scaling on the cliffs was not only one of the most physically demanding jobs on the dam, it was quite probably the most dangerous. Ropes, electric lines, live air hoses, and dangling bundles of drill steel cluttered the canyon face, forming a treacherous maze that the workers had to pick their way through. A slight miscalculation, a momentary slip, failure to hear a warning from above, momentary dizziness from heat or exhaustion, a frayed rope—all could lead to a fatal plunge. Falling objects were so dangerous that some of the scalers manufactured their own safety helmets by dipping cloth hats into tar and letting the tar harden into a tough shell. The effectiveness of this homemade headgear was demonstrated when several workers were struck hard enough for their jaws to be broken but escaped skull fracture because of the protection provided by their helmets. The contractors were sufficiently impressed to order thousands of factory-manufactured "hard-boiled hats" and to suggest strongly that men working in exposed areas wear them, making the Boulder Canyon Project one of the first hard-hat jobs in American construction history.[44]

Even with the men wearing hard hats, falling debris continued to inflict casualties. Most exasperating to Frank Crowe and his foremen was the fact that much of the deadly shower raining down on the workers from the canyon rim was composed not of rock fragments, as would be expected, but of tools dropped by careless hands. For a number of reasons, many of which had nothing to do with work, the men, no matter what

High scalers drilling on the Nevada wall of Black Canyon, April, 1933. (Bureau of Reclamation)

their particular jobs, were armed to the teeth with hand tools. The right to bear tools, regardless of need, was sacrosanct to the dam workers, and it was pointless to try to limit access to them. Crowe recognized this when at one point during the project he placed an order for an entire freight-car load of crescent wrenches. The purchasing agent who handled the order was certain that a mistake had been made: a freight-car load of wrenches would amount to a hundred gross, or 144,000 wrenches. He held up the order until he had a chance to ask Crowe about it at a meeting. Crowe told him the order was not a mistake and in response to questioning by the officials present reportedly said: "Gentlemen, I want a freight car load of crescent wrenches, for until every man on the job gets one under his bed, one in his automobile, and one in his hip pocket, I can't get any work done!" [45] And so the wrenches and other hand tools kept arriving on the rim of Black Canyon and, in defiance of repeated exhortations about job safety, kept being dropped into the gorge.

In spite of the risk, or perhaps because of it, many men were attracted to high scaling, and these individuals often performed stunts to entertain their comrades on the cliff and anyone else who might be watching in the canyon below. When the foreman wasn't looking or was busy in another part of the work area, a daring young man on a Hoover Dam trapeze would grasp the lead rope, plant his boots, and give a mighty shove, propelling himself out from the wall. There was an ongoing competition to determine who could swing the farthest from the wall, the highest over the canyon, execute the most spectacular in-air exhibition with hands and feet, and so on.

One scaler who gained a measure of fame for his exploits was 23-year-old Louis ("The Human Pendulum") Fagan of Jonesboro, Arkansas. Dangling at the end of a two-hundred-foot rope, Fagan transferred shifter Guy ("Bulldog") Bray's crew, one by one, around a large rock projection on the Arizona cliff. "This feat is accomplished," reported the *Las Vegas Age,* "by a face-to-face position in which the man who is being transferred . . . locks his legs around the waist of Fagan. Both men get a good grip on the rope and Fagan begins his dare-deviltry by kicking off at a tangent which swings the human pendulum 100 feet out into space and around the projection of the cliff to where they make the high-line cable coming up from the lower portals. He has been performing this feat of transferring men, according to his declaration, for about three weeks." [46]

As impressive as Fagan's aerial maneuvers were, the most spectacular in-air feat was generally conceded to be the Tarzan of the Apes exploit performed by scalers Oliver Cowan and Arnold Parks. One afternoon a Bureau of Reclamation engineer, Burl R. Rutledge, was trying to inspect a portion of the cliff that had just been cleared of protruding rock when he leaned out too far, slipped, and began to roll and tumble down the precipitous slope toward the river far below. Cowan, who was working some twenty-five feet beneath the spot where Rutledge had been standing, heard the engineer's muffled exclamation, looked up, and saw him begin his fall. Without hesitation, he pushed out from the wall, propelled himself horizontally through the air, swung back in to the cliff face, and snagged the falling man's leg. Arnold Parks, who had seen the accident unfolding, swung over seconds later and pinned Rutledge's upper body to the canyon wall. The two scalers then held the stunned engineer in place until a line could be rigged to haul him back to safety.[47]

During the first half-year of scaling and tunneling, the elements had conspired to lull the Black Canyon force into a false sense of comfort and security. After the blazing months of summer, the lingering desert fall, with its warm days and crisp nights, seemed almost idyllic, and the steamy squalor of River Camp and Ragtown and the scourge of heat prostration were all but forgotten. The river, too—the shifty, treacherous, rampaging Colorado—was seemingly asleep. The nervous engineers and contractors watched and waited for some sign that the Red Bull might be rousing from its torpor, but it remained somnolent, and as summer gave way to autumn they rested easier. The swift change in the appearance of Black Canyon was also comforting. Gone was the vast, overwhelming emptiness, now filled with men and machines. Gone was the menacing blankness of the sheer canyon walls, now chopped, chiseled, and blasted full of holes. Gone was the majestic, clean geometry drawn by river, cliff, and sky, now warped and cluttered by roadways, bridges, and sagging power lines. Gone was the brooding wilderness silence, now broken by the grunts, grumbles, and growls of industrial activity.

Looking down on this anthill of activity from the rim high above, the casual observer could be forgiven for concluding that man had deposed the river as undisputed master of Black Canyon. The English writer, J. B. Priestly came, saw, and was thrilled and moved. "It is like the beginning of a new world," he rhapsodized about the dam site, "a world of giant machines and titanic communal enterprises. . . . When you look down . . . and you see the men who have made it all moving far below like ants or swinging perilously in midair as they were little spiders, and you note the majestic order and rhythm of the work, you are visited by emotions that are hard to describe, if only because some of them are as new as the great Dam itself."[48] But the Colorado was also possessed of a "majestic order and rhythm," and when its challenge to the dam builders came, it was swift and stunning.

Somewhere in the valleys north and east of Black Canyon, a line of thunderstorms rolled down from the mountains and slashed the desert floor with dark, heavy sheets of rain. Newborn streams, boiling with red

End of the day shift, March 18, 1932. The workers are boarding transport trucks for the trip to Boulder City. (Kaiser Collection, Bancroft Library)

mud, raced downhill, tumbling boulders aside and uprooting mesquite trees as they rushed madly toward low ground and a rendezvous with the Colorado. The big river shouldered the flash flood, carried it downstream, and hurled it without warning into Black Canyon on the night of September 26, 1931.

The men who were working outside the tunnels that night heard the faint, breathy rushing of the flood's crest rolling toward them through the darkness. The sound broadened and deepened into a powerful liquid rumbling, and they saw the undulating white disks of reflected electric light on the river's smooth surface shatter into a thousand glistening points. The inky water began to heave and froth violently. Up the banks it came, rising with a speed the gaping men could scarcely believe, climbing an inch a minute. Within hours waves were lapping at the base of the blacksmith shop underneath the mouth of the Arizona adit and spilling onto the floor of the blower house that pumped fresh air into the narrow shaft. The ledge these buildings sat on had been twelve feet above river the day before; now the little cluster of structures appeared to be a flotilla of tin boats plowing through a chocolate-brown sea. For a time it seemed the water would rise high enough to spill into the diversion-tunnel portals, and dikes of sand-bags and broken rock were hastily thrown up. Then the flood subsided as rapidly as it had come, leaving the frantic workers both frightened and re-lieved; surely, they thought, there would be no more high water like this until next summer.[49]

They were wrong. During the first week of February, 1932, a warm, steady rain fell in the Mormon Mountains in southeastern Utah. It melted the snow that had frosted the tops of the peaks and propelled a powerful surge of water into the Virgin River. Swollen by this heavy runoff, the Vir-gin spilled into the Colorado just thirty miles upstream from the dam site. The flood barreled into Black Canyon at 3:00 P.M. on February 9. The swing-shift crews arriving at work were ordered to get all the equipment and vehicles out of the tunnels and onto higher ground, then pitch in to strengthen the dikes erected in front of the upstream portals in September.[50]

Hour after hour, the water level rose. The trestle bridge used by the muck-hauling trucks to reach the Arizona side of the river was swallowed up, and the machine shops were inundated again. By 3:00 A.M., twelve hours after the flood's arrival, the river had risen eleven feet and was chew-ing at the face of the protective dikes the workers were struggling to re-inforce. Just when it appeared that disaster could not be averted, the water suddenly retreated as before, leaving in its wake a thick coating of brown slime.

Cleanup started during the day shift on February 10 and excavation of the tunnels was resumed, but the battle with the river was not over. That afternoon Frank Crowe received word from a gauging station well up-stream that another flood was on its way toward Black Canyon. The same storm system that had engorged the Virgin had rolled on to the east, dumping heavy rains on the barren plateaus of northeastern Arizona and turning the Little Colorado into a seething torrent. If the report from the gauging station was accurate, this fresh onslaught would in all likelihood dwarf the previous two floods. Again the workers were instructed to clear the tunnels and man the dikes, but the damage caused by the preceding day's inundation hindered their preparations, particularly on the Arizona side. The washout of the trestle bridge had left the narrow wire-and-plank catwalk the only open route across the river, and although 150 workers were sent over to shore up the barricades in front of the upstream portals of tunnels No. 3 and No. 4, there was little they could do but patch and pray.

The third flood struck on February 12, a wall of water so angry and powerful that Frank Crowe feared it "would wash us right out of the canyon."[51] The foaming crest rose with stunning and terrifying speed—seventeen feet in three hours—melted the levees on the Arizona bank like sugar, and poured into the tunnels. Higher and higher the water rose until

The flash flood of February 12, 1932, spilled into the diversion tunnels. (Kaiser Collection, Bancroft Library)

its flow peaked at fifty-seven thousand cubic feet per second, the equivalent, one observer calculated, of a "tank of water a city block square and thirty-five feet deep being emptied into the canyon every sixty seconds." [52]

The flood receded the next day, and the workers returned to assess the destruction and curse the rampaging Colorado. Damage to equipment and installations proved to be minimal, but the job of cleaning up was Augean in scope: diversion tunnels No. 3 and No. 4 were awash with two to three feet of gelatinous, foul-smelling mud thick enough to clog sump pumps instantly but still thin enough to make shoveling a frustrating and agonizingly slow exercise akin to drinking coffee with a fork. Outside the tunnels the blacksmith shops and air-compressor plants, the pipes, hoses, and tunneling equipment—everything the water had touched—was covered with a yellow-brown coating of silt that had to be laboriously wiped, scraped, or washed off. It took almost a week to remove the sludge and debris and refurbish and repair equipment, and even then the cleanup was not complete; much of the mud on the tunnel floors was simply left to harden and was blasted away later when the invert section was excavated.

While drilling and dynamiting of the bench sections in tunnels No. 1 and No. 4 resumed, work began in No. 2 and No. 3 on removing the invert and trimming the jagged spurs of projecting rock left on the ceiling and walls by the earlier blasting rounds. The jumbos had been so successful in speeding up the drilling of the bench that two modified versions were built,

The concrete lining in the diversion tunnels was three feet thick, reducing their diameter to fifty feet. (Bureau of Reclamation)

one a low carriage with two wings that folded out for drilling down into the invert, the other a circular steel framework fifty feet in diameter with drilling platforms attached all around so that the drills could reach any outcrop that needed to be trimmed. These rigs worked in tandem, the trimming jumbo operating a short distance ahead of the invert jumbo; blasting was synchronized to save time.[53]

Removal of the invert and trimming proceeded rapidly, and on March 16, 1932, concrete pouring began in tunnel No. 3. The specifications required all four diversion tunnels to be lined with concrete three feet thick, reducing their diameter from fifty-six feet to fifty feet but ensuring their stability and improving their water-carrying capacity. The concrete, approximately 400,000 cubic yards of it for the 3.1 miles of tunnel, was delivered by trucks that roared back and forth from the upstream portals to the low-mix plant on the Nevada side of the river. It was mechanically troweled into place by huge, movable forms. The pouring in tunnels No. 3 and No. 4 was pushed with special urgency for it was through these two bores that the river was to be diverted first and they had to be lined and ready when the Colorado's water level began to drop in the fall.

The lining operation was closely and personally monitored by Frank Crowe, who, it seemed to the workmen, never slept and never stopped worrying about the job. One night during the graveyard shift, Marion Allen, who was now working as a concrete finisher, was debating with the foreman whether enough concrete had been poured into a form to complete a section of invert. It was 3:00 A.M., a time, remembered Allen, "when it seems that a human being is at his lowest ebb, both physically and mentally." He and the foreman were taking too much time to decide how to proceed. Suddenly a sharp, alert, commanding voice rang through the tunnel behind them: "Who is holding up this pour?" Allen and the foreman jumped with surprise, the dump man inadvertently emptied his concrete bucket into the form, and the entire crew turned to see the tall, stooped figure of the superintendent of construction stalking off down the tunnel.[54]

The gray-white concrete tubes lengthened rapidly, and the Six Companies engineering staff was encouraged that the deadline for diverting the river would be met. They were also heartened by cooperation from an unexpected source: the Colorado. After ravaging the dam site with rare winter floods, the river had remained unusually placid during the late spring and early summer months, when it was normally most rambunctious. Winter snowfall in the Rocky Mountains had been relatively light, reducing runoff and causing the water level to drop faster than in a typical year. This earlier-than-usual end to the flood season, combined with speedy progress in lining the Arizona tunnels, permitted scheduling of the initial diversion attempt for early November.

During the middle of October, preparations were made for the opening joust with the Colorado. Trucks rumbled down to the Nevada edge of the river just downstream from the upper portals and began dumping dirt and broken rock into the water. Caterpillar tractors shoved the muddy fill into place, forming a long dike that jutted out into the center of the canyon and pushed the river's flow into the curving Arizona side of the channel. The dike was extended downstream five hundred feet and then was routed back to the Nevada bank so that it had the shape of a capital C, enclosing the Nevada half of the site where the upper cofferdam would be erected and sealing the downstream work areas off from the river.

As events in the isolated world of Black Canyon moved toward the climactic moment in the eighteen-month struggle to divert the Colorado, the rest of the nation was poised on the brink of a historic political realignment. On November 4, Herbert Hoover was swept out of office and Franklin D. Roosevelt became the thirty-second president of the United States. Soon after the election Hoover announced that on the way back to Washington from his home in Palo Alto, California, he would visit the site of the dam that carried his name. His motives for making the visit were not clear. Perhaps he hoped that touring a giant public-works project that had

A load of muck is dumped into the Colorado River on November 13, 1932. The rapids in the foreground show that the diversion dam is about to break the surface. (Kaiser Collection, Bancroft Library)

been started during his administration might somehow counteract the public perception of him as a man oblivious to the plight of the nation's unemployed. Perhaps as an engineer he wished to satisfy his personal curiosity by inspecting the progress on the dam that in earlier and happier times he had done so much to make possible. In any event, his special train pulled into the nearly deserted Boulder City yard at 8:30 on the night of November 12, 1932. The president, Mrs. Hoover, and their small entourage were greeted by Interior Secretary Wilbur, Walker Young and Raymond Walter of the Bureau of Reclamation, and E. O. Wattis representing Six Companies. They were taken by Frank Crowe on a 2½-hour tour of the dam site. After the tour the thin, wan president made a short speech, reboarded his train, and departed.[55]

The next morning a hundred idling trucks, their dumpers loaded with broken rock, were lined up along the road that led to the bridge spanning the river just below the upstream portals of the diversion tunnels. The smooth, coffee-colored sheet of water slid quietly past the trestles, flowing at seven thousand cubic feet per second, far below its volume at flood but still powerful enough to be treacherous. Just hours before President

Hoover's visit, a barge carrying an electric shovel downstream had been upset in an eddy and the big piece of machinery swallowed up by the seemingly placid waters.

At 11:30 A.M. everything was finally ready, and the signal was given to begin the diversion operation.[56] Charges placed against the banks of the dirt cofferdam in front of tunnel No. 4 were fired, breaching the barrier and opening the way for water to enter. At the same time, the driver of the first truck in the long line of trucks stretching back up the steep access road gunned the engine, drove out onto the bridge, maneuvered to the downstream edge, dumped his load of rock into the water below, and raced back toward the rim and the muck pile, where an electric shovel waited to refill his truck. One after another the rest of the trucks roared onto the bridge, dropped their loads, and roared away, only to return again a few minutes later with another load.

The flow of traffic became smoother after each of the one hundred trucks had made its initial run across the bridge, and the pace of dumping quickened. Hours passed, day shift gave way to swing shift, but still the truck procession continued unabated, bombarding the river with a fresh avalanche of rock every fifteen seconds. A faint, iron-blue haze of exhaust hovered over the bridge, and the angry grinding of gears, the splashing of rock tumbling into the river, and the hoarse shouts of the men directing the traffic along the bridge echoed through the gorge.

The chill early dusk of November cloaked the canyon bottom in shadows of mauve and indigo. High above, the crags and jumbled rock slides of the rim were bathed in a rosy glow, a deceptively warm blush of rich color that flared briefly, then expired. Headlights were switched on in rapid-fire succession and their white beams stabbed through the deepening gloom, appearing and disappearing in a chaotic crisscross dance as the trucks bounced up and down the skein of roads running from rim to river.

At the bridge, banks of arc lights illuminated the dumping area and the surface of the river. Trucks emerged from the shadows along the Nevada canyon wall at the rate of four a minute, bursting into the brilliant oval of white light, where they sent their tons of rock careening down into the Colorado, raising great gouts of silver-colored spray. The water was rising noticeably now, straining to stay above the barrier of muck being hurled in its path. It sluiced over the crest of the crude diversion dam and rushed down the sloping downstream face. Upstream from the barrier and the trestle bridge, beyond the glare of the lights, a swirling pool had formed. The black water was inching up the rocky banks and lapping at the muddy roadway that led to the portals of the diversion tunnels, groping blindly for an opening, pressing to break free and surge on downstream.

The start of the graveyard shift went unnoticed on the bridge. Ton upon ton of rock went over the side and hours passed, but still the pace was maintained, one truck dumping every fifteen seconds. The gray light of dawn revealed that the pool upstream from the barrier was turbid and swollen, boiling just a few feet beneath the planks of the bridge and rising rapidly toward the mouth of tunnel No. 4. As seven o'clock approached, small groups of spectators began to congregate on the Nevada side of the canyon across the river from tunnels No. 3 and No. 4. They smoked cigarettes, rubbed their hands together, and stamped their feet, trying to ward off the early-morning chill. Several photographers arrived and set up their equipment.

The top of the diversion dam had broken through the river's surface. The muddy water backed up behind the barrier was moving purposefully now, up the last dirt embankment that separated it from the portal of tunnel No. 4. It was 7:30. The spectators on the bank and the bridge watched as the edge of the churning pool climbed the final barrier, reached its lip, and held there for a fraction of a second. Then a boisterous, frothing tongue of water raced down from the crest and plunged toward the dark tunnel inlet. In an instant the small embankment melted under the watery onslaught and with a long, bubbling sigh the Colorado gave up its struggle to flow free through its ancestral bed and poured into the smooth concrete channel that had been laid for it.

In the shadow of the towering canyon face, the onlookers stood silent, transfixed by what they had just seen. Then one of the workmen lifted his hat from his head and waved it over his head. "She's taking it, boys," he bellowed. "By God, she's taking it!" [57]

CHAPTER FOUR
Under the Eagle's Wing

The desert moon, white, fat, and luminous, hung low over the rocky spine of the River Mountains and flooded the saucer-shaped Eldorado Valley with pale light. At the northwestern lip of the valley, the lamps of Boulder City twinkled brightly, outlining a glimmering triangle on the sloping desert floor The triangle's apex lay on the crest of a gentle ridge where a silver-sided water tank loomed like a ghostly sentinel. The base of the triangle was outlined by a long row of identical white frame cottages, each one a tiny picket in a thin fence separating the small island of lights from the shadowy sweep of the surrounding desert.

Beyond the city's perimeter the moonlit hardpan stretched away for miles, barren save for the shriveled clumps of mesquite and greasewood, silent but for the faint pattering of kangaroo rats and scuttling of prowling scorpions. But inside the glowing line of lights there was noise and bustling activity; 11:30 was approaching and the graveyard shift was getting ready to go down to the dam. The center of the commotion was the dusty lot in front of the mess hall where scores of motor transports were parked in neat lines, ready to pull out for the dam site. The graveyard crew, almost completely assembled, stood in the shadows beside the empty trucks, waiting for the command to climb aboard. The men laughed, swore, and smoked cigarettes. From inside the mess hall came the low babble of conversation, punctuated by the crash and clatter of falling crockery in the kitchen, where the cooks and their helpers were already hard at work preparing the next morning's breakfast. The scent of frying onions mingled with the thick odor of sweat, motor oil, and tobacco smoke and hung stale and heavy in the night air. Across and down the street from the mess hall, the lights of the Six Companies recreation hall beckoned invitingly. Day-shift men on their way to the hall to shoot pool or play poker waved and called out to the graveyard-shift workers, who responded to the derisive greetings with jeers and obscene gestures.

All was quiet at the reservation gate, where a deputy U.S. marshal sat in the guardhouse and struggled to stay awake. Floodlights bathed the unpainted shack and several hundred feet of Boulder Highway in a brilliant glare. They also illuminated a large sign, which proclaimed in bold, black lettering: "You are entering the Boulder Canyon Project Federal Reservation. Property of the United States. You are subject to all regulations of the reservation, in part as follows. . . ."

The marshal yawned and rubbed his eyes. He did not need to read the sign; he knew its contents by heart. "The possession of, or transportation within the reservation of, intoxicating liquors, narcotics, explosives or fire arms is prohibited," he mumbled in a low, singsong monotone. Later, in the wee hours of the morning, when carloads of carousing construction workers came racing back to the reservation from the Las Vegas clubs, the Boulder Highway roadhouses, or the casino at Railroad Pass, he would be busy rousting drunks out of their automobiles and searching under backseats for stray bottles. The men who resisted or could not stand up straight would not be allowed through the gate until they were sober. He had to hand it to these dam stiffs, though: they could be dead drunk when they pulled out of Las Vegas, but the sight of the guardhouse and the big project reservation sign stiffened their spines and cleared their heads in a hurry. If they missed their shift because they had been held at the gate, there was an excellent chance that they would be fired. Leverage like that certainly simplified law enforcement, the marshal thought as he stared down the empty road and struggled to stifle another yawn.

On Wyoming Street, in the heart of Boulder City, the white-brick massif of Grace Community Church seemed to float in the moonlight like a gothic iceberg. Across the street in the parsonage, the Reverend Thomas Stevenson scratched out a draft of the sermon he would deliver on Sunday. Like all his efforts, it would be a real pulpit thumper, full of fire and brimstone. The sin and corruption of Nevada might surround the project reservation, but in Boulder City the church was going to be ahead of Satan, at least if Tom Stevenson had anything to do with it.

South of the parsonage, the residential blocks abutting the desert were dark and still. A single lamp burned in one of the bungalows below New

A double-decked transport truck with 154 men aboard about to depart for the dam site. (Bureau of Reclamation)

Mexico Street, where a young mother was humming a lullaby to a crying baby. Several houses away, a door swung open and the shadowy figure of a man emerged and slipped away up the street toward the dormitories, leaving behind the wife of another man who would soon be returning from his shift at the dam. A pair of house cats crouched on the canvas roof of a parked car, hissing and snarling at one another, but otherwise the neighborhood was hushed and peaceful.

On the northwest side of town, in front of the mess hall, the graveyard crew was now fully assembled and ready to go. On command, the workers dropped their cigarettes and scrambled aboard the trucks. Headlights snapped on, engines coughed to life, and clouds of dust and exhaust fouled the air. With a harsh grating of gears, the lead transport pulled away, followed single file by the others, each truck creaking and swaying under its human load as it made the sharp turn onto Cherry Street and headed up the hill toward the big water tank. The glowing headlights disappeared into the night, and the engines' roar grew fainter, then died away completely as the convoy cleared the ridge overlooking Hemenway Wash and began the steep descent down the other side toward Black Canyon. For a moment there was a deep silence and the moon shone bright and hard on the empty expanse of sand where the trucks had stood. Then the door to the mess hall slapped open and a string of oaths spilled out, followed by a bull cook in a stained apron, dragging an overflowing trash can.

Inside the dim, barnlike Six Companies recreation hall, the whir of ceiling fans muffled the clack of billiard balls, the rustle of newspapers, and the desultory clinking of soft-drink bottles. The atmosphere was quiet and subdued, in direct contrast to the throbbing, smoke-hazed saloons of Las Vegas, twenty-two miles miles to the northwest, where sweating, bellowing men gulped vile whiskey and flung silver dollars onto the threadbare felt of the gaming tables. Not by accident was the recreation hall such a staid and sterile place; in Las Vegas a man could drink, gamble, and whore in the time-honored tradition of the western boomtown, but within the boundaries of the Boulder Canyon Project Reservation temperance and morality were strictly enforced. The social chaperonage was as heavy handed as the U.S. marshals could make it. "Recreations [in Boulder City] shall be about the same as those that can be legally enjoyed in Peoria," a disgruntled easterner had written after coming to southern Nevada expecting to find another Goldfield or Virginia City.[1]

But for all its aspirations to midwestern normalcy and its veneer of tidy, all-American wholesomeness, Boulder City was no Peoria. It was a government-sponsored company town isolated in one of the harshest, wildest corners of the continent, a tiny outpost dedicated to a single purpose, building Hoover Dam, and organized and run as if it were a military base. Its people had come to this little patch of desert from every state in the Union and from all walks of life to shed their former identities under the watchful eyes of Uncle Sam and assume new ones at the command of the contractors. Whatever they had been before, whatever they might become later, they were now dam builders and citizens of Boulder City, U.S.A., toiling for Six Companies and living under the eagle's wing.

The federal government's decision to build a full-fledged modern city near the Hoover Dam site was made early in the advance planning for the Boulder Canyon Project;[2] it was a bold, innovative step, for men toiling on construction projects in the West traditionally had lived in camps of their own making, like Ragtown. That such camps lacked clean water or sanitary facilities of any kind had been accepted as inevitable; the hardships suffered by the men and women who lived in them was considered a necessary sacrifice or, more often, simply ignored. But Bureau of Reclamation personnel had spent enough time in and around Black Canyon to know that a rough-and-tumble construction camp of the traditional variety would be unacceptable on the Hoover Dam job. The temperature extremes endured by the geological and topographical survey teams that had explored the canyon before its selection as the dam site had made it abundantly clear that the building of the dam would be slowed, and perhaps even brought to a halt, unless adequate arrangements were made for the housing, feeding, and general care of the work force.

In a speech delivered at the Massachusetts Institute of Technology on January 9, 1931, Bureau of Reclamation Commissioner Elwood Mead explained why Boulder City had been conceived by describing the climatic hazards that awaited the men who would build the dam. "The summer wind which sweeps over the gorge from the desert feels like a blast from a furnace," he told the audience. "How to overcome this and provide for the health and welfare of the workers has had much attention. At the rim of the gorge where much of the work must be done, there is neither soil, grass, nor trees. The sun beats down on a broken surface of lava rocks. At midday they cannot be touched with the naked hand. It is bad enough as a place for men at work. It is no place for a boarding house or a sleeping porch. Comfortable living conditions had to be found elsewhere, and these are found on the summit of the divide, seven miles from the dam site. Here there is fertile soil; here winds have an unimpeded sweep from every direction; here there is also an inspiring view of deserts and lonesome gorges and lofty mountains."[3]

What Mead left unsaid, but what was well understood by the men responsible for organizing the Boulder Canyon Project, was that a general, heat-induced collapse of worker health and morale, leading to the importation of what the press euphemistically referred to as "foreign tropical labor," would be a disaster for the Bureau of Reclamation and the administration of Herbert Hoover. The presence of large numbers of blacks in Black Canyon, "with its implied confession that within the continental United States a task had been found too difficult for [white] American physique and morale to perform," was unthinkable, and so the blueprint for a modern community that would keep "3,000 or more Americans, mainly of the native or northern European stocks contented and healthy" was approved.[4] To ensure the contractors' cooperation in this undertaking, a clause was inserted in the construction contract stipulating that "the main camps of the construction employees shall be established and maintained in Boulder City," and Interior Secretary Wilbur proudly announced to the press that a model town would be built under federal auspices near Black Canyon. "Boulder City will truly be an oasis in the desert," he wrote. "The town will lack for nothing to be found in the average progressive community in the country."[5]

The ink from Wilbur's pen was hardly dry before controversy erupted over the legal status of the unbuilt town. Nevada, panicking at the thought of losing a bonanza of tax revenue, insisted that it, not the federal government, should have jurisdiction over the dam site and Boulder City. For its part, the Interior Department panicked at the thought that the Silver State, with its legalized gambling, its tolerance of prostitution, and its open contempt for Prohibition, would hold sway. "There is general agreement in administration circles that if Boulder City should be permitted to become an 'open saloon' town something in the nature of a national scandal would soon be exploded," reported the *New York Times*. "Also there is a strong feeling among all those intimately connected with the Hoover Dam project that liquor and dynamite as well as liquor and the Nevada desert climate are mixtures likely to prove fatal to both health and the dam construction job itself." [6]

For a time Las Vegas had been considered as a base for the dam workers, although its distance from the site and its Wild West reputation had raised doubts about its acceptability in the minds of the government planners. When Secretary Wilbur and Commissioner Mead came through town on an inspection trip to the dam site in the summer of 1929, the Las Vegas city fathers ordered all the gambling clubs and houses of prostitution shut down and decreed that no liquor be sold until Wilbur and Mead had finished their tour and left Nevada. This valiant attempt to impress the government with Las Vegas' sobriety and its suitability as a headquarters for the dam project failed, however, when a member of the Wilbur party, guided by several local newspapermen, entered the notorious Arizona Club on North First Street, had several drinks, and then returned to the train, where he regaled the shocked interior secretary with his tale of Las Vegas hospitality. [7]

The memory of this episode and of his pocket being picked at the silver-spike ceremony in September, 1930, undoubtedly strengthened Wilbur's determination that there would be no saloons, casinos, or bawdy houses in his model American community, and he moved swiftly to clamp federal authority on the dam site and to create a buffer zone between it and the bars and brothels of Las Vegas. Invoking the Reclamation Act of 1902 and the Boulder Canyon Project Act, he created the Boulder Canyon Project Federal Reservation, a 144-square-mile enclave that included the dam site, the lower portion of the future reservoir, the site of Boulder City, and vast stretches of open territory around the town. On May 26, 1931, a plat and description of the project reservation lands were delivered to Nevada Governor Fred Balzar in Carson City; upon his receipt of these documents, federal jurisdiction was formally established and gambling, the sale of liquor, and other practices deemed injurious to the workers and the orderly progress of work were strictly prohibited. [8]

With the legal underbrush thus cleared, and with vice formally banished from the desert wilderness around the dam site, the way was clear for the Bureau of Reclamation, working hand in hand with the contractors, to create the oasis that Secretary Wilbur had promised. On April 6, 1931, less than four weeks after receiving the contract to build Hoover Dam, Six Companies established the Boulder City Company, a wholly owned subsidiary under the direction of Felix Kahn. This company was to construct

Workers preparing to pour the foundation walls of the Six Companies executive lodge and guest house in Boulder City, August, 1931. (Bureau of Reclamation)

and rent worker housing, provide for worker feeding, organize transportation to and from the dam site, and build and operate a company store and recreation hall.[9] For its part, the Bureau of Reclamation would be responsible for providing homes for government personnel, constructing buildings to house the city government, laying out and grading city streets, and installing the water, sewer, and electrical systems.

Award-winning Denver architect S. R. DeBoer was commissioned to design the model community according to a strict set of specifications provided by the bureau. He envisioned a core of permanent buildings, including the Bureau of Reclamation and Six Companies office complex, multiple-family housing for government personnel, and a business district surrounded by an industrial zone and two large, circular, temporary residential areas for the dam workers. A block-wide green-belt forest was to separate the permanent and temporary areas, and an eighteen-hole golf course was penciled in for good measure. In a final burst of inspiration, DeBoer rotated the entire plan nine degrees east of a straight north-south axis so that streets and buildings would not bear the full brunt of direct summer sunlight.[10]

According to the original timetable for the Boulder Canyon Project, Boulder City was to be completed before work in Black Canyon began, but

Boulder City in the summer of 1931. In the foreground is a squatters' tent colony. Behind it, four-mule teams pull Fresno scrapers grading city streets. In the background stand rows of new dingbat houses, rented to Six Companies workers. A pall of white dust obscures the mountains on the horizon. (Union Pacific Railroad Collection, UNLV)

President Hoover's decision to accelerate the entire job meant that the town would have to be built hastily, even as other phases of the dam construction, involving hundreds of men with no place to live other than the rough dormitories of the River Camp or the shacks of Ragtown, proceeded through the hot summer months. When Walker Young and Frank Crowe examined DeBoer's site plan for Boulder City in the spring of 1931, they were appalled by its impracticality. They needed housing put up as quickly and cheaply as possible, and the elaborate circular residential zones and wedge-shaped building lots envisioned by the Denver architect were not only prohibitively expensive but impossible to lay out given the eroded topography of the Eldorado Valley. The notion of a green-belt forest and an eighteen-hole golf course was, under the existing conditions, ludicrous. Equipped with pencils and a roll of drawing paper, the two men drove to the Boulder City site one afternoon, and as *Las Vegas Evening Review-Journal* reporter John Cahlen looked on, they sketched their own design for the new community. DeBoer's forest, golf course, and concentric circles were thrown out and the axis of the town was rotated back so that it ran north and south. A triangular design was adopted, with the Bureau of Reclamation administration building at the apex and with traditional rectangular building lots in the residential zone. The only part of DeBoer's site plan that survived unscathed was the industrial and railroad zone, which stretched along the northwest edge of the town.[11]

A wing of one of the dormitories where unmarried Six Companies workers lived. The residents have pulled their cots onto the upper porch for hot-weather sleeping. (Bureau of Reclamation)

By the end of April, 1931, crews of carpenters, plumbers, and electricians, under the direction of Six Companies Assistant Superintendent Charlie Williams, were hard at work knocking together the first buildings to rise on the Boulder City town site: a large machine shop, a temporary mess hall, and a cluster of temporary bunkhouses. At the same time, workers for the New Mexico Construction Company, which had been awarded the $300,000 streets and sewers contract, were busy grading the first of Boulder City's wide avenues. The rail yard, finished several months before, bustled with freight cars, flatcars, and tank cars groaning with lumber, bricks, cement, wire, pipes, electrical equipment, fuel, food, and water. Trucks came and went in a steady procession, raising a towering cloud of dust that was visible miles away.

Newspaperman Pop Squires, touring the town site to write an article for his *Las Vegas Age*, was thrilled by the atmosphere of energy and vitality he found. The promise of bold new beginnings was for him embodied by a great mountain of raw lumber, which, he wrote, "gleams in the desert sunlight and smells clean and fresh." He could not have smelled the lumber for long, however, without choking on the miasma of exhaust fumes and wind-blown sand, and he commented, only partly in jest, that "the building of Boulder Dam will be as much a battle against the curse of dust as against hard rock." [12]

In June the first of eight large dormitories that would house unmarried Six Companies workers was finished. The frame building was laid out in two long, parallel wings connected in the middle by a transverse corridor. Each of the two-story wings was divided into eighty-six rooms; a large porch where the men could gather to talk, play cards, and sleep in hot weather ran the length of the wings on both floors; the showers and toilets were located in the connecting link. "The dormitories . . . are striking examples of what can be achieved . . . according to modern architectural theories, when a problem is solved simply and unostentatiously. They are direct answers, completely comfortable—and beautiful," enthused a journalist after touring one of the new buildings.[13] Another writer was less impressed: "I then went to see the great, modern Boulder City, but what a surprise! Buildings that resemble army barracks. Cheap two-story affairs. The rooms are small, the walls are of plaster board which will later become a hatchery for bed bugs."[14]

Beautiful architecture or cheaply built shacks, the Six Companies dormitories would be home to almost sixteen hundred men at the height of the Hoover Dam construction. The charge for one of the cramped, 72.5-square-foot dormitory cubicles, plus three meals in the mess hall and transportation to and from the dam site, was set at $1.65 a day, to be deducted from the worker's biweekly paycheck. For a laborer earning the minimum wage of $4.00 a day, this amounted to more than 40 percent of his gross pay; it was a stiff price but one that men who had been living in tents or lean-tos and scrounging food wherever they could find it generally were happy to pay.[15]

Just as important as furnishing housing—perhaps more important in the eyes of many of the workers—was providing satisfactory kitchen and dining facilities. Through its Boulder City Company subsidiary, Six Companies subcontracted the feeding of its employees to Anderson Brothers Supply Company. During the first months of the Boulder Canyon Project, "Ma Anderson," as the men affectionately called the boarding company, fed workers in temporary mess halls at Boulder City and the River Camp. These glorified tents were small, hot, and dirty, and because of poor sanitation and the lack of refrigeration, the food served in them was unappetizing and occasionally sickening. In July, 1931, twenty-seven men at the River Camp were stricken with ptomaine poisoning attributed to spoiled pork, and diarrhea, caused by bad food and contaminated drinking water, was a nearly universal affliction.[16] Relief was on the way, however, in the shape of a large, modern cooking-eating facility being built on the block between Cherry and Date streets on the west side of Boulder City.

From the day it was completed the new mess hall, with its two 575-man dining rooms and its giant stainless-steel kitchen and pantry, was the pride and joy of Boulder City. It was featured prominently on the tour given visiting VIPs and was a favorite subject for journalists, who seemed never to tire of writing about how much food the dam workers wolfed down each day. The quantities were, in fact, staggering: each week 12 to 13 tons of fruit and vegetables were trucked in from Southern California; 5 tons of meat

The mess hall at meal time. Its staff could serve 1,150 men at one sitting. (Kaiser Collection, Bancroft Library)

and 2.5 tons of eggs, packed in dry ice, came from Reno; 3.5 tons of flour were shipped by train from Ogden; 2 tons of potatoes arrived from Enterprise, Utah; and 1.5 tons of miscellaneous groceries were delivered from various points east and west. The demand for milk, butter, and cheese was such that Anderson Brothers purchased a 160-acre ranch in the Moapa Valley near Logandale, Nevada, some eighty miles northeast of Black Canyon, stocked it with a hundred cows, and built its own dairy, from which caravans of ice-cooled trucks departed day and night for Boulder City.[17]

The cooks, supervised by Harold Anderson himself, transformed this cornucopia of food into six thousand meals daily, served by a staff of thirty white-clad waiters, called flunkies, at half a dozen sittings. The workers ate family style at eight-man tables and were welcome to have as much of any item as they wanted. Box lunches were prepared for the men going to the dam, but if they wished to do so they could select a lunch from seven different kinds of sandwiches and an assortment of pies, cakes, and fresh fruit. Again, there was no limit on the amount of food a man could take, and some of the workers made extra-large lunches, which they shared with financially strapped friends (usually family men) who did not take their meals at the mess hall.[18]

Besides providing food, the mess hall became the main social center for the new town. The first dance in Boulder City, attended by upwards of two thousand men and women, was held there on Thanksgiving Eve, 1931;[19] thereafter, dances were a regular Saturday night event. The big building also was used for fraternal-organization meetings, Christmas pageants, and political rallies. The work of preparing the next meal never stopped, however, no matter what the event. At one political meeting Frank Crowe got up to introduce Nevada Senator Key Pittman to a large audience only to be interrupted by a tremendous crash of breaking dishes and the sound of Harold Anderson screaming curses in the kitchen. "I was supposed to make the first speech," deadpanned the construction superintendent, "but Harold just beat me to it." [20]

During the summer of 1931 the construction of single-unit housing for Six Companies' married employees began and a peculiar phenomenon was revealed: the layout of Boulder City's residential section bore an uncanny resemblance to the organization charts hanging in the Six Companies and Bureau of Reclamation offices. At the highest point in town, at the crest of the ridge overlooking both Hemenway Wash and Eldorado Valley, a large, rambling, Spanish-style hacienda with dazzling white stucco walls and a red-tile roof was built and luxuriously furnished at a cost of more than thirty thousand dollars.[21] This was the executive lodge and guest house, to be used by the Six Companies directors, their families, and guests. On a knoll west of the executive lodge, slightly lower than it in elevation but with an equally commanding view, another large, Spanish-style house was built, this one for Frank Crowe, his wife, Linnie, and their daughters, Patricia and Bettee.

Down from the executive lodge and the Crowe residence, along Denver and Colorado streets, the Bureau of Reclamation built three- and four-room houses for its senior engineers and administrators. Although far less imposing than the gleaming mansions looking down on them from the northeast, they were nevertheless substantial brick-and-tile buildings designed in the popular Hollywood Spanish style. Each of these homes had a fireplace and was furnished with the two most coveted luxury items in Boulder City: an electric refrigerator and an indoor toilet. Not far from the little Bureau of Reclamation enclave, Six Companies carpenters built fourteen five- and six-room frame houses for the assistant construction superintendents and the high-level managers and engineers.

It was some five hundred feet farther down the slope, however, south of Wyoming Street, where the desert valley flattened out, that most of the Six Companies homes were built. These rough three-, two-, and one-room frame cottages, called dingbat houses by the carpenters who hastily pounded them together, were to be rented to married laborers and then torn down when the project was finished. Six hundred thirty of them were erected in about six months' time, all identical and all ugly, built with no concern for aesthetics and with only minimal attention to construction quality.[22]

Six Companies had two pressing reasons for building the laborers' cottages so quickly and cheaply: it wanted to get its employees and their families out of the slums of Ragtown and McKeeversville and into the more comfortable and sanitary environment of Boulder City, and it wanted to start earning income from rent. The cottages, which cost a few hundred dollars each to build, were to be rented for fifteen dollars (one room), nineteen dollars (two rooms), and thirty dollars (three rooms) a month and promised to generate a tidy profit over the projected five- to six-year duration of the dam job.[23] Consequently, for the crews putting them up, the emphasis was on speed and more speed rather than painstaking craftsmanship. They were told to put up one and a half units each day and were made to understand that this figure was not an idle wish but an iron-clad quota. "There was a crew of two carpenters and a helper," recalled Carl Ballantyne, a Six Companies carpenter who had been a stockbroker before the 1929 market crash. "If we didn't finish [the quota], the whole crew was fired."[24]

The cottage interiors were depressingly spartan. In one corner stood a sink on a two-by-four frame. Across from it was a tiny three-burner gas range and oven. The bathroom, which was an enclosed portion of the screened porch and could be entered from the interior of the house, had a toilet and a shower but no sink or water heater. Except for several naked light bulbs, the rest of the interior was empty: bare wooden floor and blank, paper-thin walls with cracks and loose joints open to the wind.

In the beginning the bleakness and monotony of these long blocks of cookie-cutter cottages was overwhelming. The streets had been graded but were not yet paved, there was no shade—in fact, there was no vegetation of any kind—and an ever-shifting ocean of sand, dotted with the flotsam left behind by the carpenters, lapped at the base of every house. Yet to the men, women, and children who had spent months living in tents or cars, cooking over open fires, drinking river water from dirty cans and buckets, bathing in the muddy stream or not at all, Boulder City looked like paradise. Paint, wallpaper, furniture, landscaping could wait; for the time being, four walls and a roof were all anyone could ask. A worker summed up this feeling when he wrote of the plain little Seventh Street cottage into which he, his wife, and their daughter moved early in 1932: "To us, it was beautiful."[25]

All through the winter and spring of 1932 the Boulder City pioneers

During the fall and winter of 1931–32, Six Companies married workers and their families moved into new cottages in Boulder City. Landscaping and paved streets would come later. (Union Pacific Railroad Collection, UNLV)

occupied the new cottages. Once they had settled in, they began turning the drab shells into homes and transforming the barren blocks into neighborhoods. At first this meant making do with what was on hand. Lumber was scavenged to build makeshift cupboards and closets; scrap boards and canvas were fashioned into window-mounted swamp coolers, which doubled as iceboxes for milk and eggs; an old sheet, a blanket, or a dress became a set of curtains; discarded Hercules blasting-powder crates served as tables and chairs. Later, perhaps, there might be enough money to make a down payment on an electric refrigerator or a set of furniture at the Boulder City Company Store, but for the time being ingenuity had to substitute for cash.

After they had the interiors of their homes in reasonable order, the renters began fixing up their Sahara-like front yards. Some built little picket fences and planted grass, shrubs, and vines; others marked out walkways with stones and transplanted barrel cactus, Spanish bayonet, Joshua tree, and other native plants they found in the surrounding desert. Gradually, as the individual lots were staked out, tidied up, and landscaped, Boulder City began to look less like a refugee camp and more like a small town.

In striving to beat back the desert, the residents were following the lead of the Bureau of Reclamation, which, with the extensive and ambitious planting program it had undertaken, seemed intent on proving the benefits of reclamation and irrigation in the desert by turning its section of

Dam worker's children sit in the yard of a Six Companies cottage in Boulder City. (Kaiser Collection, Bancroft Library)

Boulder City into a giant garden. The key to this transformation in Boulder City's natural appearance was the completion of an elaborate, expensive, six-mile network of pumps and pipes that carried water from the Colorado out of Black Canyon to the Eldorado Valley. Even at a cost of half a million dollars the canyon system was a bargain compared to the alternative, a twenty-six-mile Las Vegas–Boulder City pipeline, and it was a major engineering achievement to boot. Like a pair of long, steel straws, twin intakes dipped into the Colorado just below the diversion tunnel outlets on the Nevada side and sucked the murky river water into a 200,000-gallon sedimentation clarifier, where it sat for three hours while the silt settled out. From the clarifier, squatting just a hundred feet above the surface of the river, the water was pumped almost half a mile straight up, and then to a treatment plant in Boulder City, where it was filtered and chlorinated and then piped into the 2-million-gallon steel tank that loomed above the town.[26]

In the spring of 1931, when the Bureau of Reclamation and Six Companies engineers revamped DeBoer's city plan, they eliminated his landscaped parks and play areas and recommended that cheaper, maintenance-free desert landscaping be used. The agonizingly long, torturously hot summer of 1931 showed, however, that DeBoer knew what he was doing when he proposed the planting of trees, grass, and shrubs. The midday sun blazed down on the naked sands and suffused the sky with a white-hot glare. In the afternoon the wind swept across the Eldorado Valley, whining through the mesquite and racing across the flat miles until it came to the open scar that was the Boulder City town site. Mule-drawn graders, scraping out streets and building lots, had denuded hundreds of acres, and the wind, unrestrained by the natural surface cover, dipped down, dug in deep, and rocketed clouds of dust and debris skyward, mixing them into a thick, yellowish haze. Blistering heat and whirling dust devils made the town site a miserable place for the construction crews working and camping there, and belatedly, Bureau of Reclamation officials realized that the proposed desert landscaping, which would provide no shade and precious little ground cover, could not improve the desolate environment enough to make it even moderately habitable for the thousands of people expected to live there.

Public relations had to be considered, too. Hoover Dam was a national story, and much had been made in the press about the model community rising near the dam site. Expectations were high. "In a sense, the Hoover Dam project is not only a construction job but also a sociological venture," observed a magazine writer, echoing one of Secretary Wilbur's more bombastic statements. Wrote another: "Uncle Sam [has] decided to extend a kindly but firm paternal helping hand . . . to insure for the workers a high standard of living and a maximum of comfort and general well-being."[27] With the 1932 presidential election approaching and the Hoover administration already beleaguered by shrill accusations that it was insensitive to the plight of the workingman, it simply would not do for the much ballyhooed model city to be a collection of desert shacks lost in a perpetual sandstorm, with nothing but a scattering of rock gardens and a few pitiful stands of cactus to soften the harsh contours of the surrounding landscape. Boulder City could not be just functional, it had to be a showplace, and so additional funds were earmarked for landscaping in the manner DeBoer had suggested.

The unenviable task of turning the three-hundred-acre Boulder City dust bowl into a lush oasis fell to a young landscape gardener named Wilbur W. Weed. He was working at the National Iris Garden in Beaverton, Oregon, when the Bureau of Reclamation called him to Boulder City, and the contrast between the moist, fertile soil he left behind in the Pacific Northwest and the sterile, arid sand he found in southern Nevada was dismaying. Nevertheless, he plunged into his new job in December, 1931, testing the Boulder City soil and soliciting the advice of landscapers in Phoenix, Palm Springs, and Hawthorne, California, about which plant varieties fared best in the desert environment. Early in February he placed orders for more than nine thousand trees from California and Nevada nurseries, and when the first shipment of saplings arrived at the end of the month, he embarked on one of the most ambitious and difficult landscaping programs ever undertaken.

Planting began in the bleakest part of town: the blocks along New Mexico Street where Six Companies cottages were still going up. One by one, young, fragile-looking Carolina poplars, Chinese elms, and European sycamores were set into the powdery soil and propped up with wooden stakes. The few leaves still clinging to the pencil-thin branches were quickly coated with dust, and when the winter winds gusted in off the desert, they rattled dryly as if mourning the bygone days of summer and mocking the notion that the new, green leaves of spring would ever grace this desolate place. But Weed had faith, and he kept on planting, watering, and fertilizing. Slowly working his way north through the raw residential neighborhoods to the business district, he put in row after row of Guadalupe cypress and line after line of oleander bushes. In two of the public parks, North and South Escalante Plaza, he made a grand, sentimental attempt to evoke his old home in northeastern Oregon by planting twelve hundred rose bushes in two formal gardens. Then he moved up the hill onto the grounds of the Bureau of Reclamation's administration building, which was the focal point of the entire town, visually as well as politically. Here the landscaping crews shoveled and blasted to form a flat, elevated tier in front of the building's brick facade. The broad tier was covered with topsoil and a lawn of ryegrass, bordered by Guadalupe cypress and Chinese elms, was planted. Below the tier, in a large public park called Wilbur Square, an even larger lawn was put in and curving footpaths were lined with trees and oleanders. The Bureau of Reclamation homes flanking the administration building grounds and Wilbur Square were also carefully landscaped, and the bureau, showing just how seriously it took this cosmetic campaign, decreed that if the residents did not tend their grass and shrubs satisfactorily, a yard-maintenance fee would be deducted from their paychecks.[28]

While Wilbur Weed planted and cultivated and worried about what would become of his seedlings in the spring, the settlement of Boulder City continued at a rapid pace. Among the families who came to the town that winter to begin putting down roots were the Godbeys: Tom, Erma, and their children, Tommy, Jimmy, Laura, and Ila. After abandoning Ragtown late in July, 1931, they had moved to Cowboy Bill's auto camp, located on a ranch just outside Las Vegas, where, for five dollars a month, they were able to pitch their tents under a row of shade trees and draw water from a ditch that drained the overflow from a nearby artesian well.

Tom Godbey had lost his job in the aftermath of the August strike, and because he did not have a pass to get through the new police checkpoint, he was barred from the project reservation. He walked and hitchhiked from Las Vegas to the gate every day to look for work, but there were hundreds of other men there, too, waiting for their names to be called from Labor Commissioner Leonard Blood's long list. One day after several weeks of fruitless waiting, Godbey finally decided that if he was to get a job in time to keep his wife and children from going hungry, he would have to find a way to get into the reservation, where the work was, pass or no pass.

That night, beside his campfire at Cowboy Bill's camp, he carefully soaked the label off an empty bottle of Campho-Phenique lotion; the label was almost exactly the same size and shape as the passes Six Companies issued to its employees. The next morning he caught a ride to the reservation with a group of electricians who were on their way to work. When the car stopped at the gate and the electricians held up their passes, Godbey held up his Campho-Phenique label. The marshal on duty was either near-sighted or sympathetic; he waved the car through, and Godbey was safely on the reservation.

The electricians dropped him off at the Boulder City town site, and he began going from one gang of workers to another, asking if they needed help. The foreman of one of the New Mexico Construction Company crews looked him over for a moment and asked if he were a Fresno man and could handle a four-up. The answer was no on both counts, but having come this far Godbey was not about to admit that he was no team-ster and that he had only a faint idea of what a Fresno scraper was. Mo-ments later he had a set of reins in his hands and was guiding four large mules and a big steel blade along the sandy incline of a new Boulder City street. When the shift was over, the new Fresno man passed through the gate on his way back to Las Vegas. The muscles in his arms and shoulders were throbbing and his palms were lumpy with blisters, but his spirits were high: the Campho-Phenique label was gone, replaced by an official pass identifying the bearer, Thomas Godbey, as an employee of the New Mex-ico Construction Company.[29]

About a month later the Godbey family moved from Cowboy Bill's to the project reservation and pitched their tent on a patch of open ground between the Boulder City train station and McKeeversville on the crest of Hemenway Wash. This area was known as the Railroad Y. On one side was the switching yard, where black locomotives rumbled to and fro; on the other side was the construction line, which led to the gravel pits and the aggregate-screening and concrete-mixing plants. Traffic was constant, and the little camp in the Y shook from the heavy vibration of passing trains and choked on the smoke and dust sucked up by their backwash. As if noise and soot were not bad enough, the tracks and the switching yard fas-cinated the children, and mothers had to be constantly vigilant to keep their youngsters away from this perilous playground.

In February, 1932, the Bureau of Reclamation announced that it was setting aside several vacant blocks on the east side of Boulder City for construction of private homes. Private construction had not been part of the bureau's plan for the town, but because demand for available Six Companies housing had outstripped supply, and because garbage and human waste accumulating in the Railroad Y and McKeeversville camps posed a health threat, there was a change in thinking.[30] Building a home seemed like a good idea to Erma Godbey, who wanted to move her children as far away as possible from the dangerous railroad tracks and "get into a civilized way of living," so she and her husband went to the recently opened administration building and applied for a building lot. The requirements for leasing a lot were simple: the lessee had to agree to build a house costing at least $250 that had indoor plumbing and did not have a flat boxcar or tar-paper roof. The Godbeys selected a lot on Avenue L and signed the papers.

Tom began to collect lumber and other building supplies from various sites on the reservation, stacking the scavenged material by the family camp in the Railroad Y as he picked it up. One of the original mess halls—a large canvas tent with a duckboard floor and four-foot-high plank walls— was no longer being used, and he contracted to tear it down in exchange for the boards, canvas, and wire screening. He was also able to acquire a large quantity of lumber that had been used for concrete forms in construction of the water-filter plant. While he was gathering these materials, Erma was sketching rough designs for a little four-room house on a Big Chief writing tablet. With money borrowed against an insurance policy, the couple hired a carpenter who had lost his job. Working from Erma's final sketch, he put up the shell of the house and in April, 1932, the family moved in, proud owners of the first private home in Boulder City.[31]

During the first weeks of spring, 1932, Boulder City underwent a startling transformation. Wilbur Weed's landscaping, nourished by Colorado River water and warm sunlight, burst into glorious life. The black locusts and Chinese elms shrugged off the effects of the chill winter months and put out new leaves. The oleanders grew glossy and lush and scented the air with their blossoms. Best of all, the ryegrass lawns came in thick and unblemished and covered the dusty construction scars in the parks and on the administration building grounds with swatches of rich, cool green.

In the desert surrounding Boulder City, bright daubs of color appeared on what previously had been dun expanses of sand. Small patches of delicate flowers sprang, as if by magic, from the desert floor, and withered bushes became succulent with olive-gray leaves. The breezes were gentle, warm, and caressing, and they carried the pungent scent of green mesquite across the Eldorado Valley to the fledgling town, anointing it with the sweet perfume of the desert's annual rebirth.

For a few short weeks, spring's intoxicating kiss breathed life into Boulder City. Then, as quickly and ardently as it had come, it was gone.

The warm breezes became furnace blasts, the flowers and tender green leaves shriveled, and it was summer again. In town, Wilbur Weed's grass and trees helped hold down some of the dust, but when the wind blew hard across the valley, hot sand flew everywhere. The loose seams of the dingbat houses gaped wider in the dry heat, and the wind-blown sand swirled and eddied, probing relentlessly for every crack and pinhole. The residents quickly learned to stuff newspapers or rags under the doors, around the windows, and in the keyholes, but nothing, it seemed, could stop the fine powder from finding its way in and drifting in abstract patterns on every flat surface. The dust storms were so frequent that a classification system soon developed, described by one writer as follows: "If after closing all doors and windows a pailful of sand sifted into the house it was a one-bucket storm. If two pails could be filled with the sweepings it was a two-bucket storm; and so on up to real ten-bucket twisters."[32] The storms caused other inconveniences, too. "I'd stand on the hill where I lived and look down across town during a storm," Madeline Knighten recalled. "I'd see lines hung with clean white laundry turned brown and ruined in no time by the dust."[33]

As annoying as the wind and sand were, they dwindled to insignificance when the mercury spiked upward in late May and the unbroken sunlight turned Boulder City cottages into sweatboxes. The technology of cooling was primitive in the early 1930s; air conditioning was still a novelty found primarily in movie theaters and other public buildings and not in private homes. The larger Six Companies buildings, such as the mess hall, recreation hall, and dormitories, were equipped with cooling plants and ceiling fans, but not the cottages. It was up to the individual tenants to find some way to keep comfortable during the long, torrid months between June and September, and the principles of evaporative cooling were applied in all manner of Rube Goldberg contraptions. Some would-be inventors created primitive wood and canvas swamp coolers, rigged out with electric fans that blew through burlap baffles kept wet with a watering can. Others simply ran hoses onto their roofs and let the water trickle down. Day sleepers who worked the graveyard shift swaddled themselves in wet sheets when they went to bed or paid forty cents to doze through a matinee at the Boulder Theater, which boasted an air conditioner. On especially hot nights when only the faintest breath of air stirred the midsummer stillness, the sloping lawn of Wilbur Square was dotted with the dark, restless figures of people desperately searching for coolness, solitude, and elusive sleep.[34]

The tree-shaded parks and squares in the northern part of town not only offered some refuge from the oppressive heat, they also provided an escape from another Boulder City constant: visual drudgery. The residential blocks looked so much alike that the people who lived on them sometimes had trouble finding the right one. Workers coming off the swing shift were particularly prone to losing their way in the dark and confusing one

A Boulder City dust storm, January, 1932. (Kaiser Collection, Bancroft Library)

identical cottage for another; wild stories about stumbling into the wrong house late at night—and the adventures that followed—were a staple of jocular conversation aboard the transports bound for Black Canyon. One such yarn appeared in the *Las Vegas Evening Review-Journal* under the headline "How the Builders of Boulder Made Error Sometimes." It told the story of worker J. N. Smith, who came off the swing shift at midnight and became confused while searching for his little cottage on the southern fringe of Boulder City. Smith went down a dark, quiet street to the block that bordered the desert and carefully counted off eight houses. He went up on the porch, stamped his boots to clear the soles of sand, pounded on the door with his fist, and yelled, "Woman, let me in!" The door opened a crack and the snout of a double-barreled shotgun emerged. "Run, or I'll fire," barked a strange voice. Startled, Smith whirled and fell off the porch. Regaining his feet, he sprinted away up the street until he was well out of shotgun range. When he stopped to catch his breath, he looked around and recognized his own house still farther up the street: an entire row of houses had been built since his last trip home the night before.[35]

The supposed infidelity of the night workers' wives (graveyard widows, they were called) was also a favorite subject for jokes and teasing speculation on the transports. Riding back to Boulder City one evening, concrete finisher Marion Allen noticed that the grizzled miner sitting next to him was holding a three-foot length of powder tamping stick in his gnarled hands. Allen innocently asked what the stick was for. With a straight face, the old miner replied that he always kept a powder stick by his back door; when he got to his house after a shift, he said, he knocked on the front door to let his wife know he was home, then ran to the back door, grabbed his stick, and got ready to club any man who came scrambling out buttoning his pants. There was silence for a moment in the back of the canvas-covered truck. Then a voice piped out of the dim interior, "That's right, and he's knocked out a different man every night for the past 20 years!"[36]

By late summer, 1932, the construction of Boulder City was almost complete. Six Companies crews had put up eight 172-man dormitories, one 53-man office dormitory, and more than six hundred cottages; they had built the mess hall and recreation hall; they had finished the Six Companies office building, the Boulder City Company Store, a laundry, and a twenty-bed hospital.[37] The New Mexico Construction Company had laid out and hooked up the sewer and water lines and paved nearly twenty miles of streets. The Bureau of Reclamation had spent well over a million dollars constructing the administration building and government residences and landscaping streets and parks, and privately financed structures housing various independent businesses had sprung up along Nevada Highway, the bustling main street.

Now that its boundaries were clearly established and its buildings in place, Boulder City was at a crossroads. For all its modest comforts and conveniences, it was still at heart a rough-and-tumble desert bivouac. It had sprung into being almost overnight, and although it resembled an average American town in superficial ways, it had no history or permanent institutions and was populated by people whose roots were elsewhere. Whether the isolated settlement at the edge of Black Canyon would be a temporary and impersonal, albeit well-appointed, construction camp, or develop into a true community, remained to be seen. Construction Engineer Walker Young summed up the choice the residents faced in an article that appeared in the *Las Vegas Evening Review-Journal* issue commemorating Boulder City's first year. "The government has furnished a collection of houses in an improved townsite," he wrote. "Whether these are formed into an assemblage of homes in a closely knit community depends upon the inhabitants of these homes and their attitudes toward Boulder City."[38]

The attitude of most residents soon became evident: Boulder City might be new and raw, its dust, wind, and heat punishing, its setting depressingly barren, its government-enforced isolation maddening, but it was their home. This sense of belonging grew, in part, out of the security of a job, a roof overhead, and three square meals a day at a time when many people throughout the nation could take none of these things for granted. It also grew out of the fellowship born of shared hopes and hardships and participation in a giant collective enterprise that was something new and indefinably special.

Boulder City and the Eldorado Valley. (Kaiser Collection, Bancroft Library)

﹀ The strong yearning for community, and the determination to infuse the shell of Boulder City with a sense of vitality and neighborhood spirit, led to the birth and rapid growth of a variety of institutions and organizations. By the summer of 1932 the new town had no fewer than four churches, all well attended. The largest was the Protestant interdenominational Grace Community Church, founded by one of Boulder City's most charismatic and colorful pioneers, Thomas E. ("Parson Tom") Stevenson.

Parson Tom, a Presbyterian minister from Burbank, California, had come to the dam site in October, 1931, at the request of the Home Missions Council and the Federal Council of Churches. The councils were concerned that no one was tending to the spiritual well-being of the thousands of workers congregating in remote desert encampments along the Colorado River, and they had decided to send them a pastor. A special type of individual was needed to establish a successful ministry among the construction stiffs in Black Canyon: he had to be young, single, physically strong to withstand the rigors of the harsh climate and primitive living conditions, and morally strong to shepherd his congregation away from the temptations of the flesh, which abounded in southern Nevada. Parson

Tom had all these attributes combined with a boundless, infectious, leather-lunged enthusiasm for the word and work of the Lord. His first services, held in the River Camp mess hall, were a resounding success. Word of his fiery, no-holds-barred preaching spread quickly, and as his following grew he began leading nighttime open-air revivals in Ragtown, complete with impassioned sermons, loud singing of hymns, shouting, clapping, and pulpit thumping—with a large, bootleg whiskey cask serving as the pulpit.

When the mess hall in Boulder City was completed, Parson Tom moved his services there, and whatever excitement and enthusiasm his harangues lost for not being delivered outdoors was amply compensated for by the joyful, off-key wheezing of a secondhand pump organ he had acquired. The mess hall was only a temporary home, however. The parson's congregation donated money, bought building materials, and erected the large, Gothic-style Grace Community Church on Wyoming Street. Not to be outdone, the Catholic parishioners, under the leadership of Father Hogan, built St. Andrew's Church four blocks down Wyoming Street from Grace Community. The Reverend Arthur Keane's Episcopalians completed Saint Christopher's Church at the intersection of Utah and Arizona streets in the spring of 1932, and the Mormons, under Presiding Elder Laurence Wortley, moved a small church building from Las Vegas to a vacant lot in Boulder City in the summer of 1932. Sin eventually struck down the un-quenchable, much-beloved Parson Tom: he was run over by a drunken driver while crossing the street between the parsonage and Grace Commu-nity Church on Christmas Eve, 1937. However, the religious enthusiasm he had worked so diligently and enthusiastically to spark did much to make Boulder City the wholesome, morally upright kind of town the gov-ernment planners had envisioned.[39]

Not everyone belonged to a church, of course, but there were many other community groups that helped draw the people of Boulder City to-gether. Thanks to the federally mandated hiring policy giving preference to veterans, the Boulder City American Legion Post was almost from its in-ception the largest in the state of Nevada, boasting upwards of three hun-dred members.[40] The Legionnaires were involved in many community ac-tivities, but probably their greatest contribution to the life of the town was a band that played many of the popular Saturday night dances at the mess hall and later at the Boulder City Legion Hall. A Veterans of Foreign Wars post also was organized, as was a large and active Masonic lodge under the leadership of Dr. Wales Haas, the physician in charge of the Six Companies hospital.

The stabilizing, civilizing influence of churches and fraternal organiza-tions helped set Boulder City apart from the booming, brawling construc-tion camps of the past, and so, too, did the presence of hundreds of children. The typical dam worker was in his early thirties, married, and already had, or was intent on starting, a family. During 1931 about 200 babies were born to women living in Boulder City and the various camps around the dam site; in 1932 the figure was 220, and in 1933 it was estimated that Boulder City mothers had given birth to 340 babies. By December, 1934, the number of children of preschool age living on the project reservation was thought to be between 900 and 1,100, and the number of school-age children exceeded 800.[41]

This unexpected fecundity revealed an embarrassing flaw in the government's otherwise thorough plan for creating a model American city: no money or land had been set aside for a school. Some of the dam workers' wives were former teachers, and they tried to fill the void by organizing classes in two vacant houses provided by Six Companies and in a small shack in the Railroad Y near the McKeeversville squatters camp. These facilities were hopelessly inadequate, however, and as Boulder City's population of school-age children grew by leaps and bounds, parents' pleas for the government to do something intensified.

Responding to the growing clamor, the Nevada congressional delegation and Interior Secretary Wilbur secured an appropriation of $70,000 to build, furnish, and maintain a school in Boulder City; Six Companies agreed to pay the salaries of eleven teachers, a total of $18,800, for the 1932–33 school year. Ground was broken on May 26, 1932, on a lot across the street from the municipal building, and exactly four months later Boulder City had its new school. The eight-room brick-and-tile structure was a source of great pride for the community, but almost from the day it opened it was too small to accommodate the swelling flood of pupils. For the 1933 academic year three classrooms that had been earmarked for high school students were reassigned to the elementary school, which still had to adopt a schedule of half-day sessions so that all of the 550 children enrolled could receive instruction. The older students, eventually totaling 150, had to commute by bus to Las Vegas High School.[42]

Besides omitting a school, the Boulder City plan made no provision for a public library. It had not occurred to anyone that construction workers would have the time or inclination to read. But again the Hoover Dam job proved to be different from previous construction projects. The Depression had driven many well-educated men to take any work they could find, whether it was waiting on tables, driving a truck, or wielding a shovel. A surprisingly large number of Six Companies employees were college-educated, and this fact, plus the poor radio reception in Boulder City, led to a strong demand for reading materials. In 1933, the Library of Congress agreed to lend the Bureau of Reclamation three thousand books to open a library in Boulder City, and a full-time librarian, Ruby Wyman, was hired by Six Companies to set up the facility and oversee its operation.[43] The library was located in the basement of the municipal building, and even though the cardboard circulation cards cost a dollar, use was heavy. Journalist Theo White noticed that some 75 percent of the titles were nonfiction, primarily biography and history. "They wanted it," Miss Wyman told him, "and what is more they read it—all of it." [44]

Pop Squires and A. E. Cahlen, editors of the two Las Vegas newspapers, the *Age* and the *Evening Review-Journal*, also competed for the attention of Boulder City's reading public. Cahlen dispatched a young reporter named Elton Garrett to live in Boulder City and write about its people and institutions for a special section, called the *Boulder City Journal*, that appeared in every issue of the *Review-Journal*. Garrett shared the pioneer lifestyle of the town's early residents; for a time he had no home of his own and slept on a cot behind one of the counters in the Boulder City Company Store. He wrote articles about progress on the dam, reported the news of Boulder City, and wrote a "Boys of Boulder Dam" series in which he profiled workers with interesting backgrounds or unique experiences. [45] The *Boulder City Journal* also published short columns, such as "Williamsville Town Topics" and "Boulder Briefs," that detailed births, illnesses, travel plans, club activities, and the like, and "Bunkhouse Biz," which showcased the wry, gossipy musings of C. A. ("Biz") Bisbee, a worker who lived in one of the Six Companies dormitories. These compilations of personal news were chatty and for the most part utterly inconsequential, but it was from just such mundane threads that the community fabric of Boulder City was spun.

Although churches, fraternal organizations, the school, the library, and newspapers played important roles in the life of Boulder City, the most influential and controversial community institution was the city government. Put simply, the Boulder City government was a dictatorship, benevolent for the most part, but a dictatorship nonetheless. This was the dark side of the federal bureaucracy's otherwise commendable effort to ensure that the Hoover Dam workers would have a safe, sanitary place to live. Because Boulder City was in a federal reservation managed by the Department of the Interior, the department, through the Bureau of Reclamation, exercised total control over the town; the residents had absolutely no voice in how Boulder City was run.

The fact that there would be no municipal elections, no popular representation of any kind, in their self-styled model American community undoubtedly made some people in the Interior Department uncomfortable. The tortuous justification of this policy—that the nation's voters, including the citizens of Boulder City, had elected the president, who appointed the interior secretary, who appointed the Bureau of Reclamation officials, who formed the chain of command exercising authority over the town—was not particularly convincing or comforting. Nevertheless, Secretary Wilbur was firm in his decision that a single man, a city manager whom he would select and whom he could fire at any time for any reason, would run Boulder City. The city manager would report to the Interior Department through the highest-ranking Bureau of Reclamation official at the dam site, Construction Engineer Walker Young. Wilbur's single, token concession to representation for the town's residents was the creation of a three-member advisory board, which was to include two government employees and a Six Companies representative. The board would be little more than a rubber stamp, however. Its members were chosen by Walker Young and served at his pleasure; they could advise and make suggestions, but they had no authority to enact their proposals or overrule any action taken by the city manager and the construction engineer.

The key figure in this setup was the city manager, for he would hold virtually undisputed sway over every aspect of life in Boulder City. For example, the city manager would have absolute police power: the squad of U.S. deputy marshals responsible for maintaining order on the project reservation would take their orders directly from him. Furthermore, the city manager would have absolute prosecutorial and judicial power: the cases of all Boulder City lawbreakers would be brought before him and he alone would decide how they should be disposed of. The city manager would wield absolute economic power: he would approve or disapprove applications from commercial developers who wished to operate businesses in Boulder City and thus control the degree of competition and, to a great extent, prices. Finally, the city manager would have absolute power over access to the Boulder Canyon Project Federal Reservation: he would have authority to bar or evict permanently from Boulder City anyone he deemed undesirable. The city manager would even wield some power over the otherwise almighty Big Six: if in his opinion the contractors were not maintaining living conditions meeting government standards in their dormitories and cottages, he was authorized to levy fines and require that corrective action be taken.[46] An extraordinary individual was required for this position: a highly skilled administrator, an executive who was supremely self-confident but not openly arrogant, a leader combining the attributes of firmness, impartiality, and absolute incorruptibility, an authority figure who could command total respect without being hated or feared.

The man Wilbur chose for the job was 69-year-old Sims R. Ely, a brilliant, quirky banker-businessman-bureaucrat from Berkeley, California. Ely's long résumé read like a Horatio Alger story. Born in Overton County, in north-central Tennessee, in 1862, he was educated in one-room schoolhouses there and in Kansas, where his family moved after the Civil War. After graduating from Illinois Wesleyan University Business College in 1885, Ely became the editor and publisher of the *Hutchinson* (Kansas) *Democrat,* a job he held until he was appointed special commissioner for county-seat litigation to the Kansas Supreme Court in 1892. His interest in government service piqued by his court experience, Ely went to Washington, D.C., in 1893 as secretary to Kansas Senator John Martin. Four years later he engineered an abrupt career change, becoming secretary-treasurer of the Hudson Reservoir Company and moving to the Salt River valley of central Arizona to participate in the planning of the Salt River Reclamation Project and construction of Roosevelt Dam. In 1898 he reentered government service, this time as the Arizona territorial governor's secretary and Spanish translator. He was appointed territorial auditor and bank comptroller in 1908 and chairman of the Board of Equalization, the Board of Control, and the Railway Commission in 1909. At the same time, he was owner and editor in chief of the *Arizona Republican,* a newspaper he had established in 1907, and president of the Phoenix Board of Education, a position he would hold until 1922. In 1911, Ely sold the newspaper and in 1914 was named general manager of the Valley Bank Adjustment Company. Five years later he was called back into state government by Governor Thomas Campbell, who asked him to serve on the delegation representing Arizona in the negotiations that would produce the Colorado River Compact.

The political infighting, bitter controversy, hostility, and paranoia that characterized Arizona's reaction to the interstate negotiations concerning the Colorado River left Ely weary and ready for another change. After the drafting and signing of the compact in Santa Fe in 1922, he left Phoenix for Washington, D.C., and a job as administrative assistant and chief clerk in the Department of Justice. A year later a new challenge presented itself: he was named director and treasurer of the Federal Land Bank in Berkeley, California. Secretary Wilbur found him there in 1931, ready and eager, at an age when most men were contemplating retirement, to leave his comfortable life in the misty hills of coastal California and start over in the bleak desert of southern Nevada.[47]

The new city manager arrived in Boulder City in November, 1931, startling many of the residents with his physical appearance. The town was populated almost exclusively by youthful, vigorous men and women. "It is a young man's job," Frank Crowe once remarked to a journalist. "The old can't stand it."[48] Sims Ely, on the other hand, looked every one of his sixty-nine years. His frame was thin to the point of being emaciated. A wispy fringe of gray hair spilled over his brow in wind-blown disarray; large ears sprang from his temples like granite outcrops jutting from sun-ravaged cliffs; creased, leathery skin stretched tautly over his long jaw, then sagged deeply inward to cover his gaunt cheeks. A great Roman beak of a nose emerged from a furrow in his forehead and plunged downward, pointing like a long, accusing finger at tightly pursed lips which seemed to be perpetually frozen midway between a sneer and a frown. His most striking features, however, were his eyes: they were the glittering, unblinking eyes of a bird of prey, hooded by dark, bushy eyebrows that drew up and back into bristling, accusing arcs when he focused on a face. The sharply creased tweed banker's suits he wore, the starched white cuffs and collars, the tightly knotted club ties, the neatly blocked Stetson—these only increased

Sims R. Ely, city manager of Boulder
City. (Bureau of Reclamation)

the aura of cool, angular asceticism that surrounded him and set him further apart from the rest of Boulder City's population.

\   The khaki-clad workmen and their young wives were not sure at first
what to make of this austere, unsmiling figure who always looked as if he
had just bitten into a lemon and not enjoyed it, and many hastily decided
that they did not like him. But being liked was not the point, and Secretary
Wilbur undoubtedly would have been gratified to know that during his
first weeks on the job, the city manager had succeeded in arousing feelings
of uncertainty and intimidation in the people of Boulder City. Ely himself
obviously understood and relished the Olympian role he was to play: he
was the law in Boulder City, judge, jury, and executioner. Respect, not
affection, was his goal, and power, not persuasion, his means of achieving
it. He would be critical, cantankerous, and crotchety and he would always
be right. For a brilliant, opinionated man who had spent more than forty
years cajoling recalcitrant politicians, voters, shareholders, and employees
to do his bidding, usually with less than total success, the job was a dream
come true.

Ely's first and most important task after he was installed as city manager was midwifing Boulder City's embryonic business district into healthy,
robust life. Orderly development of commerce was one of the keys to preventing Boulder City from becoming a western boom town of the tradi-

tional rip-roaring, hell-for-leather variety. Consequently, the Interior Department had announced that the number and type of businesses allowed to operate in Boulder City would be severely limited and that the "personal fitness"—including character, personality, age, and physical condition—of all would-be business operators, as well as their financial fitness and past experience, would be scrutinized and graded by government personnel before selections were made and permits issued.[49]

The purpose of establishing quotas and employing a rigid selection process, the Interior Department said, was to protect Boulder City residents from unscrupulous and exploitative commercial practices and to keep away the many small businessmen who would be wiped out by ruinous competition if Boulder City were open to all. This solicitude was laudable, but it stemmed just as much, or more, from concern about the government's image than from genuine concern about the welfare of the dam workers. Interior Department officials cringed when they thought about their much-publicized model town being overrun by hucksters and hootchy-kootchy dancers and devastated by a wave of business failures. They blanched when they thought of its becoming the target of congressional investigations and providing ammunition for scathing political attacks by members of the opposition party. The solution was to clamp tight restrictions on the businesses that were allowed to operate, thereby protecting Boulder City's citizens and simultaneously protecting the political flanks of the government agency that was their landlord.

In March, 1931, Secretary Wilbur appointed Louis C. Cramton, a Michigan lawyer and former congressman, special attorney in charge of planning Boulder City's commercial development. Cramton understood the problem and the political stakes, and stated them succinctly in a June article: "Boulder City . . . can not be the ordinary construction camp," he wrote. "The Boulder Canyon Project . . . has aroused the interest of the people of the United States to a greater degree than any other construction project . . . with the possible exception of the Panama Canal. The new town shares in this widespread interest of the country. . . . There is no doubt in my mind if the Government desired to do so it could create in Boulder City in the next year one of the most spectacular boom towns in recent history. . . . Ruin would inevitably follow . . . for the business possibilities in Boulder City are limited. Certainly the Bureau of Reclamation does not desire to have any part in such wholesale business disaster."[50]

Even as Cramton went to work devising a system for selecting the small group of businessmen who would be allowed to open stores in Boulder City, shopkeepers across the nation were reading about the new town and the dam project's giant payroll. Many of these merchants were on the edge of bankruptcy or already had gone over the brink; they dreamed of prosperity in a place insulated from the harsh economic climate of the Depression, and by the thousands they sent letters to the Bureau of Reclamation asking for information about Boulder City's commercial prospects and pleading for permission to engage in business there. Others were less constrained and more opportunistic. They simply packed up and headed for Nevada, hard on the heels of the army of would-be dam workers that had invaded the state in late 1930 and early 1931. This second wave of immigrants was made up of peddlers and prostitutes, bootleggers and bunco steerers, cardsharps and quacks—purveyors of every kind of good and service, legal and illegal—all intent on getting their piece of the Hoover Dam pie. The loose morals and wide-open business climate of Las Vegas appealed to many of the newcomers, who ventured no farther than the city limits. Others, driven by curiosity or slim pickings, made forays across the desert to the project reservation, which in the spring and summer of 1931 had not yet been sealed off by U.S. marshals, to drum up business among the small advance crews of engineers and carpenters encamped at the Boulder City town site.

One such enterprising entrepreneur was Danny, a traveling refrigerator salesman from Los Angeles, who wandered into the Boulder City camp one afternoon in May, 1931. The two hundred or so construction men then living in the camp had learned to expect the unexpected from the desert: it served up snakes, scorpions, sandstorms, and other unpleasant surprises without warning, but for a refrigerator salesman, resplendent in blue serge suit, snowy white shirt, and gleaming wingtips, to materialize out of the burning wastes was sufficiently novel to cause a stir. Danny was no desert mirage, however. He had read or heard about the new town rising in the wilderness, smelled opportunity, and followed his nose to Nevada. His salesman's instincts were sound, for in the stifling heat home refrigerators were a very necessary and desirable commodity, but his timing was terrible. There was as yet no running water, no reliable source of electricity, and only a handful of homes standing in Boulder City.

With the dream of a sales bonanza gone, Danny pondered his next move and cockily sized up the rough-looking construction stiffs who were returning to the camp after a hard day's work. "His opinion about us could be read upon his face," recalled one of those stiffs, engineer John Meursinge. "He compared us to the wild men of Borneo." For their part, the workers looked Danny over and concluded, correctly, that he was a smooth operator from California come to take advantage of them. With battle lines thus drawn, the city slicker and the desert rats entered one of the sweltering temporary bunkhouses and sat down to play a game of poker.

Outside the sun set and the stars rose, but in the bunkhouse the combatants remained fixed in their seats, eyes glued to the game. Arrayed around the table were three grimy, hirsute field engineers, naked except for shorts, sweat dripping from their arms and chests, and the neatly barbered Danny, also perspiring heavily but refusing to loosen his collar or shed his wrinkled blue suit coat. Back and forth the cards flew in the dim lantern light; lower and lower the players slid in their sweat-slick chairs, hunching their shoulders and knitting their glistening brows as the game went on. Just when it seemed the marathon contest might never break up, Danny announced that he had to answer a call of nature. The engineers reminded him that there was no running water and offered detailed directions for finding the outhouse, which was about one hundred feet behind their little dormitory. Up to this point the city-bred salesman had more than held his own with the grizzled camp dwellers, but when he stepped to the door and looked out into the inky-black night, something inside him quailed. Heat, dust, and an alien landscape he could cope with, but a trek through the pitch-dark desert to a lonely outhouse stretched his tolerance to its limits.

One of the engineers saw him hesitate and, sensing a chance to seize a psychological advantage that might lead to quick profit when the card game resumed, intoned, "Do not sit on a scorpion." Danny gritted his teeth and plunged off into the night while the engineers leered at each other and listened to his stumbling footfalls, the snapping of brush, and his muttered oaths. Presently the outhouse door creaked open and silence descended. Then, without warning, the night erupted. A flash lit up the sky, a voice yelped with surprise and fear, and the dull thump of an explosion rattled the flimsy wall of the bunkhouse. The engineers rushed outside and beheld the outhouse engulfed in flames. In the flickering red light they also saw Danny, his smoldering pants still bunched around his ankles, scampering away from the blaze in the general direction of Las Vegas. He had struck a match to make sure there were no scorpions on the toilet seat and, unaware that human waste decomposing in desert heat releases large amounts of explosive methane gas, flicked the still-burning match into the outhouse hole and detonated the entire structure.[51]

Such was the rough-and-ready commercial environment on the federal reservation during the early days of the Hoover Dam project, but there were not enough outhouses and matches in all of Nevada to fend off the plague of salesmen who would be Boulder City–bound if word went out that the gate was open and the dam payroll fair game. Consequently, within weeks of his appointment, Special Attorney Cramton announced the rules the government would apply to ensure the orderly development of Boulder City's business district. Four categories of permits to engage in business would be issued: a Class A permit would give its holder the exclusive right to operate a "public utility," such as a telephone system, garbage plant, tourist camp, or airfield, free from competition. Class B permits were for wholesale and retail stores and were limited to two per type of store. Class C permits were for operating "industries or services requiring special treatment," such as service stations, banks, bus lines, and telegraph companies, and were not limited in number. Class D "personal" permits for doctors, lawyers, and dentists, were not limited, either.[52]

The low ceiling on the number of Class B permits to be issued for wholesale and retail stores raised some eyebrows. "Government curbs trade at Hoover Dam," announced the *New York Times,* pointedly noting that the Bureau of Reclamation was warning potential applicants that "Six Companies will be operating a commissary for its employees."[53] Sensing potential controversy, Cramton moved quickly to scotch any speculation that restricting wholesale and retail businesses to two of each kind was intended to protect the contractors' company store from too much competitive pressure. "The purpose of the government in limiting the number of permits at this time is to guard against the wholesale bankruptcy which would otherwise result from the extravagant overdevelopment of the business district because of nationwide interest in the project, and plans based in many cases upon hopes rather than information," he wrote in a press release.[54] The statement also provided detailed information on how applications for permits were to be filed, how the final selection of competitive permits would be made, the cost of leasing a business lot (three hundred dollars a month on average), and a warning about the rugged climate and uncertain prospects for Boulder City after the dam had been completed.

By the cutoff date of June 30, 1931, several hundred applications had been filed, and the laborious selection process began. A three-member board sifted through the stacks of paper, evaluating the applicants' personal and financial backgrounds and grading them in order of desirability. The decision to issue or deny a permit was made by City Manager Sims Ely, who imposed his own stringent standards to make sure that Boulder City remained wholesome and that the degree of competition and the level of prices suited him. Ely was particularly concerned that unsavory characters might be issued permits to operate legitimate businesses which then could be used as fronts for bootlegging, gambling, or prostitution; he was unstinting in his investigation of applicants' moral fitness and was quick to deny a permit if he was the least bit suspicious of their motives.

Sometimes the prospective permittees made his job easy, as in the case of the application for a taxi-dance hall, which included the less-than-convincing assurance that the female dance partners for hire would all be "respectable girls on salary."[55] Sometimes the subterfuge employed by applicants to disguise their true plans was so subtle that Ely was taken in. For example, he issued a permit to a man named Frank Gotwall to operate a taxicab service, only to discover later that Gotwall had an arrangement with the Railroad Pass Casino, located just outside the reservation gate: Gotwall delivered the customers from Boulder City to the door and the casino paid their fare.[56]

Approximately one hundred Class B permits were issued in the first batch. Once these businesses had opened, Ely used his authority to grant or deny additional permits as leverage to keep them toeing the price line he drew while also seeing to it that they turned a sufficient profit to continue operating. "As matters stand," he reported in August, 1932, "there has been no business failure . . . and all establishments are doing fairly well. They, of course, are not doing so well as they would like, but all are probably doing better than they would do anywhere else just now, for here is one of the few unimpaired pay rolls of the country." [57]

Soon after the first group of independent businesses had opened in Boulder City, it became clear that the primary threat to competition would not come from the permittees' conspiring to fix high prices, as Ely had feared, but from Six Companies' Boulder City Company Store. The company store had an enormous advantage from the start: it was the first retail business to open and also the biggest, offering food, drugs, clothing, appliances, furniture, hardware, school supplies, and other sundries under one roof, as was certainly not the case with the Class B permit holders, who were generally limited to selling just one line of merchandise. The company store's competitive edge was sharpened even more by Six Companies' decision to offer its workers the option of receiving advances against their wages in scrip, which could be redeemed only at the company store. The scrip was issued in the form of coupons and coins that had an engraved likeness of the dam on one side and the words "Good for Trade Only at Boulder City Co. Stores" stamped on the other.

Anyone who had scrip could convert it to cash in a pinch, but only by selling it at a steep discount, usually 20 to 25 percent off face value. Six Companies defended payment in scrip, claiming it amounted to nothing more than offering a simplified charge account at the company commissary, but the proprietors of the independent stores saw the situation differently. They complained that because almost every wage earner in Boulder City worked for Six Companies, and because the company store offered an almost complete line of goods and services, the scrip-issuing scheme was a deliberate attempt to stifle competition and drive them out of business.

On the question of scrip, the city manager was uncharacteristically silent. Although he almost certainly disapproved of the practice because it was anticompetitive, he was not about to buck Secretary Wilbur and Commissioner Mead, who on this and other issues had allied themselves with Six Companies and made it abundantly clear that they saw nothing wrong with the contractors' turning a profit on the dam job whenever and wherever they could. The scrip controversy was destined to simmer for another two years until the Roosevelt administration took control of the Interior Department in 1933 and ordered scrip payments stopped. [58]

Along with his duties as business commissar, City Manager Ely's most important official responsibility was fostering and maintaining the wholesome living environment that was Boulder City's principal raison d'être. Given the long history of hell raising in construction camps and Nevada's notorious laxity in enforcing Prohibition, this was an almost impossible assignment, but it was one that the ascetic Ely especially relished. He was determined that demon rum was not going to disrupt the Hoover Dam project or tear apart the social fabric of Boulder City, and his struggle against it quickly took on the overtones of a holy war.

The battle against bootleggers had begun on March 24, 1931, some seven months before Ely arrived, when federal officials issued an ultimatum: the Hoover Dam job was going to be "bone dry." The announcement was timed to coincide with a raid by Prohibition agents on a prospector's cabin near the dam site. The shack was thought to be the headquarters of a large moonshine ring, and government authorities hoped that by smashing it in a blaze of publicity they could scare off any other would-be smugglers or amateur whiskey distillers who might be thinking of starting operations in or around Black Canyon. But instead of serving as a warning, the raid turned into a fiasco. Someone had tipped off the bootleggers, and when the squad of agents charged into the cabin, they found it deserted. The only contraband left behind was twenty cases of flat beer.[59]

The embarrassed federal officers regrouped under the leadership of Marshal Claude Williams, Ragtown's unofficial mayor, and began planning their second strike: a large-scale sweep of the project reservation. First, however, they had to find the bootleggers in the wild country around Black Canyon, and this was not easy. The 144-square-mile reservation was flanked on the west by the River Mountains and on the east by steep, boulder-strewn ravines leading down to the Colorado. Both areas were pocked with caves and slashed by narrow side canyons, offering hundreds of potential hideouts for whiskey peddlers, stills, and caches of bottled liquor. Williams was able to pinpoint many sites, however, by cataloging tips received from informants and then sitting with field glasses and a topographical map, patiently observing and plotting the telltale wisps of smoke from operating stills and the movement of men and mules in and out of the sheltering arroyos. Eventually his map was covered with little crosses, and in April and May, 1931, he launched a series of raids that succeeded, at least temporarily, in flushing out the bootleggers and clearing the project reservation.[60]

Meanwhile, outside the reservation, another force of Prohibition agents was trying to clean up the seamy strip between Las Vegas and Boulder City. On May 18, 1931, fifty "dry officers," led by Colonel George Seavers of San Francisco, descended on a tent and shack colony, known as Midway, on Boulder Highway. "We're going to make this place safe for Hoover Dam workers," the flamboyant Colonel Seavers announced to accompanying reporters as his men began fanning out at noon. By nightfall they had raided every known bootleggers' rendezvous within an eight-mile radius, including twenty-five speakeasies, five breweries, and three stills, and had managed to confiscate large quantities of liquor and make ninety arrests. The newspapers gleefully reported that the Las Vegas jail was "filled to overflowing," and the foray was adjudged a great success.[61]

The flow of illegal whiskey was interrupted only briefly by the spring raids, however, and as the excitement died down and the press turned its attention to other matters, bootleggers and law officers settled in for a long war of attrition. Big sweeps were made occasionally, but for the most part the Prohibition battles were small-scale affairs involving the matching of wits rather than the use of brute force. Typical of these encounters was Claude Williams' discovery and apprehension of a moonshiner in the shantytown at Railroad Pass. Williams was walking through the town one afternoon when he noticed a suspicious-looking character lolling outside a tent. The sharp-eyed marshal noticed that the man's right shoe had a deep crease worn through the middle of the sole, suggesting that he had been doing some hard shoveling. Entering the tent, Williams saw that the dirt floor had been disturbed recently; doing a little shoveling of his own, he dug up a chest full of liquor and then made his arrest.[62]

Erection of the Boulder Highway gate in August, 1931, helped police catch some of the rumrunners before they reached Boulder City, but in spite of car searches, backseat smuggling continued, and the movement of whiskey packers into the reservation via a pass in the River Mountains a few miles northeast of the gate resulted in a new name for the entire area: Bootleg Canyon.[63] This was the situation when Sims Ely entered the fray, and although it was obvious that he could not control the availability of liquor with anything less than an army, he did think he could restrict its consumption in Boulder City by imposing severe penalties for drunkenness. The rules he set were draconian: a worker caught intoxicated within the boundaries of the reservation was to be fired from his job and expelled from Boulder City. Drunken workers trying to enter the reservation after a night in Las Vegas were to be detained at the gate until they sobered up; second violations would result in incarceration and usually expulsion. Not even Six Companies senior personnel would be exempt from his edict against drinking in Boulder City. In July, 1933, at one of the Saturday night dances at the mess hall, Assistant Superintendent of Construction Woody Williams was apprehended with a flask of whiskey in his coat pocket. Even though Williams was Frank Crowe's right-hand man and a key figure on the job, Ely refused to make an exception: the Bull of the River was banished from the reservation for thirty days.[64]

To make sure that potential violators understood the consequences of being deported, the city manager made statements and issued press releases dwelling on the national unemployment rate and describing the hardship that awaited men who lost their jobs on the dam. "In the interest of the workers here, it should be understood by everybody that to be drunk on the Boulder Canyon Project Federal Reservation means expulsion from the reservation, with consequent loss of the job," he wrote in a typical broadside. "It is unfortunate that some of the men working here have very short memories concerning the terrible conditions of unemployment which prevail elsewhere. Within a short time after getting their jobs they forget all about the privations, hardships, and despair they underwent before they secured employment, and they proceed to 'blow' their money with bootleggers and toss away the jobs for which they have waited so long. If the workers on this reservation can not refrain from getting drunk, they must step aside for the sober men who are anxious for employment on this work. It is a matter of choosing between drink and the job." [65] In spite of such stern warnings and Ely's unfailingly swift and harsh punishment of violators, drinking was never fully curbed. The repeal of Prohibition in 1933 made enforcement of the continuing restrictions in Boulder City even more difficult, and arrests and evictions from the reservation continued unabated for the duration of the Hoover Dam project.

Bootleg whiskey might have been Sims Ely's special bête noir, but it was by no means the only aspect of life in Boulder City that grated on his sensibilities and provoked him into issuing warnings and taking disciplinary action. For example, he had a fetish about neatness—the city streets were swept clean by men with brooms every day but Sunday—and if he saw a front yard that had trash in it, he was likely to call the offending renter to his office in the municipal building, deliver a lecture on the importance of sanitation, and order that the garbage be picked up at once. This fastidiousness extended to matters of dress. Ely was from the old school; he always sore suits and tightly knotted ties, even in the middle of the summer, and he believed that others should follow his conservative lead. A formal dress code was never actually promulgated in Boulder City, but if the city manager saw a man or woman walking along the street in clothing that he thought was inappropriate or offensive, he would stop the individual and express his disapproval in full view of passersby.

Ely also had no qualms about intervening in domestic disputes. If he heard that a husband and wife were fighting, he would call the unhappy couple before him and try to mediate their differences. If that failed, he would on occasion act as a one-man divorce court, decreeing separation, awarding custody of children, and fixing the level of support to be paid by the father. In one instance, investigators traced to Boulder City a man who had abandoned his wife and children in another state. They asked Ely to expel him from the reservation so that he could be extradited to his home state to face charges. In typically high-handed fashion, Ely refused their request, telling them the man was a good worker who had behaved while in Boulder City. He did offer one concession, however: he would see to it that the man sent part of his wages to his family every month. [66]

With such peremptory actions as these, Ely consolidated his position in Boulder City as an unassailable autocrat. His word was law. His authority was paramount: above that of Six Companies and even above that of the courts and police agencies of other states. He used every tool at his disposal, including superior intellect, remoteness, age and appearance, and the frequent exercise of raw, arbitrary power to cow the residents of Boulder City into behaving as he wanted them to. Perhaps the ultimate symbol of his pervasive, unremitting campaign of control was the siren that blew every night precisely at nine o'clock. Ostensibly it was sounded to remind children that they should be in off the streets, but its long, piercing wail, keening across the little desert town, was really intended to warn the adults that the city manager was on his throne and that they should think twice about the possible consequence of any acts, illegal or immoral, they might be contemplating.[67]

⟍ The use of threats and crude psychological stratagems to keep Boulder City's citizens walking the straight and narrow was at least understandable if not particularly appealing, but a more sinister aspect of Ely's despotism, the surveillance and harassment of suspected labor organizers, was much harder to justify. This activity began in the wake of the August, 1931, strike when the small Boulder City police force was beefed up into a ten-man corps of so-called reservation rangers. One of the new rangers was Glenn E. ("Bud") Bodell, former Clark County chief deputy sheriff, former operator of the Boulder Club in Las Vegas, and the man behind the jailing of IWW organizer Frank Anderson and his colleagues in July, 1931. On the strength of the latter qualification, he was named police chief of Boulder City on August 24, 1931, and was instructed to prevent any recurrence of labor trouble by weeding out agitators and radicals, who, according to Frank Crowe, had infiltrated the Six Companies work force.[68]

Bodell soon held sway over the entire ranger corps because of his involvement in antilabor activities, and he further augmented his authority by forming a broad network of spies and informants who reported directly to him. In 1932 he boasted to a newspaper reporter that he had recruited thirty "undercover agents" from the ranks of the Six Companies employees and bragged that the "agitators" were marked men. "We've got 'em spotted," he said. "Give 'em rope until they blab too much—then off the reservation they go." When the reporter asked him how he felt about being called Butcher Bodell and Little Mussolini by the IWW newspaper *Industrial Worker,* he replied with a sneer: "We *do* know how to handle those guys. I guess maybe one fellow did get kind of cut up."[69]

Bodell was both physically and temperamentally suited to ruling with an iron fist. In his ranger hat and Sam Browne belt, he looked like a caricature of the head-knocking, union-busting bull of labor mythology. He was a heavyset man with a thick neck, lumpy face, and barrel chest. His legs were short and bandy, a characteristic accentuated by the riding breeches and cavalry boots he wore and by the large ivory-handled revolver that seemed to be grafted permanently to his right hip. There was nothing

Glenn E. ("Bud") Bodell, deputy U.S.
marshal, reservation ranger, and Six
Companies special investigator and
security chief. (Dennis McBride
Collection)

phony about the aura of toughness he projected, however. He had a taste
for bare-knuckles brawling, and he was not afraid to draw and use his gun
to pistol-whip or shoot any suspect foolish enough to resist arrest.[70]

With the blessing of the Bureau of Reclamation, the full cooperation
of Six Companies and City Manager Sims Ely, and the assistance of his
paid informants, Bodell cracked down hard on labor organizers and their
followers. Workers suspected of being IWW cardholders or IWW sympathizers were detained, interrogated, and summarily deported, and their
names were added to the employment agency's long black list. In the
twelve months between October, 1931, and October, 1932, the purge
reached epic proportions. According to a published Bureau of Reclamation
report, 1,056 people were labeled undesirable and expelled from the project

reservation during that period.[71] This astonishing figure, representing nearly three expulsions per day, is made even more noteworthy because Boulder City's peak population in 1932 was only 5,000.

⬩ Within months of his appointment as Boulder City police chief, Bud Bodell's power was such that he was able to start a gambling operation in the Six Companies recreation hall in open violation of the reservation regulations he was sworn to enforce. The recreation hall setup of six eight-man poker tables was modest by Nevada standards, but it was startling that it existed at all, given that circumvention of Nevada's legalized gambling had been offered as one of the reasons for building Boulder City in the first place. Bodell defended his poker game as a necessary measure to protect the workers from crooked operators who would cheat them of their money. "You can't stop gambling among these men," he said in his 1932 interview. "Until the games moved into the open, card sharps were active. I found eleven decks of marked cards. Then the company built its clubhouse and we moved the games there where we can watch them."[72] There were intimations that Bodell was doing more than just watching the games, that in fact he was drawing on his casino experience and his position in Boulder City to take a house cut, which he shared with Six Companies. Whatever the reasons for violating reservation regulations and whoever was receiving the profits, there could be no question that the poker operation was sanctioned by the Bureau of Reclamation, for Sims Ely turned a blind eye to it, even as he was busily turning down all business-permit applicants he suspected of having any interest in gambling.[73]

If there was any resentment of this twisting of the reservation rules, the forcible eviction of hundreds of workers, and the creation of a police-state atmosphere, it was not expressed loudly. In fact, many of Boulder City's citizens heartily approved of the tough antiunion law-and-order line taken by the city administration. Boulder City "represents most of our national ideas as to what a city should be," wrote one of its residents in 1932. "It has no past replete with scandals and mistakes . . . and a spirit of cheerfulness and democracy prevails. . . . Regulations are, of necessity, strictly enforced here as not only must the inhabitants of this remote little town in the desert be protected but the great dam must be forever safeguarded and eternally watched."[74] Others were less sanguine. "The average worker in Boulder City . . . was a very dissatisfied person," a construction veteran recalled of those years. "There were no facilities for entertainment, no swimming pools, no picnic grounds, nothing, nothing. Once you were there you were at the mercy of the Six Companies. The whole set-up was close to slavery."[75] But even those who hated Boulder City for its drabness and isolation and who recoiled at its company domination and repressive government rule recognized the futility of complaining. Labor Commissioner Leonard Blood's list of applicants for jobs at Hoover Dam, numbering twenty-two thousand at the close of 1932, cast a long shadow, as did the specter of Sims Ely and Bud Bodell, and it was evident that from the outside looking in, Boulder City, where everyone had a job, a full stomach, and a roof overhead, appeared to be the model town the government said it was, whatever the reality.[76]

Certainly from the Bureau of Reclamation's and Six Companies' perspective, the Boulder City experiment was a resounding success. Both took considerable pride in the role they played in the birth and development of the town, and this was understandable in light of the results that were achieved. For example, the tremendous speed and efficiency with which the construction of the dam proceeded was linked directly to Boulder City's existence—a connection starkly illustrated by the contrast between the number of heat-related casualties and the work delays during the summer of 1931, before Boulder City was finished, and the much improved health statistics and work productivity figures after its occupation.[77] Boulder City also made it possible for the government to clean out the pestilential slums of Ragtown and McKeeversville, set and maintain reasonable housing and sanitation standards, and control the flood of job seekers into the project reservation, an influx that had threatened to create anarchy in the camps on the banks of the Colorado.

In addition to promoting worker health and efficiency, Boulder City turned out to be an unqualified public-relations triumph for the bureau and the contractors. The town received a great deal of national publicity, and the descriptions were almost uniformly positive, comparing it very favorably with Las Vegas and with other construction camps at earlier projects. Six Companies quickly grasped the effect such favorable reviews had on congressmen when they voted on Boulder Canyon Project appropriations, and on the consortium's chances of being awarded contracts for future jobs. A public- and press-relations director, Norman Gallison, was hired in September, 1931. Gallison cranked out articles and press releases, shepherded reporters around Boulder City and the dam site, orchestrated photo opportunities, and arranged celebrity visits, all to great effect. The stories extolling Boulder City and the photographs and newsreel film of the gleaming mess hall, the neat dormitories, and the landscaped parks helped to foster a favorable impression of the entire project and deflect charges of unsafe working conditions and labor exploitation. The positive attitude created in the press was so strong that Six Companies and the Bureau of Reclamation could, with complete impunity, turn the Boulder Canyon Project Federal Reservation into a miniature police state, all in the name of worker health and well-being.

Of all the pages of statistics describing the housing and feeding of the dam laborers, the smug, congratulatory editorials about social responsibility, and the glowing accounts of life on the federal reservation in the Nevada desert, perhaps the most telling assessment of Boulder City was offered by a 75-year-old prospector named Jack Williams. The grizzled

Williams, a veteran of the booms at Deadwood, Cripple Creek, Pioche, and Tombstone, was interviewed by a *New York Times* reporter who found him camping in an arroyo outside Las Vegas. Asked what he thought of the model city by the dam site, Williams reflected for a moment, then spat into a creosote bush. "A feller couldn't make a real old-time Western town out of Boulder City if he tried," he said, his voice dripping with scorn. "If these fellers were after gold you might do something to stir up some life around the edges. But this crowd is only after jobs." [78]

CHAPTER FIVE
"Incessant, Monstrous Activity"

During the short, dark days of December, a chill wind gusted across Las Vegas Wash and blustered up the fan-shaped bajada at the western base of the River Mountains. It swirled in the arroyos, gathering momentum, then raced through the passes and swooped down on Boulder City. Dirty gray clouds gathered along the River Range's jagged peaks, only to be ripped away by the wind and sent hurtling across the Eldorado Valley, dragging long, curving tails of freezing rain in their wake. The raw afternoons faded quickly, and in the bitter gloom of early-winter dusk the wilderness of sand and rock became a lifeless, monochromatic moonscape. Summer, with its brilliant sunlight and shimmering heat, was an uncertain memory, something that might have been imagined so implausible did it seem now. Reality was the wind: cold, damp, and implacable.

When daylight finally expired, low clouds blotted out the stars and spat snow flurries, which melted into icy puddles on Boulder City's deserted streets. The western horizon was blank and dark, but to the east a strange yellow-green glow illuminated the underside of the clouds. This eerie, artificial sunrise came from Black Canyon, which was ablaze with thousands of brilliant electric lights. Glowing bulbs, each fitted with a burnished metal dishpan reflector, were perched on poles and strung on long, sagging lines. They snaked along the canyon's rim in long zigzag patterns, forming an incandescent crown on the brow of the cliffs. Far below, the river bottom was bathed in the harsh white glare of powerful floodlights, and tiny clusters of dancing shadows, like swarms of moths flickering around a flame, showed where the graveyard-shift crews were working.[1]

On the canyon floor the bellow of engines and the clash of steel on rock drowned out the wind's shrill whistle. Bulldozers rumbled forward, pushing up great curling mounds of dirt. A dragline bucket, its jaws agape, slid gracefully down its cable and bit into the moist earth. The stubby boom of an electric shovel executed a glittering pirouette, and a load of muck cascaded into the bed of an idling dump truck. Farther up the canyon, above the busy excavation site, the Colorado swung in a gentle curve into the face of the Arizona cliff and disappeared, swallowed up by the twin holes of the diversion tunnels.

The rerouting of the river into these tunnels on November 14, 1932, had concluded the first critical phase of the Hoover Dam construction, but the main event—pouring the towering concrete body of the dam—could not begin until the work site had been sealed off from spring and summer floods and the foundation had been drained and excavated. Dewatering and digging up more than half a mile of riverbed was a backbreaking and unglamorous task, but it was essential to guarantee that the dam's base rested, solid and secure, on a footing of dry, smooth bedrock.[2]

To ensure that the hard-won victory of turning the river out of its old channel was not lost later during the high-water months, and to prevent any seepage or backwash into areas where concrete would be poured, work was begun on two earth-and-rock cofferdams. The barriers—one five hundred feet downstream from the diversion tunnel inlets and the other just upstream from the outlets—would straddle the width of Black Canyon, bracketing the dam and powerhouse sites. The upstream structure was particularly important because it would have to hold back a lake, swollen with approximately 600,000 acre-feet of water, that would pool between it and the diversion tunnels' portals. It would also have to be sturdy enough to withstand the devastating impact of a Colorado River flash flood.[3] During the summer, when the river's mountain watershed was awash with snowmelt, a peak flow of 150,000 cubic feet per second was possible. By comparison, the February, 1932, flood that swamped Black Canyon and temporarily halted tunneling operations, had crested at a relatively modest 57,000 cubic feet per second.[4]

To withstand the battering fury of a full-fledged summer deluge, the upstream cofferdam had to be massive, and it had to be finished before the warm sunshine and balmy winds of spring began to thaw the snowpack. To get a head start, work on the big structure was begun in September, 1932, eight weeks ahead of river diversion. A C-shaped earthen dike was built from the Nevada canyon wall to the center of the river and back, enclosing half of the cofferdam site. The impounded water was pumped out, and a five-cubic-yard dragline and three electric shovels chugged down into the mudhole to begin scooping out the foundation. After the river was turned into the tunnels in November, the Arizona portion of the foundation was pumped out and excavated.

Although it was not necessary for the upper cofferdam's base to be anchored to bedrock, its footing had to be solid. The exact depth of the loose sand and gravel deposited by the Colorado over the centuries was a mystery, and as this material was dredged up day after day at the rate of 217 cubic yards per hour, the supervising engineers anxiously awaited word that a firm, compact dirt layer had been struck. Such a stratum was finally exposed 18 feet down. This meant that the cofferdam, designed to contain a flood of 200,000 cubic feet per second, would measure 98 feet from base to crest. Its length would be 480 feet from canyon wall to canyon wall, and it would be 750 feet wide at the base. These were huge dimensions for a temporary structure, but, given the tremendous power and unpredictability of the Colorado, and the disaster that would occur if the barrier were breached, the Six Companies and Bureau of Reclamation engineering staffs did not think them at all excessive.

When the foundation excavation was finished on December 5, 1932, nearly 213,000 cubic yards of river silt had been gouged out and hauled to a dumping ground near the mouth of Hemenway Wash.[5] To replace it, dry sand and gravel from pits in Hemenway Wash was carried along the construction railroad to a trestle siding in Black Canyon, unloaded into huge piles, then reloaded by electric shovel into trucks that carried the mixture to the cofferdam site.

The scene at the upper cofferdam was one of frantic turmoil. Trucks raced back and forth through a haze of dust, dropping loads of fill as Caterpillar tractors grunted around in wide circles, spreading the sand and gravel into foot-deep layers with their bulldozer blades. Other Caterpillars followed, dragging barrel-like six-ton rollers studded with iron sheep's feet to compact the layers, which were periodically sprinkled with water. At the base of the canyon walls, jackhammer operators chopped three-foot-deep parallel grooves into the rock. Reinforced concrete counterforts would fit into the grooves and extend five feet into the mounded earth of the cofferdam to prevent water from flowing between the cliff face and the dam's body. The bass rumbling of truck and tractor engines and the percussive stuttering of jackhammers filled the canyon with a heavy, vibrating rhythm punctuated by the steady cymbal-like clang of a big McKiernan-Tiery air hammer driving steel sheet piling down to bedrock at the cofferdam's upstream toe.

During December as many as four thousand truckloads of fill were deposited each day on the crest of the swiftly rising cofferdam, and on New Year's Day, 1933, the final few loads were rolled into place. In cross section the body of the completed structure looked like a low and very wide pyramid with a flattened peak. Both the sloping upstream and downstream faces were blanketed with crushed rock, and the upstream face was then paved with reinforced concrete. Most of the paving was poured during the frigid days and nights of January—the only time during the entire Hoover Dam project that fires had to be kindled to keep setting concrete from freezing.[6]

The last of the sixteen-foot-wide paving strips, stretching from the toe of the cofferdam to its flat crest, was put into place in February, 1933, and the imposing levee was complete. The conclusion of this job well before the onset of the spring flood season virtually eliminated the possibility that high water would inundate the main construction area. A smaller cofferdam was to be built just above the diversion tunnel outlets to prevent backwash from eddying up into the canyon and interfering with excavation of the powerhouse foundation. This task had to be postponed for several months, however, because crews were still at work on the cliffs above the site, prying and blasting loose rock off the face where the canyon wall outlet works were to be installed.

As they had since they began their dangerous aerial work in the summer of 1931, the high scalers put on an exciting show for tourists who gathered at Lookout Point on the Nevada rim and for workmen in the canyon below. Monkeylike, they swung over and down the cliffs at the ends of long manila ropes, hoses, pinch bars, and jackhammers in tow, rappelling across the rocks with remarkable speed and dexterity. The scalers' wage, $5.60 per day, was 40 percent higher than that given men who did similar work in the tunnels or on the canyon floor, but few of the laborers toiling in relative safety on the riverbed begrudged the extra $1.60 received by the men dangling high above. As one of them succinctly put it: "A fellow got to risk his life to make that money."[7]

A devil-may-care attitude was an integral part of the high scaler persona, standard issue like the bosun's chair and hardboiled hat, but for all their bravado, the "human flies" were acutely aware of the dangers of their work and the terrible price to be paid for even the slightest miscalculation. Accidents caused by frayed ropes, poorly tied knots, falling objects, or the slips of scalers who were clumsy, tired, or drunk resulted almost invariably in a long, fatal plunge witnessed by everyone on the cliffs and hundreds of men working below.

The death of high scaler Jack ("Salty") Russell on September 21, 1931, was perhaps the worst such incident, judged by the terrifying impression it made on the men who witnessed it. Russell had gone over the Arizona rim at the start of the day shift and was operating his jackhammer on a ledge far above the river. Workers on the rim began lowering a sling full of drill steel when, without warning, one of the long, glittering bars

A worker pauses to drink from a water bag. (Library of Congress)

slipped loose and hurtled downward. One group of scalers heard the warning shouts and maneuvered frantically to get out of the way, but Russell was not so fortunate: the plummeting missile struck him flush on the head. Without a sound he fell, bouncing like a rag doll off several promontories, four hundred feet to the riverbank. The carnage was dreadful. "Brains and blood with pieces of torn flesh and tattered clothing [were] scattered over the jagged rock. . . . Three separate sheets were required to cover the remains," said an account of the accident.[8]

꙼   The aftermath of this tragic episode was as disturbing to many of the onlookers as the accident itself. By law the body of an accident victim could not be removed until the county coroner had examined it at the scene of death. Russell's body had landed on the Arizona side of the Colorado, in Mojave County, which had its seat in Kingman, sixty-five miles to the southeast over a rough desert trace. On this day it took the coroner six hours to arrive, and for almost the entire shift workers had to pick their way around the blood-stained sheets and swarms of buzzing flies.

Beyond the question of whether the Clark County coroner or the Mojave County coroner should be called to certify the cause of death, there was another ghoulish legal twist to the placement of bodies of men killed in Black Canyon: the state of Arizona paid more compensation to accident victims or their survivors than did Nevada. If a man was maimed or killed on the job, it was to his family's financial advantage that the accident take place on the Arizona side of the Colorado. After the river had been diverted and the work crews deployed on the riverbed, the disparity in compensation rates became even more important. The border between Nevada and Arizona, which for all intents and purposes had been a 300-foot-wide river, was now a razor-thin imaginary line bisecting the exposed canyon floor. The practical implications of this change and the opportunity afforded by delay in the arrival of the coroner was not lost on the workers; when possible, the bodies of men killed in Nevada were dragged to the Arizona side of the canyon. Likewise, men who were injured near the border made every effort to get across before medical and supervisory personnel arrived on the scene. There were stories of laborers with badly broken legs gamely crawling through dirt, over rocks, and across the path of oncoming traffic in an attempt to get into Arizona.

Six Companies, which was contributing to both states' insurance funds, also was acutely aware of the rate difference between Arizona and Nevada and tried to reduce the amount of compensation it was paying to dependents of Arizona casualties. The problem was discussed at a board of directors meeting in San Francisco on August 15, 1931, and it was resolved "that so far as practical, only single men shall be employed in Arizona."[9] This long-distance effort to change the actuarial picture was not particularly successful, however. The contractors' Arizona compensation costs continued to mount, pushed upward by the men on the scene who, by

High scalers at work stripping loose rock from the Nevada wall of Black Canyon.
(Manis Collection, UNLV)

hook or by crook, managed to have their accidents in Arizona. Sometimes the reason was poverty or gambling debts: electric-shovel operator Red Wixon saw a worker who was desperate for insurance money carefully position himself in Arizona and deliberately blow his little finger off in the exhaust jet of a jackhammer. More often the reason was sympathy. On one occasion, Altus ("Tex") Nunley, who worked as a surveyor's helper, was given the job of measuring to determine in which state an accident victim had died. "My boss told me, privately, 'Make it Arizona,'" Nunley said years later. "He [the dead man] really was in Arizona, barely." [10]

Salty Russell's fate aside, high scaling was not without its special re-wards. Shared risks instilled esprit de corps, and the nature of the job allowed a measure of freedom that was denied other men who labored under the unblinking gaze of an assistant superintendent. Not that the scal-ers were left completely to their own devices. "You just could not loaf on that job," remembered Buck Blaine, who went to work on the cliffs in 1931 when he was twenty years old. "Say you took off your gloves while you were drilling to wipe the sweat off your face, they would fire you for that." [11] The difference was that the high scalers were often out of their su-pervisor's line of sight, working alone under overhangs or in shallow de-pressions in the canyon walls. This lofty independence was admired, and to a certain extent envied, by others on the Hoover Dam project. "The sur-veyors considered taking lunch with the high scalers a thrilling experi-ence," wrote John Meursinge. "It seemed that always a few square feet of level surface could be found. . . . Then the lunch pails would come down. Sitting on a narrow bench some 500 feet above the river, legs hanging down, quite often straddling a piece of steel, [they] would consume their meals and talk." The men also enjoyed setting up the surveyors' transit and training its crosshairs on the tourists' viewing platform at Lookout Point. "Every girl has her eye on a high scaler," one of the young men would an-nounce as he eyeballed the distant crowd, but to the surveyors it seemed that just the opposite was true. [12]

By February, 1933, the high scalers were almost finished with their work on the cliffs between the lower portals of the diversion tunnels and the site of the outlet works on the canyon walls, and it was safe for crews to go ahead with the building of the downstream cofferdam. Although this struc-ture did not have to be as large or water resistant as its upstream counter-part, the amount of earth moved during its construction was still huge. All told, some 300,000 cubic yards of fill were hauled into Black Canyon, trucked to the lower cofferdam site, and bulldozed into position. In lieu of concrete paying, a 54-foot-high barrier of crushed rock was erected just downstream from the body of the lower cofferdam. Not long after the rock barrier was finished, in mid-April, Bureau of Reclamation engineers in-spected both cofferdams and pronounced them satisfactory. With this official acceptance, one of the special conditions in the project specifications was met and the government assumed the risk of fighting floods and paying for cleanup of river-related damage in Black Canyon.[13]

In the Financial Center Building at California and Montgomery streets in downtown San Francisco, the offices of Six Companies pulsed with excitement and optimism. Thanks to ingenious tunneling and concrete-placing techniques, innovative use of motorized equipment, meticulous planning and coordination of tasks, and relentless driving of the labor force by Frank Crowe and his lieutenants, the four huge tunnels, the two cofferdams, and most of the high scaling had been completed more than a year ahead of schedule.

The payoff for achieving such speed was great. When Frank Crowe and the Six Companies executives prepared bid figures for the various construction units in the winter of 1931, they assigned relatively high dollar figures to units that were part of the diversion phase and relatively low figures to units that were part of the subsequent construction phases. The result was a bid that was unbalanced so that the company would receive a large portion of its payments early in the job.[14] In this way the partners hoped to recoup their original five-million-dollar investment and, if all went well, have handsome profits in hand long before the project was finished. The soundness of this strategy was revealed on August 15, 1932: the Six Companies board of directors, noting that "surplus profits exceed one and one-half million dollars," declared a $34.75-per-share dividend on their capital stock. Six months later, on February 13, 1933, as the first loads of dirt were being dumped into the base of the downstream cofferdam, another dividend was declared, this one to distribute "surplus profits exceed[ing] one and two-thirds million dollars."[15]

Burgeoning bank accounts did much to foster cordial and harmonious relations among the Six Companies directors and to keep the dam work moving smoothly. That the consortium would be torn apart by warring egos had been a major concern of the project's underwriters, but forecasts of intramural wrangling proved to be exaggerated. After a brief period of strife at the outset, management of the project had been streamlined by creation of the executive committee, which had clearly defined prerogatives and responsibilities and was composed of four men who respected one another and worked well together.

The friendship of cerebral Felix Kahn, chief custodian of financial and legal affairs, and earthy Charlie Shea, head of field construction, was typical of the tightly knit professional and personal relationships that developed among executive committee members. The two men, so dissimilar in background and temperament, discovered a shared passion for gambling, and on this odd foundation their camaraderie grew. "I used to meet [Charlie] and Felix at the [San Bernardino] station when they came up from San Francisco and drive them across the desert to the dam," said Frank Crowe.

"All the way—five hours—they'd shoot craps on the floor of the Lincoln. . . . Charlie and Felix used to say to each other, to settle an argument, 'Right or wrong, you're right, you son of a bitch.' They really felt that way toward each other."[16]

Crowe had his own reasons for appreciating Charlie Shea. It was Shea who had prevailed on the directors to give Crowe a free hand at the job site. It was also Shea who, as executive committee member responsible for construction, spent most of his time in Black Canyon and served as both conduit and buffer between Crowe and the company directors in San Francisco. The esteem in which the superintendent held the rumpled, cigar-puffing Irishman was evident when he looked back at the project and reflected on the personalities of the various directors. "Kaiser and Morrison always thought of a job in terms of draglines and steam shovels," he said. "Kahn figured in terms of money and an organization chart. But Charlie Shea always thought in terms of men. He was the kind of man who'd ask you the time not because he wanted to know, but to see what kind of watch you carried."[17]

Like Crowe, Shea disdained pretension and chafed at inactivity. He was seldom at his desk in Boulder City, preferring to don his work boots, roll up his shirt sleeves, and wade into the whirl of construction activity. One afternoon, concrete finisher Marion Allen was boarding a transport in the company of a recently hired worker named Morgan. Morgan pointed to a bedraggled laborer sitting on a nearby rock and said, "That poor old man is too tired to even get on the bus."

"I looked to see who he was talking about," wrote Allen, "and saw Mr. Shea. True, I could see what Morgan saw—a man with a decrepit old straw hat, scuffed worn boots and well-worn pants and shirt. Many of the men laughed, and someone said 'Old Charlie would appreciate having someone feel sorry for him.' Before I could explain to Morgan who Charlie Shea was, Mr. Crowe drove up. . . . When he stopped, Mr. Shea climbed in the sedan. I then explained [to Morgan] the poor old man was probably a millionaire and one of the sharpest construction men around and was part of the Six Companies. . . . When I explained all this to Morgan, he said 'I see—and that was his chauffeur.' When I said, 'No, that was Mr. Crowe, the general superintendent,' all he did was shake his head."[18]

So it went, with Frank Crowe and Charlie Shea working shoulder to shoulder, choreographing the activities in Black Canyon and inexorably moving the job further ahead of schedule. At the same time Felix Kahn was enriching the contractors with his legal acumen and his astute management of the Boulder City Company subsidiary. This organization helped Six Companies reduce its cash outlay through payroll deductions and thereby recover a substantial portion of the wages its employees received.

The Boulder City Company managed cottage and dormitory rentals; provided utilities, garbage-collection service, and transportation to and from the dam site; and operated the mess hall, the company store, and the recreation hall. Charges for rent, gas, electricity, trash pickup, transportation, and dormitory residents' board were not billed: they were deducted from the workers' paychecks, and much of the money that was left was likely to be spent at the company store and recreation hall, a transfer Six Companies encouraged by allowing its employees to draw scrip, good only for purchase at company-owned businesses, against their future earnings. Also brightening the profit picture were several hundred thousand dollars in taxes due the state of Nevada, drawing interest in the company bank account while Nevada's claim of jurisdiction over the project reservation was litigated in federal district court.

➤   The contractors were riding high, probably much higher than they had thought possible when they formed the consortium and worked up their bid in 1931. They were earning multimillion-dollar profits at a time when other construction firms were begging for work or going out of business. Their relationship with federal officials, both at the field level in Black Canyon and at the executive level in Washington, was incestuously close, and they were being hailed in the press for their role in fighting the Depression and for making Boulder City an oasis in the desert. Already these accolades and the contacts established through the Hoover Dam job were paying off in the form of contracts for new projects.[19] The national political and economic atmosphere remained stormy, however, and the tide of events, which had run so strongly in Six Companies' favor since the awarding of the Hoover Dam contract, was about to turn. Soon the partners would see their profits put in jeopardy, their association with government upset, and their public image tarnished.

➤ The first harbinger of the difficulties ahead came in the spring of 1932. In Black Canyon, thirty-four hundred men were working around the clock seven days a week in the drive to divert the Colorado, but in the rest of the country unemployment was still on the rise. Thousands of desperate veterans were preparing to march on Washington to demand immediate payment of their World War I bonuses, and a panicky Congress, casting about for measures to counteract the deepening depression, was trying to eliminate all unnecessary expenditures. One of the targets the House Appropriations Committee zeroed in on as it moved to cut the federal budget was the ten million dollars requested by the Bureau of Reclamation to keep the Boulder Canyon Project funded. The House trimmed the figure to eight million dollars, and the Senate, not to be outdone, then slashed the amount to six million.[20]

In Boulder City and San Francisco the news of Congress' one-two punch was received with shock and dismay; a 40 percent cut in appropriations would have a devastating effect on the course of the work in Black Canyon; it would force the layoff of laborers, the cancellation of equipment and supply orders, and possibly the shutdown of the entire project for six months. The reduction "will cut 1500 men from our present payroll . . . with the further possibility of our losing the year we are now ahead of schedule," Six Companies President W. A. Bechtel told reporters.[21] Henry Kaiser was dispatched to Washington to lobby Congress to restore the funds it had cut. He was joined in this effort by officials from the Bureau of Reclamation. "Those familiar with the Federal reclamation policy realize that this is about the only Federal activity that by law is placed on a paying basis, because of the requirement that the construction cost of the irrigation projects shall be returned to the Federal Treasury through the annual repayments of the water users," said an editorial in the bureau's official monthly publication, *Reclamation Era.* "Yet the Bureau of Reclamation is continually forced to combat the ill-digested views of eastern and middle western critics who see only a menace in western irrigation by the Federal Government, and overlook entirely the benefits that have resulted to the Nation as a whole in thus bringing together the landless man and the manless land."[22]

Congress refused to regurgitate its ill-digested views on the Boulder Canyon Project appropriation until Herbert Hoover intervened. The president pointed out that, instead of saving money, reducing the appropriation and blunting the project's momentum would deplete the federal treasury because the project contract would have to be extended later at additional cost and the collection of revenues from the sale of power generated by the dam would be delayed. He asked that the appropriation be increased, and in July, Congress responded by passing a relief bill authorizing the release of additional money to keep work on the dam moving ahead.[23]

A financial catastrophe for Six Companies had been narrowly averted thanks to the intercession of Herbert Hoover. The taciturn engineer from Stanford had been a staunch supporter of the Boulder Canyon Project since his tenure as secretary of commerce in the early 1920s, when he midwifed the Colorado River Compact into being. Later he had assisted in drafting the legislation authorizing construction of the dam and had helped persuade Calvin Coolidge that the bill deserved presidential support despite its high cost.[24] His own administration, primarily in the person of his lifelong friend and confidant Interior Secretary Ray Wilbur, had worked closely with Six Companies in organizing the dam construction and running Boulder City, giving the contractors the widest possible latitude in managing the job as they saw fit. Wilbur and his Bureau of Reclamation subordinates had chosen to ignore the overtime that was being worked in Black Canyon in violation of the project contract's provisions, had winked at the contractors' practice of paying workers in scrip and had acquiesced in the use of federal police to suppress union activities, investigate insurance claims, and carry out other company business. They had also allied themselves squarely with Six Companies on a number of important occasions, most notably during the August, 1931, strike, when the threat of deploying army troops and the intervention of Construction Engineer Walker Young had been decisive in ending the walkout.[25]

Hoover had been a good friend to the contractors, and they would have been pleased to see him returned to the White House for another four years. As the 1932 presidential campaign progressed, however, it became clear that both Hoover and the Republican party were going to be overwhelmingly defeated and that a new set of Interior Department appointees with a less benevolent attitude toward the Big Six's operation of the Boulder Canyon Project soon would be in power. Even though the election results were virtually a foregone conclusion, loyalty dictated that Six Companies support the man who had supported it so faithfully. On the national level this backing would not make a shred of difference and in fact might damage the company's relationship with the incoming administration, but in Boulder City a gesture could safely be made by ensuring a large Hoover plurality. Thus in the fall of 1932 a strange, awkward, and faintly comical company-sponsored campaign to reelect President Hoover got under way.

For a time it looked as if Boulder City residents would not be allowed to participate in the presidential election at all. The status of the project reservation vis-à-vis Clark County and the state of Nevada was still being litigated, and neither Six Companies nor the federal government wished to concede that Nevada had jurisdiction over Boulder City. It was decided, however, that from the standpoint of public relations and fair play it was necessary to allow the Black Canyon work force to vote in the presidential election, and so the clerk of Clark County was asked to open an office in Boulder City and register eligible residents.

Meanwhile, photographs of President Hoover materialized on office walls, Hoover literature was distributed, and Hoover signs sprouted on lampposts all over town. It was abundantly clear to Boulder City's citizens where the company's loyalty lay, and there was nothing subtle about the pressure applied to ensure a large Hoover majority on election day. On October 20, 1932, Frank Crowe received a letter from Six Companies President W. A. Bechtel extolling Hoover's efforts that summer to secure additional appropriations for the dam construction work. Wrote Bechtel: "It is interesting to note by [President Hoover's] words: 'the resultant distress to these men and their families is unthinkable' that President Hoover has ever before him and in his heart a deep understanding that employment must be provided for men so that the families of the United States may be properly taken care of. It is my feeling that your organization and your men should know exactly what the President has done for them and that, therefore, this letter should be posted in a conspicuous place."[26]

The letter was posted and also printed—conspicuously, as Bechtel wished—in a large advertisement in the *Las Vegas Evening Review-Journal*. Marching orders in hand, Crowe organized a Republican rally to

be held in the mess hall on October 24, and invitations bearing the nota-
tion "This card to be turned in at the door" were issued to Six Companies
employees. The invitations, with their thinly veiled threat of a head count,
were really summonses and the hall was packed to hear Frank Crowe's address, titled "Roosevelt," and the speech of Nevada Republican Representative at Large Samuel Arentz. The latter ardently praised Hoover and
excoriated Roosevelt, but the only heartfelt response his impassioned ora-
tory drew from the hostile crowd was a snickering laugh when he said, "If
you fellows vote for the man [Roosevelt] born with the silver spoon in his
mouth, grass will grow in the streets of Boulder City!" Wrote one of
Arentz's listeners: "Living in a sea of sand, we wondered where the grass
would come from." [27]

As election day drew nearer, the campaign for Hoover gathered momentum. A Boulder City Republican committee was formed, and the names
of its members, with Frank Crowe at the top of list, was published in the
paper, causing a scramble among the Six Companies office staff to demonstrate allegiance to their employer by joining. No form of persuasion was
left untried. The town's children were invited to Grace Community Church
to hear Parson Tom Stevenson give a talk titled "When Herbert Hoover
was a Little Boy" and to eat free ice cream. Hoover banners and posters
continued to proliferate. In this coercive atmosphere, Boulder City residents either professed support for the Republican ticket or kept their political views to themselves. It was obviously unwise to buck the company-
backed Hoover forces by openly endorsing the Democrats, but support for
Roosevelt was widespread, and as the campaign moved toward its climax,
hostility toward the Hooverites began to break into the open.

One of the assistant construction superintendents had decided to pro-
claim his support for Herbert Hoover on his automobile. This was not
easily done in the pre-bumper-sticker days of 1932, but the superintendent
hit upon the solution of pasting large, waist-to-crown portraits of the incumbent president on the covers of the spare tires attached to his car's running boards. For days Boulder City pedestrians had the eerie experience of
being fixed momentarily by the dour, lifeless gaze of the president as his
large likeness glided by on the side of a dusty automobile. Then, one night,
a Roosevelt supporter slipped into the assistant superintendent's yard and
scissored out the heads of the Hoover portraits. Rather than admit defeat,
the superintendent continued to drive about displaying his now decapi-
tated president, much to the amusement of many onlookers, who took it as
a portent of what would happen on election day. [28]

On November 8, 1932, the people of Boulder City went to the polls,
and that night a crowd gathered in the Six Companies recreation hall
where the forty-eight states had been listed on a large blackboard. As the
results came in on the Western Union wire, a running tally was chalked up
on the board. There was a murmur of excitement as the first state went to

Roosevelt, light applause after the second state also went to the challenger, and finally loud cheering as state after state went into the Roosevelt column and it became clear that a new president had been elected. Announcement of the Boulder City vote totals unleashed pandemonium; the Hoover campaigners were shocked, the previously circumspect Roosevelt backers ecstatic. Of the 2,140 voters registered in Boulder City, 2,074 had cast ballots, and of these, 1,620, or 78 percent, were for Roosevelt. In the town that was building a great monument to his name, Hoover had been trounced by a margin of more than three to one.[29]

In Black Canyon on the night of November 8 the swing shift crew was hard at work pouring the concrete lining of the diversion tunnels, drilling on the canyon walls, hauling muck to the waste dumps, operating the locomotives that plied back and forth between the Arizona gravel pit and the low-mix plant beside the river. From truck cab to locomotive to concrete jumbo form, tidings of Roosevelt's victory flashed, carried by hundreds of eager tongues. Soon everyone on the job had heard the news, everyone, that is, but the high scalers who, dangling at the end of their ropes in the blaze of the floodlights, toiled in spectacular isolation. At 11:30 P.M., when the shift was over, they would rejoin the rest of the world, but the waterboys on the canyon rim, bursting with excitement, could not wait until then to tell them what had happened. Over the side went scores of waterbags, bouncing and jerking as they were eagerly lowered to the tiny figures below. On each bag was pinned a scrawled note: "He is elected."[30]

He is elected. To the workmen of Boulder City and the nation the words were charged with hope for better times, but to the Six Companies partners they had a faintly ominous ring. Proven political allies were on the way out. New, untested officials were about to take power. For Six Companies, the Bureau of Reclamation, and everyone else involved with the Boulder Canyon Project, the question was: Who would succeed Ray Wilbur as secretary of the interior? The answer came in February, 1933. Harold L. Ickes, an obscure, reform-minded Chicago lawyer who had helped swing liberal Republican votes in the Midwest to the Roosevelt ticket, was the new president's choice.

The Ickes appointment was something of a surprise to those familiar with Washington political machinations; traditionally, the secretary of the interior had been the cabinet's token westerner, a man familiar with the important role that reclamation and the management of federal lands played in the region's economy. Roosevelt had in fact first looked to the West to find someone to head the Interior Department, offering the job to Senator Hiram Johnson of California and then to Senator Bronson Cutting of New Mexico. Rebuffed by both men, he had turned to Ickes, who eagerly accepted the interior portfolio.

The Six Companies hierarchy scrambled to find out who this little-known figure was; their initial impressions could not have been reassuring. Unlike Wilbur, a conservative Californian, Ickes was a liberal crusader with a hair-trigger temper, a vain, abrasive, confrontational man who had little patience with those who did not share his views. For the contractors the only saving grace in the Interior Department reshuffling was the retention of Elwood Mead as commissioner of the Bureau of Reclamation. It was their hope that Mead, with his experience as a civil engineer, his intimate knowledge of federal water policy, and his strong belief in the importance of the Boulder Canyon Project, could act as an intermediary between them and the new secretary of the interior.

Ickes wasted no time in becoming a player in the Boulder Canyon Project drama: on May 8, 1933, in one of his first administrative acts as interior secretary, he decreed that henceforth the dam rising in Black Canyon would be called Boulder Dam rather than Hoover Dam, the name given to it by Secretary Wilbur in September, 1930. "The name Boulder Dam is a fine, rugged, and individual name," he wrote to explain his decision. "The men who pioneered this project knew it by this name. . . . These men, together with practically all who have had any first-hand knowledge of the circumstances surrounding the building of this dam, want it called Boulder Dam and have keenly resented the attempt to change its name." He further defended his name-change order by arguing that Hoover had had nothing to do with the conception of the project, that the legislation authorizing it had been passed during Calvin Coolidge's administration, and that Wilbur had acted inappropriately when he named the dam after the president he was serving.[31]

Ickes was correct about the public confusion created by Wilbur's christening the Black Canyon structure Hoover Dam. During the planning of the project and the fight over the authorizing legislation, the dam had been referred to as Boulder, although this was due not to any formal decision or general consensus but rather to the initial belief that Boulder Canyon, not Black, was to be the construction site. After Wilbur's action in 1930 the name *Hoover Dam* was used in various official contracts and congressional appropriations, as well as by many newspapers and magazines, but references to Boulder Dam also continued to appear.[32] However, Ickes' contention that Hoover had had almost nothing to do with the shaping of the project legislation or its eventual passage by Congress was blatantly false. The nature and importance of Hoover's contribution could be debated, but it could not be denied that he had played a major role in the birthing of the Colorado River Compact and in getting the Boulder Canyon Project bill through Congress. Furthermore it was his administration that had signed the construction contracts, launched the project, and secured five appropriations for it. As for the propriety of Secretary Wilbur's naming the Black Canyon dam after the president, there was a well-established tradition of naming dams after the presidents in whose administrations they were built, as could be attested by Wilson Dam in Alabama and Roosevelt and Coolidge dams in Arizona. Thus it appeared that Ickes' renaming of Hoover Dam was a political act, a mean-spirited attempt to shred the already tattered reputation of the former president, as well as an unsubtle snub to those who had supported him.[33]

The new interior secretary also hastened to resolve the long-simmering dispute over the payment of scrip to the dam workers. Criticism of this particular Six Companies policy had risen to a fever pitch during the 1932 campaign when Nevada Senator Tasker Oddie and Representative Sam Arentz took turns slamming scrip as "un-American," an attempt to "establish a business monopoly," and "a pernicious practice which should be stopped."[34] Oddie was particularly strident in his denunciations, threatening a full congressional investigation and bitterly castigating the contractors for their greed. "The housing and feeding of the workers has been commercialized and monopolized by the contractors to such an extent that their profits are exorbitant and too large for the workers to stand," he charged, adding that a probe of the inner workings of the Hoover Dam job would "put to shame the Teapot Dome scandal."[35]

At the Interior Department the mention of Teapot Dome touched a raw and exquisitely sensitive nerve, and both Ray Wilbur and Elwood Mead leaped to the defense of the department and of Six Companies. "Standard business practice . . . used by most of the large companies," Mead said of issuing scrip.[36] "Seriously misinformed," Wilbur said of Senator Oddie, and went on to attack the Nevada "evils" of prostitution and gambling.[37] Oddie hit back and the battle was joined, with "Purity Ray," as the Nevada papers jeeringly called Wilbur, sniping at vice in the Silver State while Oddie and Arentz engaged in ever harsher criticism of the Interior Department and the way in which the Boulder Canyon Project Federal Reservation was being run. The war of words was a standoff, but for the Hoover administration, and hence for scrip, the handwriting was on the wall; on May 8, 1933, after a meeting with W. A. Bechtel, Harold Ickes announced that he had ordered Six Companies to "cease immediately" paying its workers in scrip and to redeem all outstanding scrip in cash. "I believe a man is entitled to his salary in money," said the new interior secretary, and he commanded City Manager Sims Ely to collect all the redeemed coupons and coins and destroy them.[38]

With the renaming of the dam and the suspension of scrip, Ickes had fired a double broadside across Six Companies' bow, leaving no doubt that the facilitation and favoritism of the Wilbur regime were things of the past. It was not that Ickes had deliberately set out to harass or antagonize the contractors; on the contrary, he very much wanted the Boulder Canyon Project to succeed and was prepared to cooperate with the builders in bringing it to a speedy conclusion. It was just that his principles, combined with a combative nature and a zeal for personally monitoring everything that went on under the aegis of the Interior Department, impelled him to dive headlong into controversies that his predecessor had either downplayed or overlooked altogether. It was not long after renaming the dam and terminating scrip that Ickes became embroiled in yet another longstanding conflict at Hoover Dam: the wrangling over Six Companies' failure to hire black laborers.

Resentment of discriminatory hiring practices had been festering in the black community of West Las Vegas since the Boulder Canyon Project began in late 1930. The job bonanza ballyhooed in the press had turned out to be a bust so far as Las Vegas blacks were concerned, as first the preliminary highway and railroad work, and then the dam job itself, began with a lily-white, mostly out-of-state construction force. The government's contract with Six Companies stipulated that American citizens were to be hired, with preference given to veterans; "mongolians" were the only racial group specifically prohibited from working on the project. However, when the proposal to limit hiring on the dam to American citizens was made in Congress, the *Las Vegas Evening Review-Journal* titled its report "White Labor for Dam Work Urged," and this understanding, that "American citizen" meant "white American citizen," was held widely and was put into practice by Six Companies.[39]

Black indignation rose as the number of men employed on the dam exceeded a thousand without a single black laborer on the payroll. On May 5, 1931, the Colored Citizens Labor and Protective Association of Las Vegas was formed to protest this discrimination.[40] At public meetings and in letters to the newspapers, members of the association pointed out that Las Vegas blacks were American citizens, too, and that there were many former servicemen in their number. "Is it patriotic on the part of the white community to stand by and see the eagle torn down from its lofty perch and the flag used as a dish-rag?" asked a disgruntled letter writer.[41] More pleas and remonstrances followed, but it was not until William Pickens, field secretary of the National Association for the Advancement of Colored People, came to southern Nevada to investigate the hiring situation in May, 1932, that any sort of response from the U.S. government or Six Companies was forthcoming. Shortly after Pickens' visit, Archie Cross, the federal employment director for Nevada, stated that Six Companies had not hired blacks because it was afraid of dissension at the dam site and did not want to have to erect a separate, blacks-only dormitory in Boulder City.[42] The fear of dissension apparently stemmed from a January, 1931, incident in which law officers were dispatched to prevent an outbreak of violence between white and Mexican workers on the branch railroad line being built from Bracken to Boulder City. The trouble had started when a white crew, placed under the direction of a Mexican foreman, threatened to riot if he was not replaced immediately.[43] As for the question of building a segregated dormitory in Boulder City, the contractors denied that it was an issue. Six Companies President W. A. Bechtel said he had "never heard of any refusal to employ colored people" and that he would see to it that some were hired "when and if they [have] the necessary experience."[44]

On July 7, 1932, two and a half weeks after Bechtel gave his pledge, ten black veterans were put on the Six Companies payroll.[45] "It is gratifying," rhapsodized the Las Vegas Age, "not alone to the people of African descent, but to all lovers of fair play that this question of Negro labor on Hoover Dam has been settled with justice and fairness."[46] By September fourteen more blacks had been hired, but neither the Colored Citizens Labor and Protective Association nor the NAACP was satisfied that anything had been settled, much less with justice or fairness. Not only did the number of blacks hired represent less than 1 percent of the dam work force, but the black laborers had been put to work in the Arizona gravel pits, the hottest and most remote spot in the project reservation. Furthermore, they had to travel over 30 miles from the West Las Vegas slums to the job site every day and then return via jolting Boulder Highway every night. Another recent development, which must have been particularly galling to the black leaders, was the well-publicized hiring of six Apache Indians, who were to perform skilled labor (high scaling) and who would be allowed to live in Boulder City.[47]

Harold Ickes had been at his post in the Interior Department for several months when he received reports of the hiring discrimination on the Boulder Canyon Project. He was told that the number of black workers had dropped to eleven in a total work force exceeding four thousand and that Boulder City was strictly off limits to these men. The blacks, said Roy Wilkins, assistant executive secretary of the NAACP, "were transported to and from the dam in buses separately from white workers, and on the job they were humiliated by such petty regulations as separate water buckets."[48] Ickes, a self-styled champion of civil rights, launched an investigation of black employment at the dam and presently discovered that the NAACP charges were true. He also found that because of the language of the government's contract with Six Companies he was powerless to force the hiring of black labor. He did hold sway over the administration of Boulder City, however, and so, as a goodwill gesture to the black organizations and as an unofficial reprimand of the contractors, he decreed that the handful of blacks on the payroll henceforth be allowed to live in Boulder City.[49]

➘ What with the squabbling over the name of the dam, scrip payments, and discriminatory hiring, the months following President Roosevelt's inauguration had been troublesome ones for Six Companies, but these setbacks paled into insignificance compared with the ruling that came out of U.S. District Court on February 15, 1933, opening the way for the state of Nevada to collect taxes on the contractors' property within the Boulder Canyon Project Federal Reservation.[50] Since December, 1931, Senator Oddie and Governor Balzar had been insisting that Nevada had not admitted or acquiesced to the legality of the reservation's establishment and that the state exercised jurisdiction over the reservation lands and therefore was entitled to assess Six Companies property for taxes and to collect poll taxes from Six Companies employees. The Six Companies directors had refused to pay, and they made it clear, both by their actions and their statements, that they had been promised immunity from state taxes by the Hoover administration as an inducement to bid on the dam contract. Clark County Assessor Frank DeVinney reported that W. A. Bechtel told him: "Before we [Six Companies] took this contract we had a definite understanding with Secretary Wilbur and the government that this [project reservation] would be a withdrawn area and free from taxation. It was with this understanding that we bid on the contract."[51] Understanding or not, U.S. attorneys had joined Six Companies lawyers in starting a protracted legal battle with Nevada over the general issue of state versus federal jurisdiction and the specific question of the Clark County assessor's right to collect taxes.[52]

The contractors' strategy was delay, and the arguing dragged on for months while the stakes grew higher with each passing payroll, each piece of equipment added to the company inventory, and each new building erected in Boulder City. Nevada officials seethed as the court granted postponement after postponement to Six Companies. At one point Clark County District Attorney Harley Harmon angrily proposed to Governor Balzar that state militia be sent into Black Canyon to obstruct the work by seizing the area between the low and high water marks of the river, territory that Harmon claimed still undisputedly belonged to Nevada.[53] Balzar ignored this inflammatory counsel, wisely as it turned out, for after much procrastinating the court finally ruled in favor of Nevada.

\ Even as the ink was drying on the judge's decision, Assessor DeVinney rendered the state's tax bill: a whopping $330,000, including $37,000 in employee poll taxes for the years 1931–33.[54] This was a heavy blow, but Six Companies still had an ace up its sleeve: the threat of a long and costly series of appeals that would hold off the tax collectors and possibly result in the reversal of the district court's decision. This card was put into play, and Depression-strapped Nevada, hungry for cash and unwilling to risk everything on a chancy appellate fight, agreed to negotiate a settlement. The meetings began in July and ended a month later with the state's announcement that it had slashed $148,000 from the tax bill in exchange for Six Companies' promise that it would not appeal.[55] Both sides pronounced themselves satisfied with the outcome, but it was by no means a pleasant experience for the contractors to hand over a $182,000 check while flashbulbs popped and Nevada lawmakers chortled in triumph.

Hard on the heels of the galling tax defeat came more shocking and disheartening news: Warren Bechtel was dead. In July, 1933, the Six Companies president had traveled to the U.S.S.R., at the invitation of the Soviet government, to inspect the Moscow subway system and recently completed Dnieperstroy Dam. While at the National Hotel in Moscow, he accidentally took an overdose of medicine prescribed for him by his doctor and died of the reaction.[56]

Tough, earthy, hard-driving Dad Bechtel had been one of the key figures in the creation of Six Companies. His organization's financial stake in the Hoover Dam project was the largest among the partners, and as president of the consortium, succeeding W. H. Wattis, he had played an active role in overseeing the construction and in dealing with Washington. The loss of his experience and his forceful personality did not presage the disintegration of the combine, but it would weaken it, at least temporarily. Seeking to ensure continuity, the directors hastily called a meeting and made First Vice-President E. O. Wattis the consortium's new president.[57] It would fall to 33-year-old Steve Bechtel, executive committee member in charge of administration, purchasing, and transportation, to take his father's place in the inner circle of company management.

Throughout the extended period of political, financial, and personal turmoil for Six Companies, the pace of construction in Black Canyon did not slacken. On the contrary, it accelerated, and the scope of operations broadened as the day approached when the raising of the dam proper would begin. A flood of concrete would be needed to sculpt the gargantuan arch, and any delay in its production or delivery would disrupt not just the tightly planned pouring schedule but the entire project. So it was that Frank Crowe, Charlie Shea, Walker Young, and the other engineers and superintendents involved in the day-to-day management of the job took a special interest in the preparations being made to ensure that concrete would flow in a smooth, nonstop stream from the high-mix and low-mix plants once the dam's foundation had been excavated.

Concrete, the very heart and soul of Hoover Dam, was made by mixing sand and crushed rock, called aggregate, with portland cement and water into a thick gray-white mud. Of the four ingredients, the aggregate was the most important because it would make up approximately three-quarters of the dam's mass; it had to be of high quality—free of clay, salts, and organic material—and of a consistent size.[58] For months Bureau of Reclamation prospecting parties combed the desert valleys within a fifty-mile radius of Black Canyon, locating and mapping gravel beds, quarrying into the dusty deposits with shovel and pick to determine their depth and density, and hauling samples back to Las Vegas to be examined and tested. Twenty deposits were analyzed before a decision was made: the 4.5 million cubic yards of aggregate needed to build Hoover Dam would be taken from a broad alluvial lens on the Arizona side of the river six air miles upstream from the dam site. This accumulation of silt, sand, and water-polished gravel had been layered by Colorado floods until it covered more than a hundred acres and was thirty feet deep in places. The bleak patch of desert harboring the vast gravel lode was grafted onto the project reservation by government lawyers and then turned over to the contractors for their use.

   Like so many of the tasks that were preliminary or incidental to the main job of erecting the dam, the opening of the gravel pit, the construction of the screening plant, and the laying and operating of the rail link between them constituted an engineering venture that would have been considered a triumph of the first rank had it not been dwarfed by the prodigious enterprise in Black Canyon just over the southern horizon. The limelight of national press coverage that bathed the dam in a brilliant glare did not extend into the desert where the sprawling aggregate operation was located. Further guaranteeing anonymity was its ephemeral nature: sometime in 1936, if all went according to plan, the pit, plant, and railroad would be drowned by the rising waters of the lake that they had helped create. But publicity or no publicity, the isolated complex was a remarkable construction achievement and one of the indispensable engines that drove the more visible job of concrete pouring through to an early completion.

The gravel-screening plant was situated on a flat stretch of featureless desert where Hemenway Wash bottomed out, two miles west of the river. Three branches of the contractors' railroad converged at this point, which was aptly if unimaginatively named Three-Way Junction. One fork described a tortuous V as it climbed up Hemenway Wash to a switching yard midway between Boulder City and the high-mix plant at the edge of Black Canyon; another fork meandered down to the river, crossed into Arizona by way of a wooden trestle, and ended abruptly in the dusty amphitheater of the gravel pit; the third fork wound around a stubby gray-red butte, breasted a long, steep grade to a promontory overlooking Murl Emory's boat landing, looped around the flank of Cape Horn, and skirted the canyon wall down to the low-mix plant just upstream from the dam site. The strategic advantages of building the gravel-screening plant at Three-Way Junction were obvious: equipment and supplies could be freighted in from Boulder City to the southwest and sand and gravel hauled in from the Arizona pit to the northeast while finished aggregates were dispatched to the mixing plants. The only weak spot in the transportation triad was the trestle spanning the Colorado. Because of the ever present threat of flash floods, a belt conveyor slung beneath a suspension bridge had been planned to carry aggregate over the stream. However, comparison of cost estimates revealed that the money saved by loading cars at the pit rather than at the Nevada end of a conveyor belt would pay for at least four rebuildings of a trestle, so the bridge was erected, with the addition of cable anchors to reduce damage in case of high water.[59]

During November, 1931, the first of 350 tons of structural steel was delivered to Three-Way Junction and the framework of the gravel-screening plant began to go up, stark and skeletal against the desert sky. In the weeks that followed, locomotives pulled carloads of building materials and machinery up and down the thirteen miles of track between the warehouses in Boulder City and the rapidly expanding plant. By day the clatter of hammers on steel filled the air, and by night the flashes of acetylene torches sparkled like swarming fireflies. Through the gray days of December construction proceeded full speed. Snow squalls swirled across the railroad tracks and icy winds whined through the naked girders and rattled the corrugated tin siding, but the construction crews barely had time to curse the chill as they hurried to install and test machinery and to put the finishing touches on the $450,000 structure, which began operating on January 9, 1932.[60]

↘ On the engineers' blueprints the gravel plant's layout of towers, tunnels, and bunkers was neat and orderly, and its multilayered skein of belts, chutes, and sieves formed an intricate but logical labyrinth. Viewed at the site, however, the plant was quite different—and so was the impression it created. Long lines of gravel moved up and down, back and forth, in and out in a bewildering maze of girders, pipes, and catwalks, all shrouded by dust and engulfed by a skull-splitting racket of pounding, grinding, and cracking punctuated by the long whistle blasts of arriving and departing locomotives. Making the scene even eerier was the apparent absence of humans; it was as if the complex were a giant machine with a mind and a life of its own, leading a independent, predatory existence dedicated solely to devouring and excreting the desert that surrounded it.

Crucial to the gravel plant's operation was the smooth flow of traffic along the railroad network. Washouts and derailments caused by rocks spilling from open dump cars onto the tracks were a constant problem, so a special team was kept on call day and night to make repairs and get the trains running again. When word of a derailment was received, the repair crew loaded a little gasoline-propelled railroad car, called a track shifter, with equipment and sped to the scene. The derailed locomotive was disconnected from the other cars and its front wheels were coaxed onto a platform of ties. Then the track was cut with a torch and shifted to a position in front of the platform and another locomotive pulled the derailed one completely onto the platform. The rails were then shifted again, to a position under the derailed locomotive's wheels, and the big vehicle was rolled back onto the track. When all the cars had been recovered in this fashion, the track was pushed back into its original configuration and a crew of gandy dancers tamped down the bed, rearranged the ties, and reattached the rails. Sometimes even this straightforward job was complicated by the unforgiving climatic conditions of Hemenway Wash; in summer, when daytime temperatures sometimes reached 130 degrees, the burning-hot rails could not be lifted, much less touched, until after sunset.[61]

Although railroad mishaps and equipment malfunctions were treated with deadly seriousness, some incidents proved to be so ludicrous that they instantly became part of the fund of humor that sustained the work force through the long days and weeks of unbroken drudgery. One such episode involved the corraling of a runaway train by Railroad Superintendent Tom Price. One afternoon in late January, 1932, Price had a crew working with a flatcar-mounted pile driver to reinforce the trestle that spanned the Colorado, linking the gravel pit and the screening plant. The superintendent was directing the job himself because of the importance he placed on strengthening the wooden structure before the onset of the spring flood season. Pile after pile was hammered into place and cables were strung tightly to anchor the piles to the bridge and the banks. As the afternoon wore on, Price and his crew grew more and more confident that the bridge would be able to withstand the river's hardest blows. These pleasant thoughts were interrupted by the jangling of a portable telephone that had been tapped into the wire at the end of the trestle. One of the workers went to answer it and then came running back along the bridge, face white and arms waving, screaming that Red Allen at the gravel plant had just told him a runaway train was coming down the track toward the crew.

The smug feeling of confidence in the bridge's strength evaporated instantly. The pile driver was sitting in the middle of the span, where a crash would send both it and the runaway careening into the river and destroy a large section of the trestle. Price yelled for the men to move the heavy piece of machinery, and with oaths and exhortations they put their shoulders to the front of the flatcar and tried to get it rolling. Their efforts were in vain: the big pile driver weighed too much and the car would not budge. In desperation Price looked about and spied an empty dump car sitting on a level part of the track several hundred feet from the Nevada end of the bridge. The gravel plant was seven miles away and a quick mental calculation told him that he and the crew still had several minutes before the speeding runaway reached them. If the empty dump car could be immobilized, it would act as a barrier and break the force of the runaway, perhaps enough to spare the pile driver and the bridge. The order was given and the laborers seized sledges, spikes, and cables from the flatcar and sprinted toward the dump car. They swarmed over it like a squad of aroused Lilliputians working frantically to fasten it securely to the track so that it would absorb the force of the tremendous collision they knew was imminent.

The job was done with remarkable speed, and the men dropped their tools and bounded off into the desert to get clear of the impact area. Finding safe positions, they turned and looked expectantly up the track to the brow of the hill, watching wide-eyed for the racing runaway. Nothing came. Five minutes, ten minutes, and still nothing. The men sheepishly regrouped beside the track, exchanging puzzled looks and casting nervous glances up the rails. Presently a small locomotive crested the hill and puffed down toward them. Far from rolling with runaway speed, it was barely creeping along; not only that, but aboard it was a crew clearly in control of

Six Companies built twenty miles of railroad connecting Boulder City, the Arizona gravel pit, the screening plant, and the dam site. Here a locomotive chugs through Black Canyon. (Kaiser Collection, Bancroft Library)

the situation. Eventually the locomotive stopped in front of the improvised barrier and Red Allen and several other men from the gravel plant jumped down from the cab. Without saying a word they walked around the empty dump car, staring with amazement at the thick web of cables and spikes that pinned it, alone and forlorn, on the level stretch of track. Then they began to laugh. "Where's the runaway?" barked Price. Red Allen turned to him and pointed at the dump car, trying hard to suppress the smile that kept curling the corners of his mouth. "Right there," he smirked, "and from the looks of things that's one son of a bitch that won't ever run away again!"[62]

The gravel pit, screening plant, and connecting railroad formed a world apart from the main base of operations in Black Canyon, but it was still a satellite world spinning in time with the dam site, the parameters of its orbit dictated by its mission to feed aggregate to the concrete mixers in precise quantities and sizes without letup for four years. Like the gravel-screening installation, the concrete plants were marvels of advanced technology, built to produce concrete in larger quantities, faster, and of higher quality than had ever been done before. The figures attached to every facet of the enterprise were unprecedented: production—nearly 4.5 million cubic yards, more concrete than had gone into all the other dams built by the Bureau of Reclamation before 1931; speed—3,500 cubic yards a day, enough to build more than a mile of highway twenty feet wide and ten inches thick; quality—a uniform product with compressive strength of not less than 2,500 pounds per square inch for the dam and 3,500 pounds per square inch for slabs, beams, and other reinforced members, exacting specifications made harder to achieve by the nearly one-hundred-degree difference between winter and summer air temperatures.[63]

As if these goals were not lofty enough, that ever present Black Canyon bugaboo, difficulty of access, made the challenge even more formidable. The consistency and uniformity of the concrete would be measured at the mixing plant by Bureau of Reclamation inspectors employing the slump test, which consisted of filling a conical mold with concrete, removing the mold, and measuring the amount of sag in the setting concrete. The compressive strength demanded by the bureau's specifications meant the slump had to be minimal, which in turn meant that the concrete had to be mixed very dry. Mixing dry did not pose a problem, but the concrete had to be delivered in steel buckets to the pouring site downstream; if the mixing plant was too far from the construction area, the dry concrete would take its initial set and be extremely difficult to work with. This was most likely to happen in summer, when the air temperature was above a hundred degrees. The prospect of returning buckets of set concrete from the job to the mixing plant, where they would have to be chipped out by hand, was a powerful incentive for putting the mixing plant as close to the dam as possible.

Examination of the precipitous cliffs and riverbanks of Black Canyon was discouraging to the engineers charged with building the main concrete plant. Making space for the three air-compressor installations had been difficult, and they were much smaller than the 78-by-118-foot, four-story concrete-mixing plant would be. The site surveyors followed the construction railroad farther and farther upstream until at last, three quarters of a mile above the blazes marking the position of the dam's Nevada abutment, they found a niche in the cliffs suitable for their purposes. The crescent-shaped shelf was leveled and enlarged with jackhammers and dynamite, and construction of the steel-frame structure began in November, 1931. The plant, designated low-mix, was to provide all the concrete for the diversion-tunnel linings, the powerhouse foundation, and two-thirds of the dam; when the two-thirds point was reached, a second plant on the Nevada rim, designated high-mix, would provide the concrete for the final 242 feet of the dam and the structures on its crest.

Inside the square, barnlike exterior of low-mix, aggregates were carried to storage bins from which they could be siphoned automatically into large containers called batchers. The eight batchers—one each for the five classes of aggregate, two for portland cement, and one for water—fed predetermined portions of ingredients into the revolving drum mixers. The entire process was automated and could be controlled by a single operator pushing a bank of buttons.[64]

Low-mix was finished late in February, 1932; the equipment was tested; and on March 8, 1932, the first batch of concrete was poured into the trashrack foundation at the inlet portal of tunnel No. 2. Eight days later, lining of the diversion tunnels began with a pour in the invert section of No. 3.[65] During the first year of what would be a continuous three-year run at low-mix, virtually all of the concrete produced—almost 400,000 cubic yards—went into the linings of the diversion tunnels. On any other job this would have been a prodigious output of concrete, but by the colossal standards of the Boulder Canyon Project, it was relatively modest, a mere freshet hinting at the torrent that was to follow. Before low-mix and its twin, high-mix, could be fully cranked up to disgorge this torrent, however, the jaws of Black Canyon had to be opened wide to swallow it.

On November 19, 1932, the Colorado River Board, the independent consulting body established by Congress in 1928 to advise the Bureau of Reclamation on the project's engineering aspects, gave its final approval to the dam's design, clearing the way for excavation of the abutment cuts.[66] The sides of the dam would lock into these huge V-shaped keyways, transmitting much of the crushing water pressure exerted by the reservoir to the solid rock of the canyon walls. Squads of high scalers began drilling and blasting out the wide, tapering notches in late December, starting at elevation 1232, the crest of the dam, and working down 400 feet, gradually diminishing the depth of excavation from 75 feet until the cuts warped all the way out to the cliff lines, about a third of the way down the face of the dam. From that point on down to the foundation, outcrops were chiseled away and the rock surface groomed, but heavy blasting was avoided to prevent fracturing or weakening of the lower cliffs, which would bear the brunt of the lake's force.[67]

Excavation of the abutments took four months and 185 tons of dynamite; meanwhile, on the canyon floor, a fleet of electric shovels, draglines, Caterpillar tractors, and dump trucks, supported by an army of shovel-wielding muckers, was ready to attack the deep blanket of silt and gravel deposited over countless centuries by flowing water. The riverbed, exposed when the Colorado was shunted into the Arizona diversion tunnels and sealed off from floods by the erection of the two cofferdams, presented an eerie aspect. From the water line staining the cliffs, great fans of sand and slimy gravel extended out and down over 100 feet to the center of the gorge, where stagnant pools, ringed by silt-smoothed boulders, traced the outline of what had been the river's main channel. Viewed from the canyon rim, the muddy bottom and opaque puddles looked like a suppurating gash in the earth's skin, an impression that was heightened by the dank, putrid odor welling out of the crevasse.

The task of scouring the riverbed began immediately after the Colorado was diverted in November, 1932, and was moving ahead at full speed by the beginning of 1933. Every day some twenty-two thousand cubic yards of silt and gravel were hauled out of the canyon and dumped into railroad cars to be carried upstream for disposal. As was the case during the drilling of the diversion tunnels, there was an intense push for speed by the superintendents and foremen, and fierce competition among the three shifts to see who could move the most muck. The winner of this hard-fought contest was the swing-shift crew, which on January 24, 1933, disposed of a staggering 1,841 truckloads—nearly four a minute—during its eight-hour stint.[68]

From an engineering perspective, the dredging of the dam's foundation was not nearly as interesting as the building and operating of the gravel and concrete-mixing plants, but to the ordinary citizen looking on from observation points along the Nevada rim it was infinitely more spectacular. Writer Frank Waters visited Black Canyon in 1933 and described the riverbed excavation:

> The vast chasm seemed a slit through earth and time alike. The rank smell of Mesozoic ooze and primeval muck filled the air. Thousands of pale lights, like newly lit stars, shone on the heights of the cliffs. Down below grunted and growled prehistoric monsters—great brute dinosaurs

Draglines and electric shovels removing sand and gravel from the 75-foot-deep slot in the middle of the river gorge. (Arizona Historical Society Library)

with massive bellies, with long necks like the brontosaurus, and with ar-mored hides thick as those of the stegosaurus. They were steam shovels and cranes feeding on the muck, a ton at a gulp. In a steady file, other monsters rumbled down, stopping just long enough to shift gears while their bodies were filled with a single avalanche, then racing backward without turning around. From the walls above shot beams of search-lights, playing over this vast subterranean arena. They revealed puny pygmies scurrying like ants from wall to wall; mahouts, naked to the waist, riding the heads of their mounts, standing with one foot on the running board and peering over the tops of the cabs while driving with one hand. And all this incessant, monstrous activity took place in si-lence, in jungle heat, and as if in the crepuscular darkness of a world taking shape before the dawn of man.[69]

‵ The grand panorama visible from the rim—a vision of concerted effort that inspired some observers to rhapsodize about national rebirth and the glories of collectivism—could not be appreciated by the thousands of workmen who toiled far below. Their perspective was down to earth, both literally and figuratively, and it was hard for them to share or even comprehend the emotions of those who looked down on their constricted,

Muckers in the slot cleaning the bedrock surface in preparation for the first pour of concrete, May, 1933. (Bureau of Reclamation)

dangerous world. As one worker put it in response to Waters' reference to the rank smell of Mesozoic ooze: "A beautiful piece of writing, but all I smelled . . . was Atlas blasting powder and diesel fumes from the trucks." [70] To the men on the canyon floor, the purposefulness and order apparent from on high, "the soul of America under socialism" discerned by Englishman J. B. Priestly during a brief visit to the dam site, was a chaos of racing trucks, swinging shovel booms, and exploding dynamite. Priestly's heroic "new man, the man of the future," used his shovel in exactly the same way as did the man of the past, tried to avoid being run over or struck on the head by flying debris, and anxiously counted the days to his next paycheck. [71]

The steady diet of drudgery and danger was spiced occasionally by unusual and exciting events. At the end of March the canyon was abuzz with the rumor that a crop of small gold nuggets had been uncovered in a gravel stratum thirty feet down in the riverbed. [72] It was not coincidental that this fanciful tale gained its widest circulation on April Fools' Day: there was no mother lode in Black Canyon. Treasure of a different and stranger sort was discovered, however, in the early morning hours of June

The first bucket of concrete was poured on June 6, 1933. (Bureau of Reclamation)

9, 1933, when a blast 130 feet below river level unearthed a drill bit studded with eight diamonds. A check of Bureau of Reclamation records revealed that the bit had been lost in mid-December, 1922, by Walker Young's survey team while drilling test holes in the bedrock. The bit was recovered in almost perfect condition, and the bureau placed its value at $2,500.[73]

In mid-April, 1933, the steel jaws of the dragline buckets struck bedrock at elevation 600, 40 feet beneath the river's level. From this elevation the bedrock shelved out from both canyon walls 150 feet to the middle of the canyon, then dropped sharply to form a V-shaped channel 75 feet deep. This narrow slot would make an excellent natural keyway for the dam's base, but first its filling of loose sand and gravel had to be dug out and its walls, which had been deeply carved and fluted by the whirlpools of the ancient river, had to be trimmed to create a uniform surface to which concrete could bond.[74]

The last shot in the foundation excavation, six thousand sticks of dynamite, was fired on May 31. The debris was shoveled away, and a network of drain pipes was installed to draw off spring water seeping from rock fissures deep in the erosional channel. The cracks themselves were calked with lead wool, then filled and sealed with grout so that the flow of buried springs would not eat away at the setting concrete in the dam's protruding toe.[75] Several days later, Jack Savage, the Bureau of Reclamation's chief design engineer, made a thorough inspection of the foundation and formally approved the excavation. With this endorsement and the successful completion of the plumbing job in the slot, the preparatory phase was over and concrete pouring could begin.

Carpenters had been busy hammering together the first of many hundreds of boxlike wooden forms, and on the morning of June 6, 1933, everything was ready for the first pour. Despite the landmark nature of the event, Six Companies had chosen to treat it in low-key fashion. A newsreel camera was set up on one of the bedrock shelves overlooking the footing, and a handful of still photographers was permitted on the canyon floor to record the historic moment, but otherwise the small group of witnesses was made up exclusively of Bureau of Reclamation and Six Companies personnel. At exactly 11:20 A.M., an eight-cubic-yard cylindrical steel bucket suspended from a web of steel cables appeared in the narrow strip of sky eight hundred feet overhead. It hung there momentarily, then plummeted soundlessly into the canyon, only to be brought up a few feet short of the slot's shadowy bottom by its steel tether. Frank Crowe gave a signal and two men moved forward, tripped the bucket's safety latches with their shovels, and leaped back as the bottom-dump doors clanged open, the bucket jerked upward, and a great liquid-gray mass sluiced down onto the rock. Flashbulbs exploded in a ragged volley. Hoover Dam was on the rise.[76]

CHAPTER SIX
"A Callous, Cruel Lump of Concrete"

Like some forbidding futuristic metropolis, the asymmetrical concrete columns of the dam reared up from the canyon bottom. They stabbed skyward, a phalanx of blank towers mottled by dark water stains and black shadows, but otherwise featureless, inscrutable, and cruelly huge. Here and there, pipes, pieces of lumber, cables, and bristling clumps of structural steel protruded from the tops and sides of the long oblong blocks, but the overwhelming impression was one of barrenness, bulk, and brooding power.

Along the stone parapets skirting the Nevada rim, crowds of onlookers stood for hours gazing down into the abyss, marveling at the gray-white wedge growing out of its depths. "Unreal, imaginative, supernatural," wrote Theo White. "Workers are scarcely visible. The thing loses scale. A steady hum, quiet and muffled, comes up between the walls, ripped violently by an occasional whistle from the high lines. . . . I stare at the thing trying to comprehend it, to fix it forever in mind's eye. . . . Beautiful concrete, sweating 'stiffs,' the two fuse and become inextricably one." [1]

Hundreds of feet below, the uneven checkerboard pattern of the dam's crest sprawled in a fat crescent. In one shadowy quadrant, men knocked loose the she-bolts, tie rods, braces, and pigtails holding a form panel in

The concrete columns of the dam rise from the canyon bottom like a jumble of windowless skyscrapers. (Bureau of Reclamation)

Hoover Dam on the rise, December, 1933. (Bureau of Reclamation)

place and winched it upward, uncovering a raw, freshly set block of concrete. At another spot, a gang of pipe fitters scrambled into a reassembled
form with wrenches and blowtorches to install grout tubing in the contraction joints where the flanking columns would interlock. In yet another
form, a black steel bucket hanging on the end of a cable jerked upward,
disgorging a mound of concrete, and seven puddlers, wearing knee-high
rubber boots, tramped the sticky mass into place with their feet.

From a distance the dam appeared to grow steadily, evenly, and of a
piece, but it was actually a jumble of concrete boxes rising upward in fits
and starts, a horizontal five-foot-thick layer poured first on one column,
then on another. The reason for this piecemeal, block-by-block approach
was not convenience or insufficient mixing capacity, but, rather, the need
to dissipate the tremendous chemical heat generated by the setting concrete. Bureau of Reclamation engineers had calculated that if the dam were
fashioned in a single, continuous pour, its internal temperature would rise
forty degrees while it was hardening, it would take 125 years to cool, and
the tremendous stresses created by the setting process would fracture its
body so severely that it would be rendered useless. The only answer was to

A gang of puddlers tenses as an eight-cubic-yard bucket spews out sixteen tons of concrete. (Bureau of Reclamation)

pour in individual blocks no more than five feet thick, and to honeycomb these blocks with one-inch-diameter pipe through which cool river water and then ice-cold refrigerated water could be pumped. The pipes, embedded at regular intervals throughout the length and width of the dam, would all extend into an eight-foot-wide central slot and connect with the main water lines from the cooling tower and refrigeration plant. By monitoring the rate of cooling with thermometers buried in the concrete and keeping the internal temperatures consistent from column to column, the engineers would be able to achieve even contraction and prevent severe crack-causing stresses from occurring. Once the blocks were cool and had stopped shrinking, grout could be pumped into the network of pipes to make each unit solid.

So that the finished structure would not be weakened by hairline spaces between the scores of individual concrete columns, the upstream and downstream surfaces of each block were to be scored with interlocking vertical grooves; the surfaces facing the Arizona and Nevada canyon walls would be striated with horizontal interlocking grooves. When cooling was completed, grout would be forced into these radial and circum-

Inside a form, two workers prepare for the next pour by cleaning the concrete surface around the cooling pipes. (Manis Collection, UNLV)

ferential contraction joints and into the slot from which the refrigerated water had been circulated. Thus the finished body of the dam would be monolithic, even though it was composed of many parts.[2]

The design solutions to the problems of cooling the concrete and grouting the contraction joints were relatively simple on paper, but translating them into action was extraordinarily difficult. Frank Crowe described the predicament: "We had 5,000 men in a 4,000 foot canyon. The problem, which was a problem in materials flow, was to set up the right sequence of jobs so [the workers] wouldn't kill each other off."[3] What Crowe neglected to mention was that the financial success or failure of the project for Six Companies, as well as his reputation as a builder, was riding on the outcome of the all-important concrete-pouring phase. The relatively lucrative tunneling, high scaling, river diversion, and foundation work had been finished more than a year ahead of schedule, allowing the contractors to recoup their five million dollar investment and bank substantial profits. But all could still be lost. Padding the bid figures for units in the first half of the job had meant cutting them for the second half, and it was on specification item 54, concrete in dam, that Crowe had gambled and made the deepest slash. He had committed Six Companies to pour 3.4 million cubic yards of concrete for $2.70 per cubic yard, a price $1.45, or 35 percent, below the price quoted by Arundel Corporation, the runner-up in the March, 1931, bidding, and $0.55, or 17 percent, less than the Bureau of Reclamation's estimate.[4] If he failed to deliver on this promise, if the job bogged down in a series of delays and it proved impossible to pour concrete as quickly and cheaply as he thought, Six Companies might well lose all the money it had earned so far.

The company directors undoubtedly fretted over the $2.70 figure they had agreed to, but they need not have worried, for it was in choreographing the most complicated jobs—marshaling equipment, motivating men, moving concrete faster and with less wasted effort—that the building genius of Frank Crowe shone the brightest. Not only was he certain that he could pour the concrete at the price quoted, he was convinced that he could do it at a profit and finish the task ahead of schedule, just as he had done with the earlier phases of the dam construction. The key would be mobility, swiftly transporting materials and equipment to the right place at the right time so that the laborers could perform their jobs smoothly and simultaneously like the moving parts of a well-oiled machine. In the narrow cul de sac of the dam site, such rapid, precise movement of thousands of men, tons of supplies, and mountains of concrete seemed almost impossible, but Crowe knew exactly how he was going to do it. He would fill the sky over Black Canyon with a web of cableways allowing him to pick up anything—a truck, a bucket of concrete, a crew of carpenters, a single wrench—and lower it in a matter of seconds to any point within the mile-long construction zone.

Aerial cableways were a Frank Crowe specialty; he had pioneered this method of transport at Arrowrock Dam in 1911 and perfected it at subsequent jobs during the next two decades. The Hoover Dam cableway system was the most extensive and elaborate yet devised, spanning the gorge with no fewer than ten high lines. The heart of this network, put into place in 1932–33, was a skein of five twenty-ton cableways stretching over the portion of the site that contained the spillways, intake towers, dam, powerhouse, and outlet works. The steel cables were strung between ninety-foot towers that moved upstream or downstream on railroad tracks, providing complete pinpoint coverage of the work areas below. Each cableway was made up of six separate lines. A wheeled carriage, rigged with fall blocks from which the load was suspended, ran along the track cable. Outhaul and inhaul cables moved the carriage back and forth across the canyon, the hoist cable lowered and raised the load, a dump line released it, and a button line prevented the hoist and dump lines from sagging as the carriage moved away from the head tower.

Controlling these intricate assemblages of cables and pulleys required a steady hand, sharp eyes, and a clear head. The main operator sat in lonely splendor in a shack perched at the end of a platform extending thirty feet out over the canyon. By manipulating a bank of levers and buttons, he could pick up an object and send it skimming along the line at twelve hun-

dred feet per minute, watching or listening for the directions of a signalman who, with wigwag arm movements or commands relayed by electric
bells, ordered the cargo to be stopped above its destination, lowered, and
released.

The coordination and judgment of the signalmen, and the cableway
operators' almost telepathic anticipation of commands, were remarkable.
Nowhere was their skill and exquisite teamwork more apparent than in
the handling, spotting, and landing of the eight-cubic-yard concrete buckets. An electric locomotive pulled a flatcar carrying two full buckets from
the low-mix plant down the canyon railroad to a terminus on the cliff side
beside the dam. Directed by a signalman, the cableway operator maneuvered a hook down to the loading point, where a crew scrambled onto the
flatcar and attached it to a full bucket. Another signal was given and the
load of concrete was lifted smoothly and silently into the air and swung out
over the canyon, dangling at the end of 650 feet of slender cables. At the
pouring point, another signalman positioned the bucket by relaying directions to the cableway operator, who piloted it to a point directly overhead.
With uncanny accuracy, the signalman picked out the precise point, an
eighth of a mile above, to halt the bucket and order the cableway carriage
to reverse, preventing a pendulumlike backswing. The bucket then plunged
downward until the signalman gave the washout sign, stopping it just
above the spot in the form where the pour was to be made. Two workers
tripped the safety catches on the bucket's bottom and leaped clear as, on
command, the operator hit the dump button and the bucket jumped upward, spewing out sixteen tons of wet concrete. The instant the bucket was
empty, it was hoisted skyward and whisked back toward the waiting delivery train, guided unerringly by the signalmen to a gentle landing.

There was something mesmerizing about the graceful, breathtakingly
precise flight of the big buckets. Even to construction men accustomed to
mechanical wonders the sight was riveting. "It was, indeed, fascinating to
watch bucket after bucket come gliding up to the landing point, slide into a
well on the car and be lowered into place—all with practically one motion
which represented the ultimate in concerted action and timing," wrote
Lawrence Sowles, a Six Companies engineer.[5] As the cableway operators
and signalmen gained experience and confidence, they delivered the concrete faster and faster, plucking the big buckets into the air, swinging them
into position, lowering them, dumping them, and returning them to the
waiting flat cars at a swift, relentless pace.

Any failure in the cableway system would have a dominolike effect on the
entire project, upsetting the pouring timetable, snarling rail traffic, and
raising havoc with the production schedules at the gravel-screening and
concrete-mixing plants. To prevent such a tie-up, upwards of fifty riggers
were kept busy oiling and repairing the drums, pulleys, and track carriages

A rigger working on a cableway headtower. (Library of Congress)

and ceaselessly inspecting cables and lines for wear. During the course of construction, seventeen track cables weighing fourteen to twenty-five tons each had to be replaced, an acrophobia-inspiring task carried out far above the canyon floor. Despite the riggers' heroic efforts, several cableway failures did occur. Lines that were constantly raising and lowering four-ton steel buckets brimming with sixteen tons of concrete were under tremendous strain, and when they broke, it was usually without telltale fraying.

One of the Hoover Dam project's most talked-about accidents, remarkable because it involved both a terrible death and a miraculous escape, occurred when a cable gave way during the swing shift on January 3, 1935. In form B-1 on the crest of the dam, a crew was waiting for a bucket of concrete to arrive. The puddlers were still tramping down the last pour, while signalman Ike Johnson stood on the edge of the form, relaying directions to the cableway operator high above who was swinging the fresh load down toward him. Next to Johnson, in the adjoining form, pipe fitter J. W. ("Happy") Pitts was hunched over installing cooling pipe. Spotted perfectly, the bucket floated down toward the waiting crew, but just as it was about to twitch to a stop, one of the hoist lines tore apart with a loud snap. The bucket tipped and began an uncontrolled pendulum swing, first over the form, then out over the downstream face of the dam, picking up speed as it went. Both Johnson and Pitts were knocked flying by the runaway bucket; the startled puddlers looked up just in time to see Happy cartwheel downward 150 feet and crash in a lifeless heap on one of the narrow catwalks crisscrossing the dam's face.

Meanwhile, the rampaging concrete bucket ricocheted off the arch several times, disgorged its contents, and crashed into the Nevada wall, where it shattered like an exploding bomb. Workmen from all over the site raced down to the canyon floor to search for Ike Johnson, but he was nowhere to be found. Then someone noticed a tiny flame flickering high on the curving face of the dam. With an excited shout, the men clambered up the ladders and stairs toward the light, and there they discovered Johnson, covered with a slimy coat of concrete but fully conscious, frantically lighting matches to attract attention. The lip of the bucket had struck his legs just as it was starting to swing, scooping him into the concrete and carrying him on a wild ride three-quarters of the way across the canyon before the bottom-dump doors suddenly opened and deposited him in a sticky shower on one of the catwalks. Astonishingly, bruises and severe eye irritation from the chemicals in the cement were his only injuries, and he was back at work the next night.[6]

Hook tender Kenneth Wilson was not so lucky; on February 19, 1935, a broken cable cost him his life. With the assistance of worker Jack Egan, Wilson had rigged a ten-ton pumpcrete machine for hoisting out of the canyon and them climbed on top of the load for a quick trip to the rim. Six Companies tried to discourage men from "riding the hook," but some

A skip full of workers being reeled across the canyon by a Joe McGee cableway. (Kaiser Collection, Bancroft Library)

of them could not resist the convenience and the thrill of being whisked up 800 feet and swung spiderlike across the chasm. Egan planned to go along for the ride but hopped off at the last second when he remembered a chore he had forgotten to do. Three hundred fifty feet up, the hoist line parted with a sound like a pistol shot and Wilson and the pumpcrete plummeted downward, striking the top of the lower cofferdam with such force that pieces of the machine buried themselves three feet into the hard-packed earth. "Wilson had plenty of nerve though," the shaken Egan later told a reporter. "I could see Kenny wave goodbye to us all. . . . And to think I was almost up there with him." [7]

Normally, workers were ferried in and out of the canyon by small cableways known as Joe McGees, so called because, like the Irishman of that name, they were "cantankerous little critters." [8] The men crowded onto a railed wooden platform called a skip and the Joe McGee reeled them out and dropped them down on signal from a skip tender who rode on board. Another form of canyon transportation was a contraption mordantly dubbed the monkey slide, a large platform that moved up and down on greased skids, powered by a 75-horsepower motor. Two of these inclined skips plied the upstream face of the dam, one the downstream face, and a fourth, large enough to carry fifty men, ran from the Nevada rim down to elevation 850 in the canyon.

Although it was safer than riding the hook, a trip on a Joe McGee could be a harrowing experience. Sometimes the load cables became twisted, causing the skip to spin first one way and then another as it jerked out over the canyon. On other occasions the signaling system broke down and instead of stopping in midair and dropping down to the landing platform the skip would keep going—right into the canyon wall. On New Year's Day, 1933, forty-five workers barely escaped death when their skip slammed into the Arizona cliff, knocking them from their feet and almost off the platform into the abyss. The Six Companies hospital in Boulder City refused to divulge information about the injuries to the Las Vegas press, but there were reports that many of the men on board the runaway skip had broken bones and five who were not hospitalized were so badly frightened that they had quit their jobs on the spot. [9]

Incidents like these reminded the workers that they risked injury or death every time they went down into Black Canyon: "It is no wonder that they are forced into a kind of bravado that goes with the game, that they take to minimizing the risks and the inhuman enormity of the job, chalking pet names on their clumsiest trucks, nicknaming their largest traveling crane

The dam workers called this inclined rail skip the monkey slide. (Bureau of Reclamation)

'The Flying Trapeze','" a journalist wrote after touring the dam site. "No wonder that they adopt a tone of sardonic superiority to dangers, painting in huge white letters on the door of a powder house, 'God Bless Our Home.'" [10] Bravado aside, every injury, every death, served to resurrect the nagging doubts about job safety, the disquieting questions about the trade-off between accidents and speed that had surrounded the Hoover Dam project from the beginning. Billboards with the bold black message "Death is so Permanent" were displayed prominently around the site as part of the well-publicized safety program Six Companies started in early 1932. [11] But at the same time, bonuses were being offered for rapid concrete pouring, and the spur of company-sponsored inter- and intrashift competition to see who could move and empty the most concrete buckets in an eight-hour period was being used to boost worker productivity.

The only real restraint on the headlong race to get the dam up to its final height of 726.4 feet was the contract requirement that no more than 5 feet of concrete be placed in any column or panel during a 72-hour period and that no more than 35 feet of concrete depth be placed in a column within 30 days unless specifically authorized by the Bureau of Reclamation. [12] The man responsible for monitoring the contractors' performance

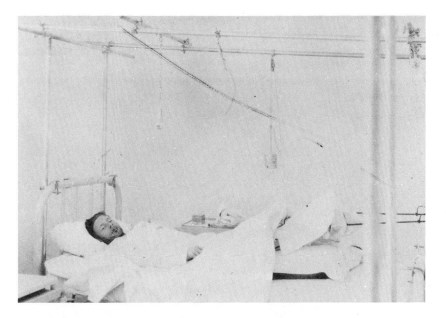

An accident victim in the orthopedic ward of the Six Companies hospital in Boulder City. (Bureau of Reclamation Collection, UNLV)

and making sure they did not exceed the specification limits in their zeal to finish the job sooner was Construction Engineer Walker Young. He and his staff of approximately 150 inspectors checked the quality of the work and acted as traffic cops enforcing a speed limit on the often impatient Six Companies crews.

Given their conflicting assignments and contrasting personalities, the potential for friction between Young and Frank Crowe seemed high: Crowe referred to Young as the Great Delayer, and Young in turn pinned the nickname Hurry Up on the hard-charging superintendent. The two men were old acquaintances, however, having met at Arrowrock twenty years earlier, and their squabbling over details of the Hoover Dam work was tinged with mutual respect and affection. "Of course we like to cry at each other and raise hell," the superintendent told a reporter in 1933. "He says my foremen are no good but he don't mean anything. . . . I'd go to hell for him." "Yes," Young agreed, "sometimes we fight with each other for the fun of it." [13]

Although the restrictions imposed by the construction contract and by Young's inspectors slowed the movement of concrete somewhat, the work pace remained frantic and the accident toll high. "We're doing our very best," Crowe had said regarding hazards, but when the choice was between getting the work done faster and making the dam site safer, there

Walker R. Young (left), the Boulder Canyon Project construction engineer, and Frank Crowe (right), Six Companies superintendent of construction, February, 1935. Young called
Crowe "Hurry Up"; Crowe called Young "the Great Delayer."
(Bureau of Reclamation)

could be little doubt where his priorities lay: "Build a dam, kill a man,"
was the cynical phrase the laborers muttered after a fatal accident.[14]

 Union activity, which might have provided some measure of protection, was virtually nil in 1933. Ever since the failed strike of August, 1931,
Frank Anderson and his IWW colleagues had lain low in Las Vegas, trying
to enlist new members and reporting on Black Canyon work conditions in
articles for *Industrial Worker,* but avoiding activities that might lead to a
direct confrontation with Clark County or federal law officers. There had
been a spate of trouble in 1932 when six Wobblies distributing union
literature on the project reservation were jailed briefly because they did not

have signed passes in their possession,[15] but by and large, aggressive organizing tactics had been avoided.

Caution was a prudent course, given the vigilance of the local police and their swift response to any challenge of their authority. On the project reservation the iron rule instituted by City Manager Sims Ely and Police Chief Bud Bodell had clamped even tighter when Bodell turned his police duties over to Deputy Marshal Warren Corley and joined Six Companies as chief special agent in charge of a group of company investigators whose job it was to contest injured workers' insurance claims and keep the lid on incipient labor unrest.[16] With both the reservation rangers and Bodell's private force arrayed against them, it appeared that the Las Vegas Wobblies had all but given up trying to challenge Six Companies directly and were content to snipe at the contractors from a safe distance. Then in August, 1933, Anderson and his followers made a bold and unexpected bid to foment a general strike on the project reservation.

What precipitated this audacious move to halt the Hoover Dam work was a mystery. Anderson may have been under pressure from the IWW leadership in Chicago and Seattle to show results in his two-year-old organizing campaign, or perhaps he merely believed that the time was ripe for action: between June 23 and July 20, 1933, there had been half a dozen fatalities in Black Canyon, the Las Vegas papers were full of stories about civil suits brought against Six Companies by workers claiming to have been poisoned by carbon monoxide gas in the diversion tunnels, and the summer heat was at its peak. Whatever the reason, Anderson made his move on August 18. At 6:30 A.M. he and seven followers burst into the mess hall, where most of the day-shift crew was just finishing breakfast, and went up and down the aisles loudly berating Six Companies and urging the workers to strike. Mimeographed sheets were handed out under the heading "I.W.W. Job Committee"; these listed the strike demands as a six-hour day with no reduction in pay, a 50 percent cut in rent for cottages in Boulder City, and a cut of sixty cents a day in the fee dormitory residents were charged for board.[17]

\ The Wobblies begged the men to stage a showdown with the contractors by remaining off shift, but their harangue failed to inspire the startled listeners. There was something faintly absurd about the early-morning hyperagitation of the would-be strike leaders, and whatever sympathy their demands aroused, it was not enough to prevent a mass exodus to the parking lot, where the transports were waiting to leave for the dam site. As the workers trooped out of the mess hall and boarded the trucks, the eight Wobblies angrily followed, and what had begun as a quixotic, if ill-advised, venture degenerated into a cruel farce. The organizers circled the trucks, first shouting at the men to get off and then cursing them as scabs and stooges. Some of the Wobblies tried to pull the drivers out of the cabs to keep them from starting the engines and there was shouting, pushing, and shoving. The disturbance caught the attention of several reservation rangers, who moved in as scuffles broke out between the drivers and the Wobblies. The workers sitting in the trucks were entertained by this melee, particularly the antics of a Wobbly who slipped on the mess-hall steps, fell and hit his face, then jumped up and ran around in circles, pointing to the bloody abrasion on his cheek and yelling, "Look what the ranger did to me!" To top it all off, Bud Bodell arrived on the scene, and in best battle-royal fashion he lowered his head and charged into the middle of the brawl with both fists swinging.

After a few minutes of furious action, order was restored, Bodell and his Wobbly opponents were separated, the truck drivers got back into their cabs, and the day shift left for the dam. Frank Crowe had been alerted to the trouble and now showed up to assess the situation. Exhibiting restraint and a fine instinct for disarming the already chagrined Wobblies, he ordered the rangers to release them, invited them into the mess hall for a free breakfast, and then had them escorted off the reservation without pressing charges.[18]

"A dismal failure," the Las Vegas press called the stillborn strike of August 18, and in fact it was the first time in the history of the IWW that the union had called a strike and workers had failed to follow.[19] That the affair had been a debacle was attested by the silence of the usually voluble Wobbly newspaper *Industrial Worker,* which printed not a word about this mortifying fiasco. For Six Companies and the Bureau of Reclamation, the episode was an unanticipated but nevertheless welcome coup, completing as it did the discrediting of Frank Anderson and his followers and terminating whatever influence they and the radical union had with the Hoover Dam workers.

It was just as well for the contractors that they scored this easy victory when they did, for much more serious and potentially costly labor trouble was brewing in Nevada state district court during the summer and fall of 1933. A Las Vegas plaintiff's attorney, Harry F. Austin, had filed half a dozen civil suits on behalf of former Six Companies employees who were seeking damages for carbon-monoxide poisoning they allegedly suffered while working in the diversion tunnels in 1931 and 1932. The gas cases, as these suits collectively were called, ran to a pattern: the plaintiffs, former tunnel men and truck drivers, alleged that Six Companies had been negligent in not safeguarding them from dangerously high concentrations of carbon monoxide, which, they claimed, had ruined their health and left them permanently disabled. Each was seeking approximately fifty thousand dollars in personal damages and twenty-five thousand dollars in punitive damages, plus the amount of lost wages since the time of incapacitation, typically amounting to several thousand dollars.[20]

The gas problem was not a new one for Six Companies. Since the

Black Canyon tunnel work began in the summer of 1931, the IWW newspapers had reported cases of carbon-monoxide poisoning and had denounced the contractors for covering up the deadly health hazard. Dr. Wales Haas, head of the Six Companies hospital in Boulder City, had been accused of deliberately diagnosing gas cases as influenza and listing pneumonia as the cause of death for men who in reality had succumbed to acute carbon-monoxide poisoning.[21] There was, of course, no proof of this, and in any event the company did not have to answer to the charges of the radical IWW. It did have to answer to the state of Nevada, however, which in November, 1931, ordered it to obey state mining safety laws that specifically forbade the use of gasoline-engine vehicles underground.

With the aid of U.S. attorneys, Six Companies had contested Nevada's order, but the controversy had not gone away, even after a three-judge federal panel ruled in Six Companies' favor. Now, with Austin's filing of six gas-related personal-injury suits, it loomed larger than ever. If a jury should rule that the contractors had compelled their employees to work in a poisonous environment, the resulting tide of bad publicity would damage their reputation and possibly cost them future contracts. Even worse, the awarding of $75,000 or more in damages to one of the plaintiffs might lead to an avalanche of similar suits; more than a thousand miners, muckers, truck drivers, and other laborers had worked in the diversion tunnels, and hundreds more were similarly employed now, drilling tunnels for the penstocks that would carry water from the intake towers to the powerhouse and canyon-wall outlet works. A black hole of financial liability, as wide and dark as one of the diversion-tunnel portals, yawned before the contractors.

At a Six Companies board-of-directors meeting in San Francisco on December 16, 1932, the question of the gas cases and how to deal with them was discussed with attorney Paul Marrin of the San Francisco law firm Thelen & Marrin. After considering and rejecting out-of-court negotiation, the directors agreed with Marrin that any gas cases filed should be fought aggressively to a court decision.[22] Just how aggressively became apparent during the spring and summer of 1933 when a strange and faintly sinister series of events culminated in the most sensational trial the city of Las Vegas had ever seen.

The affair began in late March, 1933, with the appointment of Bud Bodell as Six Companies' chief special agent. One of his first assignments was to dig into the backgrounds of the gas-case plaintiffs and spy on their current activities, in hopes of gathering evidence that would contradict their claims of disability. The detective work paid off handsomely in May when Bodell discovered that one of the plaintiffs, former truck driver Ed F. Kraus, had a Utah police record for grand theft and receiving stolen property and that he had been seen drinking and dancing in Las Vegas nightclubs since filing his lawsuit in March.

Kraus's after-hours carousing in the Fremont Street saloons and dance halls made a mockery of his complaint that he was suffering from gas-induced dizziness, weakness, and impotence. This was not good enough for the Six Companies investigators and legal team, however; they had decided to make Kraus's suit the test case, and their intention was to degrade him (and by association the other plaintiffs) so thoroughly that Austin would be cowed into dropping the other suits.

To achieve this end, Bud Bodell recruited a lowlife named John Moretti to seek out and befriend Ed Kraus, involve him in as many drinking bouts and sexual liaisons as possible, and then lure him into participating in criminal activity. Posing as a two-bit crook, Moretti went undercover in June, 1933, and tracked Kraus to Salt Lake City, where he was visiting friends. The successful completion of Moretti's mission was heralded three months later by a *Las Vegas Evening Review-Journal* headline trumpeting "Holdup Reveals Dangerous Racket!" The accompanying story told how, through the "diligent investigative efforts" of Chief Special Agent Bodell and his accomplice John Moretti, E. F. Kraus and two other men had been arrested for assaulting and robbing an old miner named Jack McEachern. Moretti, the paper reported, had also told authorities that Kraus was planning to commit insurance fraud by staging a fake accident at the dam site, to pass $100,000 in counterfeit money, to blackmail three prominent Las Vegas businessmen "through the use of attractive women," to hold up the Apache Hotel, and, last but not least, to steal sugar from the Clark County Wholesale Warehouse to make bootleg whiskey.[23]

Less than forty-eight hours after his arrest and the dramatic disclosure of his alleged criminal plotting, the other shoe dropped for Ed Kraus and his lawyer Harry Austin: Judge William Orr announced that Kraus's damage suit against Six Companies would be the first to go to trial, even though several other identical damage suits had been filed months before it. Furthermore, the date set by Orr, October 16, fell only two weeks after Kraus was scheduled to be tried on the assault and robbery charges.[24] Clearly, Six Companies was flexing its legal and political muscle in Clark County, for it beggared the imagination that the timing of Judge Orr's announcement, the Kraus case's sudden unexplained leapfrogging ahead on the court calendar, and the scheduling of the Kraus criminal and civil trials within days of each other were a coincidence. Although he was undoubtedly shaken by this turn of events, Austin was not about to back down; he managed to get the criminal trial postponed while he plunged ahead with preparations for the gas case.

On October 17 the Clark County Courthouse was packed with off-shift dam workers and other interested spectators as the 34-year-old Kraus took the stand to testify about working conditions in the diversion tunnels and their effect on his health. In response to Austin's questions, he told the jury that during the months he was employed as a Six Companies truck driver hauling muck from the tunnels he lost twenty-six pounds, suffered headaches, chills, and nausea, passed out on several occasions, and finally become so dizzy and weak that he had to quit his job and go into a hospital for two weeks. Since that time he had continued to suffer from dizziness, blurred vision, and weakness that prevented him from working and also had lost his sexual powers and sexual desire. These ailments were all caused, he claimed, by breathing carbon monoxide in the tunnels, where at times seventy trucks, engines running, were lined up to be loaded with muck, and where the exhaust fumes were so thick "you couldn't see the electric lights." [25]

Two well-known Las Vegas attorneys, Frank and Leo McNamee, were kept on retainer by Six Companies, but for this critical case they had imported a crack San Francisco trial lawyer, Jerome White, partner in the companies' firm of Heller, White, Ehrman & McAuliffe. White now launched into a relentless four-day cross-examination of Kraus, forcing him to admit that his symptoms could have been caused by heat sickness, that he had not reported his illness to the foreman or asked to be relieved, that he had not worked exclusively in the diversion tunnels during his period of employment, and that he had refused to submit to blood tests, X rays, and a spinal tap when approached by Six Companies doctors two weeks before the trial started. White also questioned Kraus sharply about his sexual activities since the alleged gassing, insinuating that he had had many affairs with women other than his wife, and pressed him about his association with John Moretti. "This is not a criminal case!" Austin repeatedly objected, but with Judge Orr's tacit approval White continued on in a prosecutorial vein, finally concluding with the dramatic warning that he was going to call John Moretti, who would testify that Kraus had gone to parties, drunk intoxicating liquor, danced, and had sexual relations with women other than his wife "as though a well man." [26]

Austin called five more witnesses. Worker O. J. Lyke testified that it was common for men to leave the tunnels because of gas and that in the late afternoon a blue cloud could be seen drifting out of the tunnel tops. Foreman John Tackie, who had been named co-defendant with Six Companies, admitted that he knew little about carbon monoxide and that the tunnel lights had been obscured sometimes by dust and smoke in the air. Mrs. Kraus claimed that her husband had lost his sex drive after being gassed: "Male members of the audience craned to catch her every word, and women buried their faces in their hands," reported the *Review-Journal*. Finally, Dr. Walter Koebig of Los Angeles and Dr. J. N. Van Meter of Las Vegas testified that Kraus's symptoms were consistent with carbon-monoxide inhalation and that their examinations of him led them to believe that he was suffering from an "extreme case of carbon monoxide poisoning." [27]

The trial, which had already run longer than any other in Las Vegas history, took on a circus atmosphere when Jerome White began to present Six Companies' case. As the courthouse crowd gawked and reporters scribbled frantically, the dapper San Francisco lawyer announced that he would show that Kraus had planned "to get evidence and manufacture it if necessary" to frame his suit against Six Companies; that, contrary to his claim of impotence, he had maintained a "love nest" in an alley behind Las Vegas' El Portal Theater, where he regularly had sex with a woman known only as Merle; and that while "drinking to excess" with Moretti he had said, "I can look pretty sick for $76,000."[28]

A number of workers testified that the air quality in the tunnels was acceptable, although they admitted when cross-examined by Austin that they, too, had suffered headaches and nausea and that they had been required to come and testify by the contractors. C. J. Seymour, a Six Companies safety engineer, said that his tests had showed the air in the tunnels to be unfit to breathe "only three times," and although he had seen many ill men come staggering from the portals, he had attributed this to the heat rather than gas. Prominent San Francisco physician H. M. Behneman swore that Kraus's ailments were not caused by carbon monoxide but by abdominal ptosis (sagging of internal organs) resulting from his "tall and angular" build. Charlie Shea took the stand to say that he had been in the tunnels from January to July, 1932, and had never been affected by gas.

For two weeks White called witness after witness to testify about working conditions in the tunnels, while both public and press waited impatiently for the promised revelations about Kraus's drinking and dancing sprees, sexual peccadilloes, and other illicit activities. They were not disappointed. "Attractive Spanish Senoritas Cop Spotlight in Kraus Trial Today!" screamed an *Evening Review-Journal* headline on November 13, 1933, proclaiming the surprise court appearance of the giggling, gum-chewing, provocatively clad Lopez sisters of Los Angeles and heralding the lurid climax of the damage suit against Six Companies. In their testimony Mary and Betty Lopez described themselves as "old friends" of Special Agent Moretti and told how, at his instigation, they had arranged a "party" for Ed Kraus. At this affair the allegedly disabled and impotent Kraus had drunk a case of beer, played the piano, performed a tap dance, jumped into a swimming pool, and then departed for a hotel in the company of a woman named June. On another occasion, Betty told the gaping jury, Kraus had taken her to a hotel and had sex with her twice in one night. On cross-examination, both women denied that they were prostitutes or that they had been paid to appear in court.

John Moretti was the last witness for Six Companies. For the better part of a week, he poured out an incredible tale about the three months he had spent undercover with Kraus, enthralling spectators with intimate details of the plaintiff's escapades at the Red Rooster, the Willows, Lorenzi's Resort, and other Las Vegas nightspots; his drinking, gambling, and criminal scheming; and his fornicating with the mysterious Merle on a cot beneath a fig tree in back of El Portal Theater. When this performance was over, the original questions about carbon-monoxide poisoning and Six Companies negligence had been all but forgotten despite Austin's efforts to refocus the proceedings. Much to his surprise, however—and to the discomfiture of the Six Companies legal team—the jurors remained deadlocked after seven days of deliberations, and on November 29, Judge Orr dismissed them and declared a mistrial.[29]

A week later, Kraus's criminal trial got under way; this time it was Six Companies' turn to be cast in the role of villain as testimony strongly suggested that Special Agent Moretti had planned the assault and robbery to discredit Kraus before his gas suit went to court. It was also revealed that Jack McEachern, the victim of the assault, was a longtime acquaintance of Bud Bodell and that the day after the crime Bodell had installed him in a comfortable job in Boulder City. Moretti and Bodell had conspired to frame Kraus on orders from Six Companies, Kraus's lawyer contended, and the jury apparently agreed, for they found him innocent of the assault and robbery charges.[30]

So it was that the first of the gas cases ended in a standoff. Six Companies had played the hardest of hardball, but had failed to administer the crushing defeat that would discourage future plaintiffs. Austin had bungled in taking a disreputable hoodlum as a client, but somehow he had convinced several of the jurors that Kraus had a legitimate claim. Each side conducted a postmortem, reviewed its performance, and vowed to do better next time.

Next time came on February 18, 1935, when the case of *Jack Norman v. Six Companies, Inc., Woody Williams, and Tom Regan* went to trial in Judge Orr's courtroom. Austin had screened his client more carefully for this contest: Norman was older than Kraus and was not a hell raiser. He had been a shift boss, or shifter, in the diversion tunnels, a position of some responsibility, had a wife and several children, and clearly bore the outward signs of a wasting illness that Austin intended to link to carbon-monoxide poisoning. Otherwise, Austin used the same strategy he had employed in the Kraus case, calling former tunnel workers to describe working conditions, physicians to testify that Norman had been poisoned, and Norman himself to tell about the pain and financial hardship he had suffered since falling ill and losing his job.[31]

Six Companies had changed its approach, too, although the nature of its revamped strategy was not clear at first to many courtroom observers. After calling fifty witnesses in a sensational, all-out effort to discredit Ed Kraus, Jerome White and the other Six Companies lawyers treated Norman quite gently, contradicting his claims but refraining from direct attacks on his character or veracity. The atmosphere of the trial was subdued and businesslike, a sharp contrast to the sideshow aura that had surrounded the first gas case. Whatever the outcome, it seemed to onlookers that the proceedings had been fair and reasoned, adjectives that could not have been applied to the Kraus carnival. Thus, when the eruption came, it was all the more shocking for being unexpected.

The trial went to the jury early on the afternoon of March 16. After leaving the courtroom, many of the participants headed for nearby saloons to hash things over and await the verdict. Shortly after 5:00 P.M., Harry Austin entered the bar in the lobby of the Apache Hotel, where he found Six Companies Assistant Superintendent of Construction Woody Williams already well in his cups. The two men had several drinks together and were chatting amicably about the trial when Williams suddenly blurted out: "We [Six Companies] have at least three men on that jury that we are sure of and perhaps four or five. You fellows may think you are going to get a verdict but you haven't a chance in the world." [32]

Surprised and disturbed, Austin left the bar and went in search of his co-counsel, attorney Clifton Hildebrand, who, it turned out, had an upsetting story of his own to tell. He, too, had gone to the Apache Hotel bar earlier that afternoon and had been buttonholed by James Begley, one of Bud Bodell's subordinates on the Six Companies investigative squad. Begley was drunk and boisterous, and he gleefully boasted to Hildebrand that Six Companies would owe the winning of the Norman case to him and Bud Bodell because they "knew how to fix juries." They and five others had been working day and night on the gas suit with an unlimited expense account, and although lawyer White strongly disapproved of tampering, he "would not have had a chance if it hadn't been for the activities of Begley and Bodell." Begley went on to say that he knew of at least five jurors who were in Six Companies' pocket, and when Hildebrand said in disgust that he didn't doubt it, the smirking investigator replied, "Well, what the hell, you would do the same thing if you could." [33]

On March 18 the jury in the Norman case voted nine to three in favor of Six Companies, and Austin and Hildebrand immediately began an investigation, convinced by their encounters with Woody Williams and James Begley that the contractors had engaged in jury tampering. Three days later, when they believed they had compiled enough information to confirm their suspicions, they petitioned Judge Orr for a new trial, claiming that there had been jury irregularities. [34] In support of their contention, Austin and Hildebrand filed nine affidavits, including statements by the jurors who had voted against Six Companies, all of whom said that at least three of their fellow jurors were plants with a fixed opinion in favor of the defendant before the trial ever started. Perhaps the most damning remarks were made by Guy Melett, who had been drawn on the panel of potential jurors for the Norman case but not actually selected. He told his friend John L. Russell that juror H. E. Hazard "was fixed. He got at least $5000." Melett also said that jurors Arthur Mosbach and Dave Stewart "did not get any more than $500" and that he knew all this because "I myself was offered $1000." At this point in the conversation juror Stewart had come up, and when he was questioned, he acknowledged that he had been paid off. Would he dare to divulge this information publicly? "If I do, somebody will shoot me in the guts," he replied.[35]

Statements like these proved nothing, but they did raise unsettling questions about the Norman trial and suggested that further inquiries might be in order. However, Austin's and Hildebrand's plea fell on deaf ears. Judge Orr had shown where his sympathies lay when he pushed Ed Kraus's case ahead on the court calendar just hours after Six Companies had engineered his arrest. Now, to no one's surprise, he brushed off the affidavits and intimations of irregularities and refused to consider a new trial. Austin's response was equally predictable. On May 22, 1935, less than three weeks after Orr's ruling, he filed nine new gas-related civil suits.[36]

It appeared that Bud Bodell's Six Companies investigators had engaged in pimping, criminal conspiracy, intimidation of witnesses, and jury tampering in their effort to dispose of the gas cases and protect the company's treasury and business reputation. Their tactics had been ruthless, their methods heavy handed, but instead of scaring Austin off, they had only goaded him on. Now the higher-ups had to decide whether the game was worth the candle. With its money and clout, Six Companies clearly could win the battles, but it might lose the war if Austin persisted, as he showed every sign of doing, in a long campaign of attrition. On August 6, 1935, the *Review-Journal* reported that forty-eight gas suits seeking a total of $4.8 million in damages were pending against Six Companies in Nevada courts. With the number of cases growing so swiftly, and the Hoover Dam job rapidly winding down, the contractors decided it was time to call off Bud Bodell and instruct their lawyers to negotiate with Austin. In January, 1936, an out-of-court settlement was reached, and checks for undisclosed amounts were distributed by Six Companies to fifty gas-suit plaintiffs.[37]

So ended the four-year battle over carbon monoxide's effects on workers in the Hoover Dam tunnels. No negligence had been shown on the part of Six Companies, no proof positive had been presented showing that laborers had been injured by company-sponsored work conditions, and yet the ambiguous resolution of the gas cases left a lingering bad taste, for this was the one instance in which charges that Six Companies "held life cheap" on the job at Hoover Dam rang true. All the other complaints and accusations could be countered: one man's exploitation and job speedup was another's efficiency and hard work. But the simple fact that it was dangerous to run scores of gasoline-engine trucks day in and day out half a mile inside poorly ventilated tunnels where hundreds of men were working could not be denied. If common sense did not suggest that eight hours of breathing concentrated exhaust fumes mixed with the dust and effluvia of dynamite explosions was a serious health hazard, and if the splitting head-aches, dizziness, and fainting spells commonly suffered by tunnels crews did not indicate a problem, then ample literature was available from the U.S. Bureau of Mines, not to mention the Arizona and Nevada mining safety laws, to spell it out.

Six Companies knew that carbon monoxide was a potentially lethal hazard, but to protect a $300,000 investment in gasoline-fueled trucks and avoid a costly delay while electric motors were installed, it was willing to circumvent state legal restrictions on the use of internal-combustion equip-ment underground and put its employees at risk. Then, when the bill for this decision was rendered in the form of a long casualty list and at least half a hundred lawsuits, the company tried to cover up and evade responsi-bility in a manner that was at best odious and at worst criminal. How many men died, how many were severely injured, how many were crippled by carbon monoxide at Hoover Dam will never be known for sure, but the figure was high, certainly much higher than it should have been. This un-necessary, easily preventable disaster, caused by corner cutting and exacer-bated by selfish concern for corporate image, was not only a tragedy for the victims but also a sad chapter for Six Companies, for it was the single major blemish on a safety record that was otherwise commendable, given the inherent dangers of the job site and the work performed there.

In Black Canyon during the summer and fall of 1933, concrete was flowing in a rising tide, not only into the base of the dam but also into the lining of the so-called Glory Holes, the two most precipitous and spectacular of the sixty-odd tunnels drilled during the project. The seventy-foot-wide Glory Holes opened at the bottom of the troughlike spillways, then swooped downward at a forty-five-degree angle through the cliffs, gradually taper-ing to a diameter of fifty feet before intersecting the lower part of the outer diversion tunnels, which would carry the runoff around the dam. Excavat-ing these steep, telescoping bores on the exact grade and to the precise di-mensions required by the plans had been difficult, and lining them with concrete proved to be even harder. The transitions in diameter gave the carpenters fits when they began assembling the forms, and the slope meant that all materials had to be winched down into the tunnels on cable cars.

When the forms were finally prepared and in place, a two-cubic-yard bucket of concrete was strapped to a cable car and sent down to be poured. It arrived without falling off the little trolley, but the pouring crew found that the concrete had already taken its set; the distance between the open-ing of the inclined tunnel and the pouring site was too great and the cable car's rate of descent too slow for the concrete to be workable by the time it reached the form. Obviously, a different transportation system was called for. The cable-car tracks were torn out and a pipeline was laid down the slope. The idea was to feed the concrete into the pipe and let it slide down to the pouring site. The only problem with this approach was that the temperature in the tunnels was 120 degrees, and try as they might the workers could not keep the inside of the pipe cool and wet. The first three buckets of concrete were emptied into the pipe, but when the gluey gray mass hit the hot metal, it hardened almost instantly into a solid plug. The men poured in water and hammered on the outside of the pipe, but all that trickled down to the form was a pile of pebbles.

⟍   Frank Crowe was famous in construction circles for his ability to overcome the most baffling obstacles, and he now exercised that talent to find a way to get workable concrete into the Glory Holes. Typically, he employed a new piece of equipment, a portable agitator, to solve the problem. The tracks were reinstalled, and the mixer, mounted on a steel frame so that it could be picked up and put down by the high line, traveled up and down on a cable car between the tunnel opening and the pour site.[38] This innovation proved successful and the lining of the Glory Holes proceeded steadily thereafter, but not without a noteworthy accident. During the graveyard shift on September 19, 1933, three carpenters, supervised by foreman B. F. Partain, were trying to knock a form free so that it could be pulled into position for the next pour. Normally the she-bolts holding the form to the tunnel sides were loosened and the men stood clear while the big panel was hauled up the steep incline, but this time the carpenters were careless. After pulling the bolts, they stayed on the form as it began to inch upward, dragged by a harness of groaning cables. Suddenly a piece of the form was torn loose by one of the cables and in rapid-fire succession the other lines gave way, snapping or ripping free under the uneven strain until the last tether holding the heavy assemblage of lumber and steel in place was gone. With nothing to restrain it, the form began to slide down the tunnel, slowly at first and then faster and faster. One of the carpenters grabbed a cable and yanked himself off, but the other three men—Partain, Armstrong, and McPhee—could only hang on for dear life as the form rocketed down the raise, careening toward the opening a hundred feet below where the Glory Hole funneled into the diversion tunnel. The horrified crew members standing above in the tangle of broken cables heard a tremendous splash as the form struck the river water rushing through the diversion tunnel. Someone had the presence of mind to telephone a warning to the men working downstream in the vicinity of the outlet, and Woody Williams and a rescue team were standing by when the battered form popped out of the portal like a cork from a bottle and fetched up against a sandbar. The bedraggled carpenters were still clinging to the side. One man had a fractured back and a concussion, another had broken ribs, and the third was so pummeled and swollen that he was unable to walk, but somehow all had survived the terrible quarter-mile roller coaster ride down the throat of the Glory Hole and through the Arizona cliff.[39]

If the oversized inclined-spillway tunnels were known as Glory Holes, the narrow penstock tunnels deserved a considerably more ignominious name. These dusty, cramped bores, eight on each side of the river, branched off from the inner diversion tunnels and the thirty-foot-diameter penstock headers and angled down to the powerhouse. From the intake towers, water would be delivered through this tunnel network to spin the turbines and generate electricity. Excess flow would be shunted past the power-house via the penstock headers and discharged into the tailrace from the valve houses of the canyon-wall outlet works 180 feet above the surface of the river. Unlike the concrete-lined spillway tunnels, the penstock headers and tunnels would, as their names implied, be fitted with steel pipes to carry water through and around various parts of the dam. In all, nearly three miles of pipe had to be fabricated and installed, a job so big and re-quiring such specialized skill and equipment that the Bureau of Reclama-tion decided to solicit bids and write a contract for it separately from the main job. The low bid, just under $11 million, was submitted by the Bab-cock & Wilcox Company of Barberton, Ohio, and on July 9, 1932, the firm was awarded the contract for building and installing the penstocks.[40]

Like almost everything else at Hoover Dam, the piping was to be huge, some of it thirty feet in diameter. There was not a truck or a railroad flatcar in the country big enough to haul such a load from Ohio to Nevada, so Babcock & Wilcox built a complete steel-fabrication plant at the dam site, shipping all the necessary machinery—including a furnace, a bending mill, a vertical rolling mill, and X-ray gear for checking welds—across the country from Barberton at a cost of nearly $600,000. By the spring of 1933 more than one hundred employees were ready to begin transforming 44,000 tons of steel plate into sections of pipe. The next problem was how to get the monstrous pieces from the plant to the canyon rim and then down six hundred feet to the openings of the construction adits. Six Companies came up with the first part of the solution by designing a Caterpillar-tractor-drawn steel trailer that could carry the 170-ton sec-tions along the one and a half miles of winding road between the fabrica-tion mill and the Nevada cliff edge; the second part of the solution was completion of the huge government cableway, the largest ever rigged, strong enough to lower penstock sections and the turbines for the power-house, and the one cableway that would become a permanent part of the Hoover Dam complex. In 1934 the first gargantuan piece of pipe was eased into the trailer, pulled at a snail's pace to the rim, slung from the gov-ernment cableway, and dropped down into Black Canyon, beginning a yearlong push to graft together the steel arteries that would pump water from the reservoir behind the dam to the dynamos in the powerhouse at its base.[41]

Penstock headers, spillway tunnels, powerhouse foundation, intake tow-ers, canyon-wall outlet works—all the pieces of Hoover Dam—were com-

A thirty-foot-diameter section of penstock pipe, with a man aboard, is lowered into the canyon by the government cableway. (Bureau of Reclamation)

ing together by 1934, the progress of each piece synchronized so that it would not delay work on any other piece. It was "the same problem as the bride's dinner: getting everything done at the right time," said a Reclamation official. At the beginning of each shift some fifteen hundred men descended on Black Canyon, took up their tools and attacked their jobs,

A section of penstock pipe arrives at the mouth of an adit leading to the interior of one of the diversion tunnels. (Bureau of Reclamation)

driving the work along at a pace that would have the project finished as much as two years ahead of the deadline set by the government.

"There is a stark and uncompromising efficiency which one feels at every point of contact with the dam work," wrote journalist Duncan Aikman after a visit to the site in 1933. "Everywhere men move fast, throw all

Another bucket of concrete is emptied into the body of Hoover Dam. (Kaiser Collection, Bancroft Library)

their power of muscle and machinery into what they are doing, waste no time in workmanly sociabilities on the job. There is no gayety about the scene, no sense of men colorfully enjoying their work. Instead, a kind of surly determination broods over their labor. . . . By now the sheer, nervous drive of the task and the ruthless hiring and firing policy of the Six Companies contracting organization, which leaves only the supremely efficient on the job, have achieved a coordination that is practically flawless." [42]

Experience through ceaseless repetition was the key to the seemingly effortless proficiency exhibited by the work crews: in June, 1932, the first month of pouring, 25,000 cubic yards of concrete were delivered to the dam site and put into the forms. By August the figure had jumped to 149,000 cubic yards. Two months later it had passed 200,000, and in March, 1934, more than 262,000 cubic yards—1,100 buckets a day, one every 78 seconds—gushed out of the mixing plants and into Black Canyon. [43]

The tall gray columns of the dam rose inexorably, merging to form an immense upside-down wedge, a concrete behemoth thrusting upward like some bizarre geological formation. So unworldly did it look from the vantage point of the Nevada rim, and so intimidating was the impression it made, that tourists, who came by the thousands to stand on the edge

Puddlers tramped concrete into place in the forms with shovels and their feet.
(Library of Congress)

of Black Canyon and stare, felt compelled to humanize it by concocting Bunyanesque stories about its builders and the hair-raising dangers they faced. It was from this deep-set need to make the impersonal personal, to endow the lifeless block with a soul, that the myth of men buried alive in the concrete sprang. The idea of workers forever entombed in the giant structure they had helped build was so irresistibly poetic, so deliciously macabre, that it became the basis for the most enduring legend of Hoover Dam, an article of faith for millions of visitors who down through the years would insist, despite the firm denials of tour guides, Bureau of Reclamation engineers, and historians, that the great arch was not only a dam but a sarcophagus.[44]

The closest anyone came to being buried in the dam was when the wall of a form collapsed during the graveyard shift on November 8, 1933. A bucket of concrete had just been dumped into the nearly finished block when the bolts and braces holding one of the form panels in place gave way and the whole sticky mass slumped free. Normally such a break would have caused only moderate inconvenience—several hours of cleanup and repair work—but as fate would have it the collapsed panel flanked the slot, the central eight-foot-wide cut housing the pipes that connected the water lines from the cooling and refrigeration plants with the tubing interspersed throughout the body of the dam. The concrete, more than a hundred tons of it, slid like a lava flow over the edge of the slot and plummeted straight down, tearing out pipes and platforms as it went. W. A. Jameson and W. H. Hammond happened to be walking through the slot on a scaffold, carrying an eighteen-foot timber on their shoulders, when the avalanche began. Hammond heard the crashing, ripping sound, dropped the timber, and scrambled back just in time to see the concrete flash by, inches in front of his face. A second later he was on his knees with a severely sprained back, peering over the edge of the shattered scaffold, his sense of relief dissolving into horror as he realized that Jameson had been swept to his death one hundred feet below.

After the rain of debris had subsided, searchers rushed to the bottom of the slot to retrieve the body, but all they found was an ugly mound of wet concrete studded with jagged fragments of lumber and pipe. It was 3:30 in the morning when the digging began; at 7:30 the following night the shattered corpse of Jameson was finally recovered. He was the only man ever buried in Hoover Dam, and he was interred for just sixteen hours.[45]

Although brief, searing moments of terror like those resulting from the Jameson accident burned themselves into the memories of all who lived through them, they came only rarely, and for nearly all the men in Black Canyon the work of building the dam eventually devolved into a monotonous albeit hazardous routine. Once every twenty-four hours they piled into the transports in Boulder City and made the jolting, swaying eight-

Swing-shift workers lining up at the paymaster's window of the Six Companies office building in Boulder City, January, 1932. (Bureau of Reclamation Collection, UNLV)

mile journey to the canyon rim. After crowding aboard the monkey slides and Joe McGee skips, they were lowered swiftly to the top of the dam or to the canyon floor, where they performed their assigned tasks for eight hours, breaking only for lunch. When the shift was over, they returned to the rim, got back on the transports, rode up through the loops and hairpin turns to Boulder City, and dispersed to their cottages and dormitories. Entertainment consisted of dinner, perhaps a movie at the Boulder Theater or a game of pool at the Six Companies recreation hall, and then sleep to ease the ache of muscles fatigued by pushing, pulling, digging, and stomping, and to relieve the stress of constant heat, danger, and boredom.

↘   Weekends provided no respite from this unremitting grind, for the work of raising the dam continued on Saturday and Sunday. Only the approach of payday, on the tenth and twenty-fifth of every month, and the prospect of pocketing two weeks' wages, offered any real surcease from the tedium. The passing time could seem interminable, but when payday finally did come, it wrought a startling transformation. Boulder City shrugged off its torpor; the men lined up for their money, clean shirts and trousers were broken out, and a long caravan of cars passed through the reservation gate and rolled northwest toward the bars, gambling halls, and fleshpots that lay at the far end of Boulder Highway.

The population of Las Vegas had more than doubled between 1930

HIGHWAY LAS VEGAS - BOULDER CITY, NEVADA
FRASHERS FOTO-POMONA, CALIF.

Boulder Highway, also known as the Widowmaker, linked Boulder City with Las Vegas twenty-eight miles to the northwest. (Kaiser Collection, Bancroft Library)

and 1934, and it seemed that every one of the ten thousand citizens, native and newcomer alike, was hustling to get his or her piece of the Hoover Dam payroll. Banks urged the workers to save their hard-earned dollars, playing on fears of continuing Depression and unemployment when the Boulder Canyon Project was finished. "If a dog could talk he'd tell you that he buries bones when he gets them so that he'll have food in reserve when there is no more to be had," proclaimed a First State Bank newspaper ad. Stores competed vigorously to lure business away from the shops in Boulder City, offering a wide variety of merchandise at sharp discounts. But saving and shopping were not what most of the men had in mind as they roared into town on payday, wallets fat with greenbacks and throats clogged with road dust. Las Vegas had always been a wide-open town where women and whiskey were cheap, but the onset of the Boulder Canyon Project and the arrival of thousands of dam workers had transformed the once casual, easygoing business of selling sin into a booming industry. "They must do something else there," wrote an astounded visitor, "but if so it is not discernible to the innocent tourist's eye." [46]

For the married or the faint of heart, milder diversions were available. Sixty cents bought admission to the Airdome, "Glorifying the Talking Screen," or the El Portal, "Nevada's finest theater." Establishments like the Meadows Hotel, the Apache Hotel, the Golden Camel, the Willows, and the Lorenzi's Resort offered drinks and dancing in an atmosphere that was reasonably civilized. The real action, however, was in the teeming, frenetic

The bar at the Boulder Club in Las Vegas, April, 1935. (Davis Collection, UNLV)

casinos—the Nevada Bar, the Boulder Club, the Las Vegas Club, the Rainbow Club, the Northern Club, and dozens of others—that lined the downtown streets, each striving to outdo the others in creating an atmosphere of blind, boisterous revelry.

"The liquor is vile and no one trusts the wheels; but all drink and play furiously," wrote journalist Theo White. "By ten-thirty things are well underway. By two everyone is drunk and begging for more. Rooms, as large as small auditoriums, are packed to bursting with sweating inebriates fighting for the edge of the gambling tables. Barkeeps frantically pull corks and pour liquor into the howling mob at the bar. In some corner a dreary, anaemic 'professor' mechanically bangs out worn tunes on a shaky piano, encouraged by philanthropic mugs of beer. Then one gets a headache and is certain that the lungs cannot take another breath of smoke-fouled air. It seems a wild, wild dream of a delirious mind." [47]

Waiting in the wings when the pleasures of drinking and gambling paled were hundreds of women eager to help the dam workers spend what was left of their pay. "Las Vegas painted ladies go in for raspberry-colored sailor pants, coral-tinted blouses and high-heeled slippers," a visitor reported. "Their faces, beneath the metallic-looking orange rouge which is universal in the Southwest, are haggard and burned like the faces of all desert people. Every second or third man you meet has had about three drinks too many, and is glad of it. The ladies, as always, drink less or hold it better, and obligingly help keep their gen'l'men friends from rolling into the gutters."[48]

The action centered on infamous Block 16, the once drowsy and decaying red-light district, which had been rejuvenated into what one observer called an "unforgettable 'skidway,'" a seedy but bustling strip of cribs and houses with such names as Blue Heaven and Ye Bull Pen Inn, a place where the "flotsam and jetsam are herded together for the purpose of satisfying the sex appetites of the men who are building the Hoover Dam."[49] A continuous parade of prostitutes and potential customers passed up and down this thoroughfare, laughing and shouting, plucking at one another's sleeves, pairing off and disappearing into the dark doorways for their brief liaisons. "Many a young worker is going to remember Hoover Dam, not as a triumph of engineering skill, but as the place where he contracted venereal disease," commented River Camp waiter Victor Castle after a tour of the Block,[50] but fear of infection did little or nothing to keep the men away from their "home away from home," as one laborer waggishly called it.

Prices, not prophylactics, were the main concern of the skidway habitués: although the prostitutes' standard two-dollar charge was by no means prohibitive, it could cut deeply into a bankroll already depleted by too much roulette and red-eye. So, during the first year of the dam construction, the cash-strapped men hit upon a ruse for getting a 75 percent discount from the denizens of Block 16. Each worker was issued an identification badge when he hired on with Six Companies, an item that had to be presented at the paymaster's window to receive wages and cost fifty cents to replace. The workers told the women that the badges were worth two dollars and were redeemable in cash at the Six Companies office in Boulder City. Badges became legal tender on Block 16 and business boomed until one day a pair of well-turned-out ladies arrived on Frank Crowe's doorstep, lugging a carton brimming with the little cards and demanding payment. As gently as possible the superintendent explained to the agitated women that the badges were worth only half a dollar, not two, and that Six Companies was under no obligation to redeem them. However, to show that he was a gentleman and to prevent more hard feelings, he promised that henceforth the cost to the workers of replacing a lost badge would be two dollars, the same as the going price for Block 16 entertainment. This new and rather unique floating standard of value for identification badges brought an abrupt end to the workers' scam, and cash on the pillowcase became the rule, although Six Companies scrip was also accepted in some houses.[51]

When payday night finally passed and the sun poked up over the horizon, Las Vegas seemed to collapse in on itself like a deflating balloon. The brilliant yellows and reds of a thousand neon tubes were replaced by the dull, dirty hues of the desert. The throngs that had filled the sidewalks dwindled to a scattering of drunks staggering aimlessly or snoring on the pavement. The scent of whiskey and perfume that had hung sweet and heavy in the night air was supplanted by the sour smell of drying beer spills and the odor of stale tobacco smoke. On Boulder Highway, which had earned the nicknames Widowmaker and Shambles Alley on mornings like this, cars full of half-sober day-shift workers careened toward the reservation gate, racing to beat the clock back to the mess-hall parking lot and the transports waiting to leave for Black Canyon.[52]

For most of the men a twelve-hour binge in Las Vegas was enough to release the tension built up during two weeks of work, but for others whose appetites were more constant, and for those who wanted gambling action, a drink, or feminine company on short notice, there was an alternative. It was grimier and more basic than Las Vegas, yet correspondingly cheaper and more convenient. This was the strip of bars and bordellos stretching along Boulder Highway from the settlement of Midway, half the distance to Las Vegas, all the way back to the very gates of the project reservation. According to Clark County officials, there were at least 125 "girl entertainers" working in the various "resorts" along this thirteen-mile stretch of highway, as well as scores of drinking and gambling dens clustered in impromptu communities that sported names like Four Mile, Sunset, Oklahoma City, and Pitcher.[53] The busiest and most notorious of these roadside hamlets was the one closest to Boulder City. It was called Railroad Pass, or more simply the Pass, and consisted of a cluster of buildings and tents situated at the place where the highway and Union Pacific tracks cut through the range of low hills separating Las Vegas from the project reservation.

The Pass had started out as just another squatters' camp in the early days of the Hoover Dam job, but its character changed overnight when an enterprising promoter arrived on the scene and built a casino almost within eyesight of the reservation boundary. Interestingly enough, the first load of building materials delivered by the Union Pacific when the branch line from Bracken to Boulder City opened in late January, 1931, was used not for construction of the dam or the dam workers' community but to erect this pleasure palace. The Railroad Pass Club, where "Gayety is the Password," opened on August 1, 1931, boasting twenty-five ceiling fans, a five-piece orchestra, "cold, refreshing drinks," and the latest in gaming equipment. Not mentioned in the advertising, but also a prominent attraction, was the clutch of little cabins on the rocky slope behind the casino where the prostitutes plied their trade.

On the north (Las Vegas) side of the Pass, surrounded by yet another paupers' bivouac, stood the Texas Acres, probably the toughest joint on Boulder Highway. Beatings, knifings, and shootings were regular occurrences in this establishment, which, because of its vicious atmosphere and wicked reputation, had drawn the attention and condemnation of even the most hardened state and federal law-enforcement officials. In 1933 it was renamed the Boulder Inn Club, and a year later its owner, Pete Pansey, announced that he was closing it temporarily for remodeling. Shortly after the closing, Walker Young, Sims Ely, and other project reservation officials were aghast to discover that the remodeling had consisted of jacking the saloon up on skids and hauling it to a new location on government land next to Boulder Highway. As if this were not brazen enough, Pansey also had put the prostitutes' cribs on the federal land in plain view of passing traffic and of a nearby schoolhouse.

The normally mild-mannered Young was livid. He wrote an angry letter to his Interior Department superiors demanding that action be taken to close the operation.[54] They agreed, and in June, 1934, government lawyers were dispatched to tell Pansey to move or face a lawsuit. Business was good in the new location and the recalcitrant tavern keeper wanted to fight the eviction, but several months after the government's suit was filed, a pair of events spelled the end of the Boulder Inn Club. First, in July, 1934, the public was outraged by a *Las Vegas Evening Review-Journal* report that school-age boys were being served liquor and allowed to gamble in the establishment. This was followed on April 9, 1935, by a particularly brutal gun battle in the bar. Bowing to political pressure, the Clark County Liquor Board suspended the Boulder Inn Club's license on April 12, and Pansey was finally forced to cart his building off to a new location near Midway. The roughest saloon was gone from Railroad Pass, but the other Pass businesses carried on, offering Boulder City residents "the essence, minus the variety, of Las Vegas without the necessity of a long trip."[55]

When the last dollar had been spent and a lingering hangover was all that was left to recall an evening of revelry, the dam reasserted its claim to the workers' lives, drawing them back to its gray-white flanks, back to the danger and drudgery of construction work. "Their existence is . . . centered about the callous, cruel lump of concrete at the bottom of the canyon," observed journalist Theo White. "To them it assumes a personality and they are devoted to it. When they have surveyed the result, the pride of the man who designed it will be no greater than that of anyone who has driven a single nail in its making. In each there is a feeling of ownership, but I think they have never paused to calculate the cost in terms of humanity."[56]

➤ On July 20, 1934, Boulder Canyon Project employment reached its peak with 5,251 men engaged in some form of construction-related activity, either in Boulder City or Black Canyon.[57] Concrete pouring continued at a rapid clip, pushing the job farther and farther ahead of schedule. Frank Crowe, Charlie Shea, Walker Young, Felix Kahn, and a gang of mud-spattered workmen gathered around a concrete bucket on January 7,

1934, and had their picture taken. A sign hanging from the bucket announced that it held the one-millionth cubic yard to be put into the dam, just seven months after the start of pouring. Less than three months later the photographer was back, this time to memorialize the record daily pour of 10,350 cubic yards, and on June 1, 1934, he returned yet again to record for posterity the placement of the two-millionth cubic yard of concrete. The dam proper was now nearly two-thirds complete, and it looked less and less like the uneven hodgepodge of individual towers and more and more like a solid, finished structure.

Work on other parts of the dam complex was moving ahead rapidly also. The lining of the inclined spillway tunnels was finished in March, 1934, and erection of the eight 2.5-ton steel drum gates that would control water flow into them began. The clumps of steel rods that crowned the four intake towers were approaching their final height of 395 feet as concrete sluiced down through a series of chutes to fill the forms shaping the fluted, gracefully tapering pillars. The powerhouse foundation had been excavated, and placement of structural steel and pouring of concrete were under way.

"Men are carpentering, digging, chopping, sliding, swinging on ropes, shouting and signaling everywhere," wrote a canyon-rim sightseer. "Out of their swarming comes a light, steady roar, half muffled by the vastness of the setting, but punctuated from moment to moment by its own inner emphasis—brake-squealings, cable-creakings, horn-honkings, whistles, hammer-poundings; the cluckings, purrings, grindings, snufflings of machinery merged with the desert wind into an open-air orchestration of modern power industry."[58] The great spectacle of construction, touted by the press and the Six Companies public-relations division, and the attractions of nearby Las Vegas, drew ever larger crowds of tourists into southern Nevada. During 1934, 266,436 visitors were checked through the project reservation gate to see Boulder City and drive to observation points on the edge of Black Canyon. Union Pacific was running excursion trains to Boulder City almost weekly, and on March 9, 1934, the railroad's special, ultramodern streamlined train, carrying celebrities, members of the press, and company and government officials, actually glided into the canyon and up to the face of the dam on the contractors' construction line. The most ambitious public-relations event took place on the night of October 20, 1934, when ten trainloads of Shriners from Utah and California held a dramatic candlelight ceremonial on the crest of the upstream cofferdam.[59]

Hoover Dam's popularity as a travel destination was reflected by the expansion and reexpansion of the Boulder Dam Hotel on Arizona Street in Boulder City. No one had anticipated the need for such a facility when the town was built in 1931—the notion of the baking patch of desert becoming a tourist mecca seemed patently absurd. But by the summer of

Structural steel bristling from the top of one of the four intake towers, July, 1934. (Bureau of Reclamation)

1933 it was clear that a hotel was necessary, and Sims Ely issued a permit and lot lease to P. S. Webb, owner of Boulder City Builder's Supply Store, to construct a deluxe hostelry complete with oak-paneled lobby and private tiled baths in every room. Three and a half months after ground was broken, the two-story Dutch Colonial–style hotel opened its doors and welcomed a celebrity-studded clientele. The Hollywood film colony had discovered Boulder City as a weekend hideaway: Bette Davis, Ronald Colman, John Wayne, Clark Gable, Carole Lombard, and Henry Fonda were a few of the names that appeared in the register. The visits of these screen stars to Boulder City and the dam site conferred upon the already famous project an aura of glamor and excitement that attracted even more visitors. Barely six months after the hotel opened, an eighteen-room wing was added, giving it fifty-one rooms; a year later, a dining room and thirty more guest rooms were built.[60]

❧ Ironically, the surge in Boulder City tourism in 1934 coincided almost exactly with the beginning of a gradual winding down of the Hoover Dam project. After hitting a peak in July, the payroll began a steady decline as hundreds of workers departed for western Arizona, where Six Companies was gearing up to build Parker Dam, the second link in the chain of Colorado River control works, and to Oregon and Washington, where work on Bonneville and Grand Coulee dams would soon begin. On November 3, 1934, the low-mix plant cranked out its last few yards of concrete and then shut down after a continuous run of more than two years. It would be dismantled and shipped piece by piece to the Parker Dam site while the high-mix plant provided the concrete needed to finish the job.

On December 5, photographers snapped away as the three-millionth cubic yard was deposited in the body of the dam. It was duly noted in the papers that the Bureau of Reclamation's bid specifications had called for concrete pouring to *start* on December 4, 1934, and that instead the structure was 92 percent complete.[61] Sixty-three days later, with almost no fanfare, cableway 6 lowered a concrete bucket to a shallow form on column K and the block was topped out, the first to be brought to the dam's final height of 726.4 feet. Five more columns were nearly finished, needing only a two-foot pour to complete the roadway that would run along the dam's crest.[62] Rapid progress was also being made in the construction of the powerhouse and canyon-wall outlet works, where scaffolding was in place and the valve house floors were being poured.

For more than two years the Colorado had flowed through the diversion tunnels, forced from its ancestral course but still rolling unimpeded toward the sea. Now, as the wall of concrete wedged into its former bed was grouted and cured into a solid, impenetrable block, preparations were made to stopper the tunnel outlets and corral the river once and for all. As the low-water season began in the fall of 1934, cofferdams were mounded into place in front of the portals of the inner tunnels, Nos. 2 and 3, and concrete plugs were dovetailed into the fifty-foot bores, sealing them permanently. Next, another temporary cofferdam was erected in front of tunnel No. 1, forcing the river to use No. 4, the outer tunnel on the Arizona side, as its sole route downstream. When the interior of No. 1 was dry, the third concrete plug was poured, this one perforated with four six-foot-

The dam approaching its final height of 726.4 feet. An eight-foot-wide slot, housing cooling pipes, runs through the center of the dam's body and catwalks crisscross its downstream face. (Kaiser Collection, Bancroft Library)

diameter holes, each fitted with a valve so that it could be opened or closed on command. January 31, 1935, was the Colorado's last day of freedom; the next morning, a steel bulkhead gate weighing more than one thousand tons was lowered, closing the portal of tunnel No. 4 and forcing the river into No. 1 and through the honeycombed plug. By manipulating the four valves, water sufficient to meet the irrigation needs of Imperial Valley, two hundred miles downstream, was allowed to pass through while the remainder began to pool behind the dam, forming the murky nucleus of a lake that eventually would cover 210 square miles.[63]

With the start of water impoundment on February 1, 1935, the topping out of the dam on February 6, and the elimination of graveyard shifts a few days later, the Hoover Dam project entered the home stretch. The number of men employed had dropped below four thousand, and those still on the job were looking ahead with a mixture of eagerness and trepidation to the day when they would quit southern Nevada and go elsewhere in search of work. Similarly, the Six Companies board of directors was in a period of transition. Seventy-eight-year-old E. O. Wattis had succumbed to a heart attack at his home in Ogden on February 3, 1934, and Harry Morrison had taken his place as president of the consortium.[64] The Boulder Canyon Project, which for four years had consumed most of the directors' energies, was now just one of a number of jobs they were working on. The group had bid on and won the contract to build Parker Dam and was preparing to bid on the contracts for Bonneville and Grand Coulee dams, and several of the constituent companies were involved in building the Golden Gate and San Francisco–Oakland Bay bridges. The construction challenges posed by Hoover Dam had been overcome, the problems of organizing and controlling the huge labor force had been solved, and all that remained, or so the executives thought, was to finish, clean up, count their profits, and move on.

This scenario proved to be too optimistic. On the morning of February 26, 1935, U. S. Attorney Edward P. Carville, accompanied by a squad of U.S. marshals, entered the Six Companies office building in Boulder City, seized all the time records and payroll checks for the period between March 16, 1931, and December 31, 1934, loaded them into a truck, and hauled them off to be audited by federal agents. Carville announced that, acting on the instructions of Attorney General Homer Cummings, he was launching a sweeping investigation of eight-hour-law violations alleged in an affidavit by one John F. Wagner, a disgruntled former employee of Six Companies.[65]

Whatever complacency the Six Companies directors had felt about the Hoover Dam project was shattered by this startling development. The construction contract they had signed with the government had incorporated sections 321 through 324 of the United States Code, which forbade any laborer to work more than eight hours in a calendar day and fixed a penalty of five dollars for each violation, "to be withheld for the use and benefit of the government."[66] Wagner, who had worked as a typist and assistant auditor in Six Companies' Boulder City office from July, 1931, to April, 1934, claimed in his affidavit that the company had kept two sets of books, one for regular time and one for "emergency time." Bureau of Reclamation auditors were not shown the second set of books, he said, which detailed at least sixty thousand violations of the overtime provision, and as a result the government had made forty-four payments to the contractors without withholding anything for overtime penalties. By his calculation, Six Companies had defrauded the government of more than $300,000, and he charged that Frank Crowe, Administrative Manager J. F. Reis, Office Engineer A. H. Ayers, Chief Auditor E. R. Baker, and Paymaster F. M. Zeller were all criminally liable.

Wagner had joined the Bureau of Reclamation as an auditor after losing his job with Six Companies, and it was in this capacity that he compiled his accusations and dumped them into the lap of Harold Ickes. The interior secretary had not participated directly in the management of the Boulder Canyon Project since his initial involvement in changing the dam's name, abolishing scrip, and pushing for increased black employment; however, his interest in the undertaking had remained undiminished. He had kept the job funded, allotting $38 million from the Public Works Administration budget in 1934 when the rapid depletion of previous congressional appropriations threatened to interrupt the construction schedule. The project was politically popular, and Ickes wanted it to proceed smoothly to a successful conclusion and serve as an example of the economic and social benefits that would flow from other large public-power developments, such as Grand Coulee, that he was championing.[67] He did not want to delay or upset the progress on Hoover Dam by picking a fight with Six Companies over eight-hour-law violations, but the charges could not be ignored, and he reluctantly dispatched a special investigator to look into the matter.

The Six Companies executives were outraged by the timing and tone of the federal investigation. Overtime had been worked in Black Canyon from the outset with the knowledge of the government's top field representative, Construction Engineer Walker Young. It had been the contractors' understanding that this overtime was to be exempt from penalties because it had been required by various emergencies caused by the hazardous nature of the construction work, an exemption clearly spelled out in the language of the overtime law. The extra work also had helped move the project ahead of schedule, saving the government millions of dollars. For accusations of overtime violations and the threat of stiff penalties to be leveled now, with the job almost done, was hypocritical and smacked of persecution. On behalf of the other directors, Henry Kaiser fired off a lengthy telegram to Secretary Ickes, stoutly defending Six Companies payroll practices, denying that there had been any secrecy or dishonesty on the part of the contractors, and suggesting that inferences of irregularities were a direct reflection on the integrity of the hundreds of Interior Department officials overseeing the construction. The overtime that had been authorized had been worked because of emergencies and was not subject to penalty, Kaiser claimed, adding that "a majority of the government representatives understand the emergency situation involved in constructing a project of the magnitude and nature of this dam."[68]

Kaiser had been made chairman of the Six Companies executive committee in 1931 "in recognition of his unique gifts for making men work together," a polite way of saying that his fast-talking, grandstanding style suited him well for dealing with the politicians and bureaucrats who controlled the Boulder Canyon Project purse strings. The Shoreham Hotel in Washington, D.C., had become his home away from home, headquarters for his efforts to keep the work funded and payments flowing into Six

Companies accounts. He had become a sophisticated lobbyist adept at political persuasion and skilled at cultivating influential decision makers. So successful were his efforts at promoting Hoover Dam that the press increasingly gave him sole credit for creating Six Companies and running the job, plaudits that he did not discourage. The publicity that Kaiser received, and his tendency to refer to "my dam" when talking about Hoover, had begun to irk some of his colleagues in the consortium, but now his egotism would be forgiven if he could use his contacts and influence to counter the embarrassing and potentially costly overtime charges.[69]

Kaiser attacked the controversy head on, contending that the Hoover Dam job had been a continuous emergency and therefore was not subject to the restrictions of the overtime law. A writer was commissioned to prepare a narrative history of the dam construction, highlighting the floods, windstorms, and assorted mishaps that had made the extra hours of work necessary, and suede-bound copies were presented to members of Congress and important government officials. More hard-hitting literature, outlining Six Companies' legal defense against the overtime charges and leveling countercharges against the government, also was produced and circulated. Never one to shy away from the spotlight, Kaiser made himself available for radio and newspaper interviews in which he described the Boulder Canyon Project as a series of obstacles overcome by the heroic round-the-clock efforts of the contractors' employees, and claimed that by speeding up the pace of work Six Companies had fostered unemployment relief at a critical time while saving the public treasury several million dollars.[70] The message delivered by the books, pamphlets, articles, and broadcasts was the same: the Interior Department's levying of fines for overtime violations was not only unjustified but downright un-American. Kaiser also kept up what a miffed Secretary Ickes referred to as a "telegraphic bombardment," flooding the Interior Department with wires supporting the Six Companies position.[71] By late December, when the board of directors authorized him to negotiate a settlement of the overtime matter, Kaiser's public-relations blitz had had the desired effect. On February 17, 1936, it was quietly announced that the contractors and the government had agreed on a fine of $100,000. A potentially damaging scandal had been short-circuited and the size of the penalty reduced by two-thirds with a masterful piece of lobbying.[72]

At about the same time the overtime controversy was flaring up, Six Companies found itself embroiled in another unanticipated conflict, this one with its work force. The crushing defeat suffered by IWW leader Frank Anderson and his Wobbly followers in August, 1933, and the continuing efforts of the reservation rangers and the contractors' own private investigators to purge the Hoover Dam employment rolls of union organizers, seemed to have ensured a trouble-free conclusion to the Boulder Canyon

Project, at least as far as labor relations were concerned. But public and governmental attitudes toward unions had changed dramatically since 1932, a transformation reflected by the Norris-LaGuardia Act, which guaranteed workers full freedom of association without interference from their employers, and by the National Industrial Recovery Act, which legalized the right to organize. These two laws made it much harder for Six Companies to engage in out-and-out suppression of labor activities, and opened the door for unions to organize at Hoover Dam.

The change in labor's fortunes was demonstrated dramatically on March 29, 1934, when more than one thousand Hoover Dam workers gathered in the mess hall to listen to representatives of the United Metal Workers and the Carpenters and Joiners of America outline plans for establishing locals in Boulder City. Three months later the American Federation of Labor granted the first charter for a Boulder City labor organization, and in short order the various crafts represented in Black Canyon, including metalworkers, carpenters, machinists, truck drivers, and electrical workers, had their own union locals. A central labor council composed of representatives from each local was formed soon after.[73]

Although now prohibited by law from openly fighting this development, Six Companies did not passively acquiesce in the unionization of its work crews. Because unemployed men were not permitted to live in Boulder City, the termination slip was a powerful weapon, and in at least some instances the contractors used it to try to remove union officials from the project reservation. The union leaders were onto the reservation rules, however, and they used their knowledge and the new legal protections provided by the National Industrial Recovery Act to frustrate company efforts to evict them. Ragnald Fyhen, one of the principal organizers of the machinists' local, was fired from his job shortly after being elected secretary-treasurer of the new Boulder City Central Labor Council. "It [had become] clear to me, as we progressed with the organization, that as soon as the company found who was behind the activities, I would be layed [sic] off," he wrote thirty-two years later. Wisely, he had planned for this possibility. "I took out an agency, as salesman with the Mutual Benefit Health and Life of Omaha, and became a sales representative in my spare time. A few months later . . . I was layed [sic] off by the Six Company [sic] and no reason was given."

The day after his name was struck from the payroll, Fyhen was escorted to Sims Ely's office by a Six Companies special agent who asked the city manager to banish the labor organizer from the reservation because he was no longer employed. "Mr. Ely asked the representative what I had done," recalled Fyhen. "The representative stated that I had started to organize a union. Mr Ely asked what union it was—the IWWs or the A.F. of L. The representative informed Mr. Ely that it was the A.F. of L. Mr. Ely then informed him that this was permissible everywhere and he brought out the information sheet that showed I was gainfully employed as an insurance sales representative. . . . The [special agent] left Mr. Ely's office and [his] language . . . was not the most pleasant."[74]

Union men like Fyhen, D. M. Leigh, Les Parker, and others proved to be smarter and tougher opponents for Six Companies than Frank Anderson and his followers had ever been, and they were aided considerably in their organizational efforts and dealings with management by the political and economic changes that had taken place since the beginning of the Boulder Canyon Project. The powerful leverage the contractors had enjoyed in 1931 and 1932, when the construction industry was dying and dozens of men were competing desperately for every available job in Black Canyon, had been weakened by the Roosevelt administration's commitment to public works as an unemployment-relief measure. New Deal construction projects big and small were starting up across the West; the pay scale on these government projects was higher than the one at Hoover Dam; and this disparity, plus the fast-approaching conclusion of the job, made the Black Canyon workers increasingly restive.

The labor force's growing assertiveness and the contractors' concomitant loss of authority showed itself in a variety of ways. Some observers thought they detected a slowing of the work pace and increasing caution on the part of the crews: "There was no more record pour [of concrete] on the Boulder Canyon Project and the men didn't run over themselves to get there first on any job like it was before," was the way Ragnald Fyhen described the atmosphere at the dam site in 1935. "The boys no longer ran after the boss, there seemed to be a more independent spirit."[75]

A more overt demonstration of labor's new strength was the elaborate ceremony held in Black Canyon on May 30, 1935, to dedicate a plaque honoring the workmen who had lost their lives during the construction of Hoover Dam. The idea of memorializing the project's dead originated with the Boulder City Central Labor Council and was enthusiastically endorsed by Nevada Senator Pat McCarran, who persuaded Interior Secretary Ickes to approve the memorial service and the installation of a permanent monument on the canyon wall near the Nevada abutment of the dam. Six Companies went along with the idea, but the directors could not have relished having the spotlight cast so glaringly on the job's casualty figures.

After prayers, songs, and brief remarks by various government and labor officials, the bronze plaque was unveiled. It read: "They Labored That Millions Might See a Brighter Day—In Memory of Our Fellowmen Who Lost Their Lives on the Construction of This Dam" and carried the insignia of the American Federation of Labor. Senator McCarran was the featured speaker, and he paid a rousing tribute to the laborers, both living and dead, who had contributed to the advancement of the project. "You may garner gold, you may hold it in store, and worship it as you will, but the greatest treasure a nation ever held to its bosom is the worker," he proclaimed. "Labor has gone through many battles, and will go through many more." [76]

McCarran's last statement proved uncannily prophetic as tensions that had been simmering for months boiled over during the summer of 1935. On May 29 a worker named Eugene Schaver fell and hit his head on the job. He was taken unconscious to the Six Companies hospital in Boulder City, where he died the next day, a few hours after the dedication of labor's memorial. The cause of death was listed as spinal meningitis, a non-job-related and therefore uncompensable illness. When Schaver's coworkers learned this, their sorrow turned to outrage. Lars Matson, a friend of the deceased, and several other workers who had witnessed the accident, went to Clark County officials and angrily demanded an investigation. The coroner's office performed an autopsy and confirmed the spinal meningitis diagnosis. "The investigation . . . proves the thoroughness and integrity of the Boulder City Hospital, which in past months has been in question," said Dr. R. O. Schofield, who had succeeded Dr. Wales Haas as head of the facility in 1933; but others suspected that it was a whitewash and remained convinced that the company was trying to avoid making compensation payments to the dependents of dead and injured workers. [77]

Bitterness about perceived mistreatment of dam casualties and their families was only one ingredient in the witches' brew of discontent bubbling in the Black Canyon caldron. The grinding seven-days-a-week work schedule and a wage scale that was lower than those at other PWA-funded construction projects were major sources of resentment, exacerbated by the workers' uncertainty about what lay in store for them once the dam was done, and the insecurity of not knowing when the layoff ax would fall. The departure of seasoned assistant superintendents for other jobs and their replacement by less-experienced men affected discipline and morale adversely. Marion Allen recorded an exchange between one of the greenhorn superintendents, whom he described as being "of the new type . . . who were coming up at that time as efficiency experts," and an old-line foreman: *Superintendent:* "I have been watching your carpenters and they are spending an average of fifteen minutes each in the outhouse." *Foreman:* "Good! You watch the outhouse and I'll watch the job and we'll get the powerhouse built!" [78] Such petty animosities were allowed to fester because Frank Crowe was spending more and more time at the Parker Dam site and was not available to rein in overeager superintendents or to ride herd on balky crews. It was during one of his absences that matters finally came to a head in Black Canyon.

On July 12, 1935, 265 carpenters and 70 steelworkers helping to build the powerhouse walked off the job when they learned that the day-shift schedule had been changed and they would have to report to work a half-hour earlier than before. [79] For the past four years the day shift had started at 7:30 A.M. and ended at 3:30 P.M., with the crew taking a half hour, on company time, to eat lunch. Now, because the 11:30 P.M.–7:30 A.M. graveyard shift had been eliminated, Six Companies management could extend the day and swing shifts to eight and a half hours and make the men eat lunch on their own time. The net result of this change was a wage cut of approximately 6.5 percent, and the workers were furious, not only because they had thought the wage scale too low to start with, but also because they believed this scheme had been hatched by one of the new assistant superintendents behind Frank Crowe's back. Their attitude was summed up by Marion Allen, who wrote of the episode: "Had Mr. Crowe said, 'Boys, we're not doing so good so it would help if you'd work a half hour longer to put in a full eight hours,' I think the whole crew would have kept on working, but [instead] . . . . It was like cutting the wages of crews which had given everything they had to the company for nearly five years, and not being stupid, the men realized the company was making money, so why *cut* their wages?" [80]

The Boulder City Central Labor Council called a mass meeting on the night of July 12, and some five hundred workers attended it to discuss strike strategy and vote on a list of demands. The next day, Ragnald Fyhen, secretary-treasurer of the council, released a statement to the Las Vegas press: "The lengthening of the working time on the Boulder Dam project by the contractor caused the touching off of the spark of unrest that had been smoldering in the minds of the workmen for over a year." The strikers were demanding an immediate return to the old shift schedule, a 25 percent pay raise, a six-day workweek, and elimination of the hiring blacklist allegedly kept by Six Companies. [81]

Not all the craft unions chose to join the walkout, and there was even some dissension in the ranks of the carpenters' local, which formed the backbone of the strike effort. Laurence Wortley, a carpenter foreman and presiding elder of Boulder City's Mormon church, refused to abide by the union's strike vote and led a group of ten form strippers back to work, braving the wrath of the pickets who surrounded the transport loading area. Enough men did choose to stay off the job, however—between one-third and one-half of the 1,770-man payroll—to slow construction to a crawl and bring work on the powerhouse and canyon-wall outlet works, the most important parts of the dam complex not yet completed, to a virtual standstill. [82]

Frank Crowe returned to Boulder City on July 15 and immediately rescinded the order extending the day shift, but it was too late to halt the walkout. The Central Labor Council announced that "a strike condition now exists" and stood firm on its demands for higher pay, a shorter workweek, and abolition of the hiring blacklist. [83] Crowe was equally firm in insisting that a pay raise was out of the question, and he pointed out that a lengthy shutdown would cost the strikers much more than they could possibly gain, because the project was nearly finished. This attempt to dissuade the Central Labor Council from pushing the confrontation was to no avail; the strike, once started, seemed to take on a life of its own. The union leaders were anxious to flex their new organizational muscle, and the workers wanted a vacation, something many of them had not had since the last strike almost four years earlier.

Boulder City was quiet, half empty, and the mood among the Six Companies managers, the federal authorities, and the strikers who remained in town was calm. There was none of the hostility or crackling tension that had marked the walkout in 1931, no talk of calling in troops or of forcibly clearing the reservation, no move to fire the entire work force and then rehire only loyal employees. The rules of the game—and the stakes—had changed. Organized labor had gained new rights and recognition, and no longer would the federal government openly engage in strike breaking; Six Companies had won the fight with the Colorado, earning handsome profits in the bargain, and it could afford to be patient and wait out the strikers rather than force a quick showdown.

On July 16, Edward Fitzgerald, a Labor Department commissioner from Los Angeles, arrived in Boulder City to try to get negotiations started between the contractors and the Central Labor Council. He offered to serve as an arbitrator, a proposal that was accepted by both sides, and on the nineteenth, Frank Crowe representing management, L. A. Parker, representing the unions, and the Reverend C. H. Sloan of Las Vegas, who was appointed to serve as a neutral third party by Nevada Governor Richard Kirman, met with Fitzgerald for eight hours.[84]

At first good progress was made: the strikers dropped their demand concerning the hiring blacklist, which Six Companies denied existed, and the contractors said they would accept a six-day week, although they contended that most workers would prefer to continue earning the extra day's pay. On the question of raising the wage scale there was no compromise, however, and neither two more days of meetings nor a decision by the machinists' union to join the walkout on July 21, bringing the number of men off the job to eight hundred, produced any movement. The talks were deadlocked, and on July 22, Frank Crowe announced he would no longer negotiate without the authority of the company directors. With that, arbitrator Fitzgerald gave up and went back to California. It was clear that the contractors were not going to offer a nickel more in wages and that they were confident that as soon as the workers had finished spending the last of their paychecks the strike would collapse.

This was, in fact, precisely what happened. The laborers drifted back to the project reservation, pleased with the extended holiday but anxious to return to work and start earning money again. "The duration of the strike was one of the most pleasant times I can remember at Boulder City," wrote Marion Allen. "We had plenty of time to go to [the Hoover Dam reservoir] where we could swim, or we picnicked almost every day. . . .

The dam was nearly finished in July, 1935. (Bureau of Reclamation)

[Then] as suddenly as the strike started it was over. . . . One of the main reasons for this . . . was the fact that the pantry was getting empty." [86]

Exactly two weeks—one pay period—after it had started, the walk-out ended when the striking craft unions voted to go back to work at the old wage scale. The Central Labor Council passed a resolution stating that it would make every effort to get an additional PWA fund allocation from the Interior Department to bring pay on the Hoover Dam job up to the level of other government-sponsored projects, but this was just a fig leaf. No additional funds would be allocated, and as Six Companies predicted, the rank and file chose not to go on a six-day work schedule. The strike was not a total failure from labor's standpoint, however: the walkout had frustrated Six Companies' bid to extend the day and swing shifts by half an hour, and throughout the two-week affair the unions had comported themselves well, negotiating on an equal footing with the contractors, preventing violent confrontations between picketers and nonstrikers, and avoiding any provocations or irresponsible behavior that might have given Six Companies a pretext to dismiss the strike organizers and break the unions.

Although better organized and managed than the 1931 strike, the 1935 walkout had none of the fiery emotion of that earlier confrontation, and its impact on the course of construction was negligible. The big jobs were all but finished when the crews returned to work on July 26, 1935. The last of the trashracks had been bolted into place on the flanks of the intake towers, and the fast-rising waters of the reservoir were lapping at their foundations and at the base of the dam itself. The penstocks that would feed the first two generators to be put in service were in place, and others were almost ready. Railroad cars loaded with powerhouse machinery had been pulling into Boulder City since June, and the equipment was being lowered into Black Canyon and installed. Much of the work left for the dwindling crews was of a cosmetic or cleanup variety: grinding and patching concrete surfaces, placing tile in the galleries that ran through the dam from the elevator shafts to the powerhouse, tearing down temporary buildings, dismantling machinery, and preparing trucks, cranes, and other vehicles and tools for shipment south to Parker or north to Bonneville.

More and more men received their layoff slips, packed up their belongings, and hit the road, following the heavy equipment to new locales and, they hoped, new jobs. Fewer and fewer cars were seen on the streets in Boulder City, row upon row of tables sat empty in the mess hall at mealtime, and delicate drifts of sand, undisturbed by broom or shovel, eddied across abandoned walks and sifted against the walls of shuttered cottages. The Hoover Dam adventure was drawing to a close.

CHAPTER SEVEN
"Twentieth-Century Marvel"

Early on the morning of September 30, 1935, the normally sleepy downtown blocks of Boulder City were abuzz with activity. Women in their Sunday-best dresses, men in pressed trousers and white shirts, well-scrubbed boys and combed-and-curled girls moved along the sidewalks, pausing under the arched porticos to greet friends or to stare at the passing throng. The traffic on Nevada Highway, running through the business district, was nearly bumper to bumper, and more cars kept coming, funneling through the gate at the edge of the project reservation, creeping through town, then heading over the crest of the valley and down toward the dam. Anywhere there was a bit of empty shoulder along the road linking Boulder City and Black Canyon, automobiles were pulled over to the side. People sat or stood close by, fanning themselves, fiddling with Brownie cameras, chatting, waiting patiently. At the observation niches overlooking the dam, the tightly packed spectators pressed against the restraining masonry walls, and near the Nevada abutment, surrounding a bunting-draped platform, the biggest crowd of all, numbering in the thousands, milled about. Before the day was over, twelve thousand people would line the cliffs, cover the abutment area, and spill onto the dam's crest.[1]

The event was the dedication of Hoover Dam, the official culmination of four and a half years of prodigious labor. Like the opening of the Brooklyn Bridge a half-century earlier, completion of the dam was cause for public celebration and self-congratulation, not only in the Southwest but throughout the United States. Just as the span of granite and steel thrown across New York's East River had been heralded as a nineteenth-century "wonder of science," an "astounding exhibition of the power of man to change nature," so had the vaulting arch of concrete, stark and gleaming in its desert gorge, been seized upon as a symbol of twentieth-century America's ability to shape a new and better world with technology. The "Great Pyramid of the American Desert, the Ninth Symphony of our day," one writer called it. "A visual symphony written in steel and concrete—the terms of our mathematical and machine-age culture—it is inexpressibly beautiful of line, magnificently original, strong, simple, and majestic as the greatest works of art of all time and all peoples, and as eloquently expressive of our own as anything ever achieved." [2]

For a nation deeply wounded by the Great Depression, the symbolism was doubly important, linking as it did the traditional American free-enterprise values of private initiative, ingenuity, and risk taking with the more recent New Deal policy of government planning, funding, and supervision of public-works projects.

Hoover Dam was a triumph of individuals: of visionaries like Arthur Powell Davis, Herbert Hoover, and Phil Swing, who helped create the legislative blueprint for developing the Colorado River; of entrepreneurs like Warren Bechtel, Harry Morrison, and Henry Kaiser, whose financial acumen, organizational talent, and gambler's luck forged the managerial team needed to run the job; of engineers like Walker Young and Frank Crowe, whose resourcefulness and leadership overcame all obstacles and transformed paper plans into action; and of workers—miners, muckers, high scalers, and all the others—whose sweat and blood literally made a dream reality. But Hoover Dam was also a collective national triumph, a stunning example of what private industry, government, and labor, working together, could accomplish for the betterment of all.

If after five years of the Depression, with all its hardships and rending uncertainties, there was doubt about the country's capacity to recover and build for the future, the dam offered powerful reassurance, tangible affirmation that the American dream of limitless possibility still lived. Twelve thousand people had come in person, and millions more would tune in on a nationwide radio hookup, to celebrate that dream and participate in a rite of national self-renewal.

At 9:30 A.M., at the Union Pacific Station in Las Vegas, the doors of a private railroad car opened and the president of the United States was helped to the platform and then into an open touring automobile. Accompanied by his wife, his personal aide, several Secret Service men, and Construction Engineer Walker Young, he rode down Fremont Street, through the outskirts of the city, and southeast into the desert, heading toward the rugged escarpment of the River Mountains. One thousand six hundred and fifty-three days earlier, a tall, stoop-shouldered engineer had made this same journey, but under vastly different circumstances. On that spring morning in 1931 the lawn of Union Pacific Park had been populated by ragged scarecrows, not the well-dressed citizens to whom the president smiled and waved. Fremont Street had been dusty, run down, and dreary, a pale shadow of the bustling thoroughfare along which the presidential limousine glided. The edge of town had been marked by tents and shacks, not neat new construction, and the smooth ribbon of pavement angling up the desert slope had been a rutted and treacherous desert track. Beyond Railroad Pass, where the desolate Eldorado Valley had stretched empty to the horizon, Boulder City sat on its irrigated green island, and at the foot of Hemenway Wash, where the Colorado had plowed its muddy course, a glittering sheet of water covered the sand flats. In the four and a half years since Frank Crowe first made his way along this route, the geography, both physical and human, had been changed almost beyond recognition.

Shortly after 10:30 the motorcade arrived at the dam, and while the thousands of onlookers applauded and snapped pictures, the president's car slowly crossed the crest into Arizona and then came back to the speaker's platform on the Nevada side. Already the dais was crowded with an impressive array of officials: Harry Hopkins and Interior Secretary Ickes from the presidential entourage; Elwood Mead, Raymond Walter, and Jack Savage of the Bureau of Reclamation; Senators Key Pittman and Pat McCarran, Representative James Scrugham, and Governor Richard Kirman representing Nevada; the governors of California, Arizona, Utah, Wyoming, and New Mexico representing five other Colorado Basin states; Harry Morrison, Steve and Kenneth Bechtel, Felix Kahn, Charlie Shea, and Frank Crowe of Six Companies; and in the back, Ragnald Fyhen and James Farndale of the Boulder City and Las Vegas central labor councils.

"Gee, this is magnificent," the president exclaimed with a broad smile as he was seated, and then Harold Ickes stepped forward to deliver the first address.[3]

"Pridefully, man acclaims his conquest of nature," began the pudgy, bespectacled secretary, but his words, amplified by the microphone, were garbled and then drowned out by thunderous echoes off the canyon walls. The horns of the public-address system were turned off and the secretary continued, although many in the audience now had a hard time understanding what he was saying. The subject of his speech was conservation and the importance of developing the nation's natural resources for public rather than individual benefit. "Here behind this massive dam is slowly accumulating a rich deposit of wealth greater than all the mines of the West have ever produced," he said, "wealth to be drawn upon for all time to come for the renewed life and continued benefit of generations of Americans."

Having spoken eloquently of the dam's purpose and of the role it would play in the nation's future, Ickes turned to politics and his personal obsession with the structure's name. He repeated *Boulder Dam* again and again—five times in a half-minute passage—and finally concluded his address with a thinly veiled dig at Herbert Hoover: "This great engineering achievement should not carry the name of any living man but, on the contrary, should be baptized with a designation as bold and characteristic and imagination stirring as the dam itself."[4]

Key Pittman followed Ickes, and then, finally, it was the president's turn to speak. Franklin Delano Roosevelt, his dark double-breasted suit buttoned in spite of the ninety-degree heat, put down the pages of his speech, gripped the edge of the podium, and looked out at the upturned faces of the crowd. "This morning I came, I saw, and I was conquered as everyone will be who sees for the first time this great feat of mankind," he ad-libbed, and his listeners smiled at the sound of the patrician eastern accent familiar from so many radio broadcasts. More amusement was provided by Secretary Ickes, who dropped a handful of coins on the floor of the platform and then blushed beet red while the press section snickered at his discomfiture.[5] But the president did not seem to notice the disruption behind him. He was gazing across the lip of the dam at the cerulean waters of the new lake miraculously growing in the midst of a wilderness of sand and rock. Then his eyes dropped to the text in front of him, and his powerful voice rolled out over the crowd: "Ten years ago the place where we are gathered was an unpeopled, forbidding desert. In the bottom of a gloomy canyon, whose precipitous walls rose to a height of more than 1,000 feet, flowed a turbulent, dangerous river. The mountains on either side were difficult of access, with neither road nor rail, and their rocks were protected by neither trees nor grass from the blazing heat of the sun. The site of Boulder City was a cactus-covered waste. The transformation wrought here is a twentieth-century marvel."

The crowd stirred and applauded politely, but for those who had been in Black Canyon from the start, the president's introductory remarks had special meaning. Looking at the spreading lake, they could easily visualize the Colorado as it had been when they first glimpsed it: a dirty red-brown ribbon snaking along between the cliffs. The blink of an eye removed all the roads and footpaths, the cables and transmission lines that crisscrossed the canyon, and once again the walls were the soaring, unbroken barriers of 1931, awaiting the bite of steel and the shock of dynamite. Somewhere beyond the looming bluff of Cape Horn, under a fast-deepening blanket of water, lay the remains of Ragtown and the River Camp, cleansed out of existence, but not out of memory.

"All these dimensions are superlative," the president continued, refer-

President Franklin D. Roosevelt delivering the dedicatory address at Hoover Dam, September 30, 1935. (Bureau of Reclamation)

ring to the size of the dam and to the vast quantities of water and electric power it would harness. "When we behold them it is fitting that we pay tribute to the genius of their designers. We recognize also the energy, resourcefulness, and zeal of the builders. . . . But especially we express our gratitude to the thousands of workers who gave brain and brawn to work of construction."

❧ Almost directly across from the rostrum, three-quarters of the way up the Arizona wall, was the spot where Jack Russell had slipped from his scaler's chair and tumbled to his death. To the left of that point, somewhere in the middle of the dam, was the place where W. A. Jameson had been buried under a hundred tons of concrete. Deep in the breast of the Nevada cliff, six hundred feet under the president's shoes, was the tunnel section where Carl Bennett had been killed by a premature explosion and Bert Lynch had been crushed by a rockslide. It was hard for the workers at the ceremony to look at any part of the dam site and not recall moments of fear, horror, sadness, and anger. But it was also hard for them to look about and not feel a surging pride of accomplishment, a dawning realization that the long days of discomfort, drudgery, and danger had been great days, that their labor at Hoover Dam was an experience that would loom ever larger with the passing of time until it became an integral part of their identity, one of the touchstones of their lives. Most had come here thinking only of making a living; now, for the first time, they began to sense that they had made history, too.

The president's voice rose as he began his peroration. "This is an engineering victory of the first order—another great achievement of American resourcefulness, skill, and determination." He drew a breath and looked squarely into the faces of his audience as if trying to establish eye contact with each individual. The pitch of his voice fell again, and his words were measured and sincere. "That is why I have the right once more to congratulate you who have created Boulder Dam and on behalf of the nation to say to you, 'Well done.'" [6]

"It's one thing to build a great public work," Felix Kahn observed several years after the conclusion of the Hoover Dam job. "It's something else to get a government bureau to admit it's finished. Unless you can saw the main job off at a reasonable point, they'll have you adding power equipment, transmission lines, roads, and other extras the rest of your life." [7] This was exactly the predicament Six Companies found itself in during the last three months of 1935. President Roosevelt had made his dedicatory speech, marking, in his words, the "official completion" of the dam, the press had extended its editorial congratulations to everyone involved, and after nearly five years, the bright spotlight of national publicity had finally dimmed. But the Hoover Dam project was not over yet for the contractors and would not be until the Interior Department agreed that they had fulfilled the contract and released them from further obligation.

Meanwhile, some five hundred workers remained in Black Canyon, finishing the powerhouse roof, installing penstocks, generators, circuit breakers, transformers, and other electrical equipment, and cleaning up the debris that littered the dam site. In military parlance, this was the mopping-up phase of the construction campaign, but just as foot soldiers knew the deadly hazard of such operations, so, too, did the dam workers recognize the dangers still present in Black Canyon. Not wanting to be the last to be crippled or killed on a job that was almost finished, they performed their tasks in a cautious, perfunctory fashion.

Generator being installed in the powerhouse, July, 1936. (Bureau of Reclamation)

But someone did have to be the last casualty, and the identity of that person and the circumstances of his death gave even confirmed skeptics pause to think about predestination and the vagaries of fate.

On December 20, 1922, J. G. Tierney, a member of one of the Bureau of Reclamation drilling crews exploring Boulder and Black canyons, had been swept off a barge and drowned in the Colorado, the second fatality of the Boulder Canyon Project. Exactly thirteen years later, on December 20,

1935, Tierney's son Patrick, a 25-year-old electrician's helper working on the top of one of the intake towers, slipped and fell 325 feet to his death. He was the last man to lose his life during the construction of Hoover Dam.[8]

Even as the rearguard of workers was carefully putting the finishing touches on the dam the structure was beginning to do the work for which it had been designed. The closing of the bulkhead gate at the portal of tunnel No. 4 on February 1, 1935, had put the Colorado River permanently under control, and within months the changes were being felt hundreds of miles away from Black Canyon. A flood peaking at more than 50,000 cubic feet per second on June 30, 1935, was intercepted by the dam and reduced to 14,900 cubic feet per second, sparing Yuma and the Imperial Valley a costly battle with rampaging high water. Later that summer, when the river level dropped precipitously, promising a repeat of the drought that had inflicted ten million dollars' worth of crop losses in 1934, the disaster was prevented by controlled water releases from the new reservoir, which already held nearly four million acre-feet of runoff.[9]

During the waning days of 1935, as farmers in the irrigated valleys along the Lower Colorado harvested their fruits and vegetables and sang the praises of river regulation, Henry Kaiser went to Washington to begin his final lobbying effort of the Hoover Dam project. Six Companies was anxious to saw the job off, as Felix Kahn had put it, to negotiate a final agreement on payments, penalties, and settlement of claims, to set a date for termination of the contract, and to collect the two and a half million dollars the government was withholding as a guarantee of completion.[10]

Kaiser convinced the Interior Department that negotiations for terminating the contract should begin, but the process was tragically interrupted when Elwood Mead, commissioner of the Bureau of Reclamation, died of a heart attack on January 26, 1936. Although not as well known to the public as some of the other men associated with Hoover Dam, Mead had been a strong and influential advocate of the project in both the Hoover and Roosevelt administrations. His passing was mourned by all who had worked with him, and it was deemed entirely fitting by everyone associated with the Boulder Canyon Project that the huge reservoir created by Hoover Dam be named Lake Mead in his honor.[11]

Not long after Elwood Mead's death, Six Companies and the federal government reached final agreement on fulfillment of the construction contract. Disputes that had arisen during the course of the project, including disagreements about the exact number of units of work done, the unit price at which certain classifications of work were to be paid, and the size of the penalty assessed for violations of the overtime law, were resolved, and Six Companies resigned any further claims against the government. The plant and machinery still in place in Black Canyon were to be turned over to the Bureau of Reclamation for use without charge until October 1, 1936, so that installation of powerhouse equipment and various small tasks could

The Arizona intake towers, with Lake Mead rising behind them. (Manis Collection, UNLV)

be finished. For its part, the bureau agreed to transfer Six Companies' remaining work crews to its payroll and to excuse the contractors from responsibility for the one major contract item not completed: placement of a diversion tunnel plug. The water for Imperial Valley irrigation was being bypassed through valves in this tunnel, and it would have to remain open until the powerhouse went into operation.[12]

On the morning of February 29, 1936, Ralph Lowry, who five months earlier had succeeded Walker Young as construction engineer, met Frank Crowe on the crest of the dam. While several reporters stood by and a lone newsreel camera recorded the scene for posterity, Crowe shook Lowry's hand and announced: "As representative of Six Companies, Inc., builders of [this] dam, I am very happy to turn over the job for your acceptance." The next day in Washington, Secretary Ickes formally accepted the dam and powerhouse on behalf of the government, terminating the contract and ending construction exactly two years, one month, and twenty-eight days ahead of schedule.[13]

In San Francisco, Six Companies banked its final payments and closed the books on Hoover Dam. The contractors' gross earnings totaled approximately $54.7 million, $5.7 million more than the original bid amount because of design changes in the penstock tunnels and powerhouse made by the Bureau of Reclamation while the work was under way. After deductions by the bureau for items ranging from provision of electric power to penalties for overtime violations, actual cash payments came to approximately $51.6 million. Of this amount, between 20 and 35 percent— $10.4 million to $18 million—was profit, prorated among the consortium members (excluding Warren Brothers, Sydney Ehrman, and Graeme MacDonald, whose shares had been bought out by the Bechtel-Kaiser unit) according to their capital contributions.[14] The gamble into which the partners had entered with such trepidation at the Engineers Club in 1931 had paid off spectacularly; in five short years thay had won wealth and fame far exceeding their fondest hopes and expectations.

For the principal author of Six Companies' success in Black Canyon, completion of Hoover Dam was a bittersweet triumph. Frank Crowe had performed brilliantly in an extraordinarily demanding job, rising to every challenge and proving beyond a shadow of a doubt that he was the finest field engineer in the world. In the process he had become wealthy: besides an $18,000-a-year salary, he had earned a bonus of 2.5 percent of profits— between $260,000 and $350,000—a princely sum in the middle of the Depression and enough on which to retire comfortably if he so chose.[15] He was also in a position to capitalize on his new celebrity status by becoming a professor of engineering or a highly paid consultant.

But neither retirement nor consulting held any appeal for the man the workers called Hurry Up. *Las Vegas Evening Review-Journal* writer Al

Cahlen interviewed Crowe on March 3, 1936, two days after the government's acceptance of the dam and the conclusion of the general superintendent's responsibilities. He expected to find the engineer in a happy, relaxed mood, savoring his accomplishments and looking forward to a well-deserved vacation. Instead he found a man anxious and depressed, beset by doubts and misgivings and desperate to get back to work as soon as possible. "I feel like hell," Crowe told Cahlen. "I'm looking for a job and want to go right on building dams as long as I live. Somehow I can't imagine myself in a big city skyscraper acting as a consulting engineer. . . . I'm going to keep my feet in the dirt someway—that's where I'm happiest. . . . I've got to find a dam to build somewhere."[16]

Finding work was not Frank Crowe's real concern. He already had a new dam to build for Six Companies—Parker, 155 miles downstream from Black Canyon—and when that was finished, he could have his pick of the big New Deal construction projects planned for the Columbia River Valley and the Central Valley of California. The real problem was more fundamental: how to fill a gaping void in his life. Hoover Dam had been his obsession for ten years; in his mind it had loomed as the ultimate challenge of his career, the focal point of all his plans and ambitions. From the day in 1925 when he left the Bureau of Reclamation and joined Morrison-Knudsen, all his efforts had been concentrated on preparing himself for the task of directing its construction. Now he had realized his lifelong ambition, but he was only fifty-four and the demons that had driven him from construction site to construction site until he finally reached Black Canyon and the pinnacle of his profession would not let him rest. Frank Crowe stood alone at the apex of the Great Pyramid of the American Desert, unwilling to rest on his laurels, looking desperately for a new job worthy of his talents, realizing that there would never be another challenge that could measure up to Hoover Dam.

To his credit, Crowe did not let the knowledge that Hoover Dam was his crowning achievement affect his performance on subsequent projects. He brought Parker Dam in ahead of schedule in 1938, directed construction of two small dams, Copper Basin and Gene Wash, that were part of the Colorado Aqueduct system, then turned his attention to the plans for Shasta Dam, the key structure in Northern California's huge Central Valley Project. Shasta would rival Hoover in size, and Pacific Constructors, the contracting organization that had won the right to build it, asked Crowe to be the general superintendent. He accepted and spent the next six years working on the big dam, guiding the project to a successful and timely conclusion in spite of manpower and material shortages brought on by World War II.[17]

When Shasta Dam was finished in 1944, Frank Crowe was sixty-two years old and had been engaged in stressful, physically demanding work for almost four decades. During that period he had had virtually no time off, not even weekends, and the continuous strain had taken its toll. In 1945, when the War Department asked him to organize and direct all the reconstruction work in the U.S. Zone of Occupation in Germany, he was eager to accept, but his doctor ordered him to forgo the assignment because of his failing health. Reluctantly, Crowe turned the job down and retired to a cattle ranch he had bought in Redding, California, not far from the Shasta Dam site. He died there on February 26, 1946.

Obituaries described Crowe's long and distinguished career in heavy construction and lauded his ability as a field engineer and builder. Eulogists in Las Vegas and Boulder City spoke of his devotion to family, friends, and protégés, of his fairmindedness, of his sense of humor, and of his absolute integrity.[18] But in the end Frank Crowe was remembered for one thing: he was the man who built Hoover Dam. Those closest to him knew that he had never hoped for anything more.

The Hoover Dam adventure was the pivotal experience in the lives and careers of other men as well. Walker Young, the government engineer who led the surveying and diamond-drilling expeditions in Boulder and Black canyons in the early 1920s and then became the project's construction engineer, seemed destined to rise to the top of the Bureau of Reclamation, thanks to his role in the Hoover Dam construction. But Young was a Republican, reaching the peak of his career at the same time the Democratic party was beginning a twenty-year ascendency in the executive branch, and when Reclamation Commissioner Elwood Mead died in 1936, Young had to swallow his frustration as one of his Boulder Canyon Project subordinates, Office Engineer John C. Page, was tapped for the top slot.

\ Although denied the ultimate prize by politics, Young did not languish after leaving Black Canyon. Between 1935 and 1940 he was construction engineer and then supervising engineer for the Central Valley Project in California. In 1940 he returned to the bureau's Denver office, where he served as assistant chief engineer and then chief engineer before retiring in 1948, at the age of sixty-three, to become a consulting engineer and president of Thompson Pipe & Steel Company of Denver.[19] Looking back on his long, eventful career in government service, Young recognized Hoover Dam as the peak, and in his later years he made it a point to share his recollections of the project. In a letter written in 1972 he summed up, in characteristically modest fashion, his thoughts on who the real heroes of Hoover Dam were: "One day near the end of the job when we were together in the canyon, Frank [Crowe] said to me 'You know, Brig, these dam jobs are hell to get started but once the construction stiffs know what is to be done there is no longer need for a Construction Engineer or Superintendent of Construction.' To illustrate what he meant there is the rule, 'If you want to know how to use a shovel you had better talk with the man on the little end of a number two muck stick.' I think what Hurry-Up Crowe said

was a fine tribute to all the workers at Hoover Dam, regardless of position
or rank. They were the ones deserving of credit to a large degree for the
successful completion of a difficult job."[20] Young died in a California nurs-
ing home in 1982; he was ninety-seven years old.

For Sims Ely, the government-appointed dictator of Boulder City, ruling the dam workers' town was the capstone on a rich and varied career in
public service. Ely was sixty-six and eligible for retirement when Hoover
Dam construction ended, but he found the role of benevolent despot so
enjoyable that he agreed to stay on in Boulder City, even though it was
widely anticipated that the community would quickly become a ghost
town. This did not happen, and Ely continued to serve as city manager,
directing town affairs in his inimitable fashion, until forced by failing
health to resign in 1941. He remained active in retirement, writing a book
on Arizona's legendary Lost Dutchman Mine. He was at work on another,
this one dealing with his experiences in the West, when he passed away in a
Rockville, Maryland, sanitarium at the age of ninety-two.[21]

For Frank Crowe, Walker Young, and Sims Ely, Hoover Dam was the great
achievement of their professional lives, but for another group of men
closely associated with the undertaking—the Six Companies partners—
the project proved to be only a single step, albeit a huge one, from their
past as small regional contractors to their future as industrial- and heavy-
construction giants.

Four of the key figures in the founding of the consortium, W. H. and
E. O. Wattis, W. A. Bechtel, and Alan MacDonald, did not live to see the
dam finished, but the others reaped the profits and the bountiful harvest of
new business opportunities their work in Black Canyon had created. With
a building organization second to none, large stocks of modern equipment,
and the confidence gained from successfully fulfilling the biggest construction contract let by the U.S. government up to that time, they tackled other
government projects and grew even richer and more powerful. Through
the late 1930s they continued to concentrate on earth-moving jobs, but as
hydroelectric power generated at dams they had built transformed the industrial potential of the West, they started looking for ways to diversify
into less risky and more permanent forms of business.

Steve Bechtel led the way in 1937 when he formed the Bechtel-McCone-
Parsons Corporation, an oil-refinery engineering firm that was the first of
its kind on the Pacific Coast. Right behind him came Henry Kaiser, pursu-
ing his longtime vision of a western industrial empire. Kaiser's first move
was to challenge the powerful cement combine that had monopolized construction jobs west of the Rockies by submitting identical bids on government projects. Although his company had no cement-making experience,
Kaiser competed against the combine for the big Shasta Dam order, under-
cutting it by 11 percent and winning the contract. He then set up the Per-

manente Corporation, built a huge cement plant in California's Santa Clara Valley, and almost overnight became a dominant figure in the lucrative cement industry.[22]

Another new venture on which the former Six Companies partners embarked was shipbuilding. In 1939, in conjunction with Todd Shipyards of New York, the consortium members formed the Seattle-Tacoma Shipbuilding Corporation and won a U.S. Maritime Commission contract to build cargo ships. With Hitler on the march in Europe and a big naval buildup under way in the United States, the timing of this entry into maritime construction was superb. In December, 1939, Seattle-Tacoma received a $160 million order from Britain to build sixty ships; this windfall was followed shortly by the beginning of the huge U.S. Liberty-ship program. Within three years, Kaiser was running seven shipyards, Bechtel two, and Morrison-Knudsen and Pacific Bridge one each; the Kaiser and Bechtel yards together were employing more than 243,000 workers and turning out ships at the rate of one every 10.3 hours; and the value of the contracts awarded to Six Companies corporations topped $3 billion—six times the amount of the Hoover Dam contract, which had been of unprecedented size just ten years before.[23]

America's entry into the war created a bonanza of construction work of the type at which the Six Companies members excelled. Morrison-Knudsen crews built a vast network of bomb shelters and underground oil, water, and ammunition depots in the Hawaiian Islands; Bechtel workers constructed pipelines and oil refineries in the Caribbean, the Persian Gulf, and the Subarctic, as well a huge airplane-modification plant in Birmingham, Alabama; a Pacific Bridge underwater team raised sunken battleships from the floor of Pearl Harbor; MacDonald & Kahn and J. F. Shea employees erected military bases and port facilities on the California coast; and Utah Construction men graded runways on South Pacific islands and highways across the Canadian and Alaskan tundra.[24]

Meanwhile, Henry Kaiser was moving into heavy industry, using connections he had established during the Hoover Dam project to secure federal loans to build new factories. The Permanente magnesium plant went up in the Santa Clara Valley in 1941, and the following year Kaiser began building a giant steel mill in Fontana outside Los Angeles. Fittingly, the construction and operation of the sprawling Fontana complex was made possible by ample hydroelectric power crackling from generators in the Hoover Dam powerhouse two hundred miles to the east.[25]

Kaiser's business interests and philosophy had been diverging from those of his Six Companies partners for some time, and his decision to go into steel led to a final parting. Although they had broken out of their old earth-moving mold, Morrison-Knudsen, Utah, Shea, MacDonald & Kahn, Bechtel, and Pacific Bridge still thought of themselves as builders, whereas Kaiser now saw himself as an industrialist. Already he was looking ahead to the postwar era and hatching grandiose schemes for manufacturing automobiles, helicopters, prefabricated housing, and other consumer products. His more cautious colleagues looked at his magnesium-producing venture, which was losing large sums of money, and, fearing that the Fontana steel mill might also be a white elephant, decided not to participate in the new enterprise. Felix Kahn spoke for the others when he said of the decision: "[Steel was] a business for which we had neither the training nor disposition. We decided to stay as we were—contractors looking for business all over the world, ready to build anything to somebody else's design and specifications."[26]

Thus, with the exception of shipbuilding operations, the Six Companies confederation dissolved and the constituent organizations went their separate ways, participating from time to time in small-scale joint ventures but never again pulling together as a whole. After the breakup the fortunes of some of the partners began to wane. The J. F. Shea Company consisted primarily of doughty Charlie Shea, and when the happy-go-lucky Irishman died of cancer in late 1941, it was a severe blow to his firm. Mac-Donald & Kahn suffered a similar loss when Felix Kahn passed away in 1958.[27]

The stars of other former Six Companies members waxed bright, however. Thanks to his wartime industrial record and his tireless self-promotion, Henry Kaiser became a national hero and was mentioned as a possible running mate for Franklin Roosevelt in 1944. After the war he added automobiles, aerospace components, chemicals, health care, real estate, and television and radio to his interests in steel, aluminum, and cement, building the corporate empire he had always dreamed of.[28]

Morrison-Knudsen, Bechtel, and Utah—the companies that had decided to persist in their original character—also prospered. Large-scale postwar reconstruction, and the emergence of new Third World nations eager to exploit their raw materials and build modern transportation and industrial infrastructures, created marvelous opportunities for experienced contractors. Harry Morrison, the Bechtel brothers, and the executives at Utah filled the bill, constructing roads, airports, harbor facilities, mines, mills, factories, pipelines, refineries, and reactors all over the globe.

Before his death in 1971, Harry Morrison had built Morrison-Knudsen into one of the biggest contracting companies in the world. Utah Construction, which in 1931 had been forced to pledge the Wattises' sheep herd to secure the Hoover Dam bid bond, metamorphosed into Utah International, a major natural-resources concern operating coal and copper mines in North and South America and Australia. Even more successful were Kenneth, Warren, and Steve Bechtel; they turned their father's firm into an engineering and construction colossus, the wealthiest and most powerful of all the old Six Companies organizations.[29]

The assets of Kaiser, Bechtel, Morrison-Knudsen, and Utah mounted into the billions and their interests and influence spanned continents, yet each of these corporate giants traced its roots to a desert canyon on the Arizona-Nevada border. For all of them Hoover Dam was the key, the job that boosted them to prominence. As Steve Bechtel, Sr., the last surviving member of the Six Companies inner circle, put it in a 1984 interview: "Coming at the time it did, [Hoover Dam] was very important. . . . It put us in a very prime position . . . as being [regarded as] big-time thinkers, real thinkers." [30]

"There's something peculiarly satisfying about building a great dam," Frank Crowe said in 1943. "You know what you build will stand for centuries." [31]

\ Hoover Dam has stood now for more than half a century, while all around it cities and states—in fact, an entire region—have been transformed by the revolution it sparked. Water and power, dispensed at the push of a button, have turned the "profitless locality" of the arid Southwest into America's new technological and agricultural promised land, ushering in an era of material wealth and physical comfort that was undreamed of, even by the visionaries who conceived the development of the Colorado River.

This transformation did not occur overnight, nor was it all directly linked to Hoover Dam, but just as the buried steel and concrete footings of a skyscraper hold up the glass aeries a thousand feet above, so did the concrete wedge in Black Canyon undergird the elaborate network of reservoirs and aqueducts, substations and transmission lines, that nourished the Southwest's oasis civilization into being.

The flood protection provided by Hoover made possible the construction of Parker Dam and the creation of Lake Havasu, the reservoir from which the cities of Southern California and Arizona draw their share of the Colorado's flow. The Colorado River Aqueduct, an incredible system of tunnels, canals, siphons, and pumping plants stretching across 240 miles of mountains and deserts to the Pacific coast, was built by the Los Angeles Metropolitan Water District with Hoover Dam power and began delivering water to Los Angeles and surrounding communities in the spring of 1941.

Work on the third phase of the Boulder Canyon Project—the All-American Canal and Imperial Dam—began in 1933. An immense 650-ton dragline chewed its way through the sandy waste of the Walking Hills eighty miles into the heart of Imperial Valley while at the same time Imperial Dam, a low concrete barrier that raised the river level twenty-three feet, and a series of desilting basins were built eighteen miles north of Yuma. On October 13, 1940, water coursed down the bed of the new canal and the long-standing dream of Imperial Valley farmers—an assured irrigation supply beyond the reach of Mexican officialdom—was finally realized.

Power generation, the Boulder Canyon Project's third major function, began at Hoover Dam when the first of four generating units installed in the Nevada wing of the powerhouse in 1936–37 was turned on. All of the Nevada units, plus one Arizona unit, were operating by 1937, producing electricity for Los Angeles, Glendale, Burbank, and Pasadena, California, as well as for Boulder City and Las Vegas. Two more generators went on line in 1938, powering the Metropolitan Water District's Colorado River Aqueduct project, and another pair was installed in the Arizona wing in 1939 to supply energy for the Southern California Edison Company. With nine turbines cranking out over more than seven hundred thousand kilowatts by the end of 1939, the powerhouse in Black Canyon was the world's largest hydroelectric facility.[32]

Hoover Dam's immediate impact was substantial: water and power for the Los Angeles metropolitan area, water and flood protection for the fertile agricultural lands of Southern California and Arizona. But even more important was its long-term impact on the way westerners thought about their region and its future.

Since its settlement in the nineteenth century the West had been an economic colony of the East, exporting minerals, timber, petroleum, and other raw materials to eastern factories and importing finished manufactured goods. Lack of electric power and reliable supplies of water stunted industrial development, discouraged immigration, and kept the region commercially and politically weak.[33] In this atmosphere of dependency and stagnation, Hoover Dam came as a revelation; in one stroke it freed Southern California from its economic fetters and made possible virtually unlimited growth. In Northern California, in Oregon and Washington, in Idaho and Montana, in Colorado and western Nebraska, farmers, businessmen, and politicians studied this example of large-scale reclamation, saw the economic potential it had unlocked, and began clamoring for similar projects of their own.

No less inspired by Hoover Dam were President Roosevelt and Interior Secretary Ickes, who saw it as a compelling demonstration of the social and economic benefits they believed would flow from centralized resource planning, public-power development, and federally funded public-works projects. The political popularity of the Hoover Dam undertaking was not lost on them, either. They embraced multipurpose water projects as the central element of their economic and natural-resources policy for the West: Grand Coulee, Bonneville, Shasta, Hungry Horse, Fort Peck, Granby, and scores of smaller New Deal dams, canals, tunnels, and power plants, stretching from the hundredth meridian to the Pacific shore and from Canada to Mexico, were all children of Hoover Dam.[34]

Less than a decade after the last bucket of concrete was poured in Black Canyon, an extensive water and power network was in place in the West, ready to support rapid expansion of agricultural production and operation of new factories. All that was needed was a catalyst to ignite an explosion of industrial growth, and the Japanese attack on Pearl Harbor provided it. Overnight America was at war in the Pacific and in desperate need of guns, ships, and airplanes that eastern and midwestern plants, already committed to supplying the European theater, could not produce in sufficient quantities. With its new power grid the West was prepared to mobilize, and in a remarkably short time shipyards, aircraft plants, and other war industries were up and running, churning out the flood of weapons that would ultimately crush the Japanese.

❮ Hoover Dam played a vital role in this speedy defense buildup. The electricity it generated flowed to Los Angeles, where it powered steel and aluminum mills and the Douglas, Lockheed, and North American aircraft plants, which built approximately sixty-two thousand fighters, bombers, and cargo planes—a fifth of the nation's entire aircraft production—between 1941 and 1945.[35] When the war was over, Hoover Dam power and water continued to sustain the aerospace industry and to attract new plants, fostering a robust cycle of growth in Los Angeles and surrounding communities that in the span of two decades turned Southern California into one of the nation's leading centers of population and commerce.

The postwar economic explosion in the Los Angeles metropolitan area was not surprising, given its climate and coastal location; the same could not be said for the boom that took place in southern Nevada. When the Boulder Canyon Project's benefits were apportioned, few anticipated that the Silver State would become a major market for Hoover Dam water and electricity. Reno, the only urban center of any size in Nevada in the late 1920s, was too far to the north, and the south was deemed too barren and hot to attract industry or significant numbers of settlers. Las Vegas, it was thought, would enjoy a modest spurt of growth while construction went on in Black Canyon, then slip back into its former desert somnolence.

The authors of this scenario could not foresee the effects of three important developments, however: the legalization of gambling in Nevada in 1931, the beginning of the New Deal in 1933, and the emergence of Hoover Dam as a tourist attraction in 1934 and 1935. In 1931, as the project began, dozens of casinos and clubs opened to cater to dam workers, and the profitability of gaming was demonstrated dramatically. Starting in 1933, the city received a much-needed facelift, courtesy of the New Deal, as government-paid work crews paved city streets, laid sewer lines, landscaped parks, and constructed new public buildings. Then came the wave of tourists—an astonishing three-quarters of a million visitors in 1934–35 to see the dam and visit Las Vegas—and the revelation that Six Companies workers were not the only ones who liked to gamble.[36]

A steady flow of tourists continued to come to southern Nevada even after work in Black Canyon was finished, and in 1941 Las Vegas confounded the forecasters who had said it would never house heavy industry when it attracted Basic Magnesium Industries with the promise of abundant water and power from Hoover Dam. Where the squatters' community of Midway had stood on Boulder Highway, a sprawling light-metals complex was built and a new town called Henderson sprang up. Shortly after Basic Magnesium located in southern Nevada, the U.S. Army Air Force established a large training school and gunnery range in the area, pumping even more defense dollars into the Las Vegas economy.[37]

After the war, power from Black Canyon kept the gaming industry expanding by running the air conditioners and neon displays of resorts that sprouted up and down the Las Vegas Strip and by supporting the population surge in Southern California, the area from which Las Vegas casinos drew many of their customers. Water was added to the growth formula in 1971 when the first stage of a project to pump a portion of Lake Mead over the River Mountains and into the Las Vegas Valley was constructed. The second stage was built in 1982, completing a system that can deliver 299,000 acre-feet a year, freeing southern Nevada from its dependence on diminishing supplies of groundwater.[38] Today, seemingly against all odds, the dusty desert village of five thousand souls that Frank Crowe saw in 1931 has become one of the great gambling and entertainment meccas of the world, a thriving city with a metropolitan population exceeding half a million.

During the late 1930s and early 1940s, while Las Vegas was drawing on the gaming and tourism experience of the Hoover Dam construction years to invent its modern persona, Boulder City, the town the dam had given birth to, was struggling to survive. From its peak in 1934–35, the city's population dropped sharply as Six Companies workers pulled out and businesses closed; by 1940 only twenty-six hundred people remained, nearly all of them employees or dependents of the Bureau of Reclamation, the Los Angeles Bureau of Power and Light, and Southern California Edison.[39]

The town contracted physically, too. Before it left Boulder City, Six Companies was required by its contract with the government to remove all the temporary buildings it had erected, a designation that covered most of the structures in town. A deal was struck between the contractors and the Wattis-Decker Company of Salt Lake City in February, 1936, with Wattis-Decker agreeing to pay Six Companies $15 for one-room cottages, $25 for two-room cottages, $50 for three-room cottages, and $750 for large buildings, such as dormitories in exchange for the right to sell the salvaged lumber, wiring, plumbing, light fixtures, and other construction materials. Demolition began on March 5, 1936, and continued for the next two years. The government purchased two of the dormitories to house Civilian Conservation Corps workers who were building facilities for the recently established Boulder Dam National Recreation Area, and a number of Six Companies cottages were resold, some to a tourist-camp operator who jacked them up and hauled them off to Utah and others to people who were staying on in Boulder City. But everything else—hundreds of cottages, half a dozen dormitories, the mess hall, the recreation hall, and many other structures—was razed, returning the southern half of the city to a state closely resembling its predam desolation.[40]

Just when it seemed that Boulder City would dwindle into a tiny, isolated maintenance station for the dam, the opening of the magnesium plant in Henderson and the beginning of Wold War II gave it a reprieve. Housing shortages in Henderson and Las Vegas pushed defense workers into the town, and McKeeversville, the tent and shack encampment established at the crest of Hemenway Wash in 1930, was reoccupied by about sixty families. The army also moved into Boulder City, building Camp Williston to quarter troops manning the checkpoints and gun emplacements that guarded Hoover Dam, a prime strategic target, from attacks by Japanese aircraft or saboteurs.[41]

Having barely survived the postconstruction decade with the help of a temporary, war-related population surge, Boulder City looked for some way to ensure a more prosperous and stable future. Although the autocratic Sims Ely had departed in 1941, the town was still governed by a handful of unelected federal officials, and it was evident, even to the Bureau of Reclamation, that economic and social stagnation would continue as long as the government ruled. In 1949, Harry Reining, Jr., a professor of public administration and political science at the University of Southern California, was commissioned by the bureau to prepare a plan for instituting self-rule in Boulder City. Following his recommendations, the secretary of the interior issued an order in 1951 administratively separating Boulder City from the Boulder Canyon Project. This preliminary step was followed by passage of legislation authorizing disposal of all government property in the community not needed for continuing Bureau of Reclamation activities and permitting the establishment of self-government. On October 28, 1959, the municipality of Boulder City was incorporated under Nevada law, and on January 4, 1960, it was officially separated from the U.S. government, ending thirty years of federal administration.[42]

Boulder City was now free to chart a new course, but its people chose to go on observing many of the strait-laced regulations inherited from the Sims Ely era. Gambling, which was making Las Vegas rich, continued to be prohibited, and the sale of hard liquor was forbidden until 1969. Ironically, these conservative restrictions, which might have been expected to suppress growth, actually attracted new residents, who found Boulder City's insularity and wholesomeness appealing; the community's population swelled so rapidly that in 1979 its citizens passed an ordinance limiting future expansion.[43]

Modern Boulder City, home to approximately ten thousand people, has a look of permanence and prosperity that would startle the band of engineers and construction workers who settled there in the summer and fall of 1931. Within the city limits the desert has been defeated; Wilbur Weed's trees are tall and full, his hedges and shrubs are thick and well rooted, and there is nothing left to suggest that within living memory these lush, shady acres were barren, wind-blown sand. Some of the original buildings are still standing—the hospital, the administration building, Grace Community Church, the Boulder Dam Hotel, the municipal building, the elementary school, a number of enlarged and renovated Six Companies cottages—but their stark appearance has been softned by time and mature landscaping. On the eastern, southern, and western fringes of town, blocks of new ranch-style houses push the edge of the desert farther and farther back, and motels, shopping centers, fast-food restaurants, and drive-in banks line Boulder Highway where it enters the community.

After more than half a century, Boulder City has finally fulfilled the destiny that Ray Lyman Wilbur, Elwood Mead, and other government officials so fervently desired for it: it has become a truly typical American small town.

The dam itself has changed very little since the day in 1936 when Frank Crowe presented it to the government. The concrete in its body is still slowly curing, growing stronger with each passing year. Faint brown, beige, and white water stains streak the once-unblemished downstream face, but the effect is not unpleasing, adding a measure of warmth and contrast to the otherwise stark and austere facade.

Alterations have been made in the powerhouse: four generators were installed during World War II, three more in 1952, and one in 1961, completing the power plant and bringing the dam's nameplate generating capacity to 1,344,800 kilowatts. In the early 1980s a program to uprate the plant was begun; as part of this program, the original cast-steel turbine runners were replaced with more efficient, lower-maintenance stainless-steel runners, and new, thinner insulation was installed in the generators, increasing their capacity to more than 2 million kilowatts.[44]

Diversion-tunnel repairs and modifications were made necessary by the most dramatic event in the dam's working life: the flood of 1983. It began in late spring when snowstorms in the Rocky Mountains were followed by a heat wave and heavy rains. Runoff between April and July was 210 percent of normal, and the Colorado River reservoirs filled rapidly. On June 6, 1983, Lake Powell behind Glen Canyon Dam was full, and emergency water releases began through its spillways. By late June ninety thousand cubic feet of water per second was rushing through the Grand Canyon toward Lake Mead, and it was apparent that Hoover Dam's flood-control capacity would be put to a severe test.

Late in the evening of July 2, the dam's giant steel drum gates were lifted and a foaming wave crashed into the spillway channels and was

Water pouring into the Arizona spillway, July, 1983. (Bureau of Reclamation)

sucked into the diversion tunnels. A cloud of mist rose into the hot desert night, and after a hiatus of fifty-two years, the wild freight-train roar of the Colorado River was heard again in Black Canyon.

The flood crested on July 24 when more than fifty thousand cubic feet of water per second was discharged through the diversion tunnels; the spill was continued until September 6, when the drum gates were closed. Millions of dollars' worth of damage occurred in the river corridor downstream from Hoover Dam, but the destruction was only a fraction of what it would have been if Lake Mead had not cut the peak from the flood.[45]

At Glen Canyon Dam, built in the late 1950s, the emergency discharges virtually destroyed the diversion tunnels' linings. Cavitation—the forceful collapse of partial vacuums in fast-moving water—chewed through the concrete as if it were cardboard and gouged large holes in the tunnel floors. Inspection of the Hoover Dam diversion tunnels revealed a much different outcome: the water, racing along at speeds up to 120 miles per hour had caused slight pitting and abrading, but the concrete linings were still intact—a tribute to the design and construction expertise of the men who had built them fifty years earlier. In spite of the minor nature of the tunnel damage, the Bureau of Reclamation decided in 1985 to patch

At night, the Hoover Dam intake towers glow like giant candles. (Bureau of Reclamation)

the concrete and install air slots to prevent cavitation during spillway discharges.[46]

The flood of 1983 demonstrated the fundamental soundness of Hoover Dam's design and the high quality of its construction. Frank Crowe and his work crews not only built fast, they built well; with proper maintenance the dam's life should be indefinite.

Although it is no longer the tallest dam in the world (Nourek Dam in the Soviet Union is 314 feet higher), Hoover Dam is still the most famous. The number of sightseers touring the dam and power plant has risen steadily since World War II, with yearly visitor totals now approaching three-quarters of a million. On busy summer days, as many as three thousand people come to see the huge structure, walk across its crest, and ride elevators down into its body to see the tunnels and generators.

The experience, especially the first glimpse of the dam from one of the hairpin turns in the road zigzagging down through the red and black cliffs, seldom fails to elicit a visceral response. The sight is unearthly, particularly at night, when recessed lamps illuminate the expanse of concrete and the

Hoover Dam. (Bureau of Reclamation)

tailrace below in a blaze of dazzling golden light. Confronting this spectacle in the midst of emptiness and desolation first provokes fear, then
wonderment, and finally a sense of awe and pride in man's skill in bending the forces of nature to his purpose. In the shadow of Hoover Dam one
feels that the future is limitless, that no obstacle is insurmountable, that we

have in our grasp the power to achieve anything if we can but summon the will.

Some try to suppress this feeling, thinking that it is naïve to be enthralled by an engineering marvel, dangerous to be seduced by twentieth-century technology, which cynics say is untrustworthy, exploitative, and destructive of the environment and the human spirit. But in the clear desert light of Black Canyon, guilt about the deeds of the past and doubt about the promise of the future shrivel. The romance of the engineer still lives in the graceful lines and brute strength of Hoover Dam. The courage of the construction worker is written in concrete and steel across the face of the towering cliffs, and a generation's belief in the destiny of the next is proclaimed by the deep-throated hum of the generators. Let poet May Sarton describe it:

> But here among hills bare and red,
> A violent precipice, a dizzy white curve falls,
> Hundreds of feet through rock to the deep canyon-bed;
> A beauty sheer and clean and without error
> It stands with the created sapphire lake behind it,
> It stands, a work of man as noble as the hills,
> And it is faith as well as water it spills.
>
> Not built on terror like the empty pyramid,
> Not built to conquer but to illuminate a world:
> It is the human answer to a human need,
> Power in absolute control, freed as a gift,
> A pure creative act, God when the world was born!
> It proves that we have built for life and built for love
> And when we all are dead, this dam will stand and give.[47]

Notes

CHAPTER 1. A RIVER AND A DREAM

1. Details of the executive meetings at St. Francis Hospital from Six Companies Corporate Records (unpublished minutes, memoranda, financial statements, etc.; hereafter cited as SCCR), Vol. 1, 1931; "The Earth Movers I," *Fortune* 28 (August, 1943): 99–107, 210–214; Neill C. Wilson and Frank J. Taylor, *The Earth Changers,* pp. 20–21. For a description of W. H. Wattis on his death bed, see "Damn Big Dam," *Time Magazine* 21 (March 23, 1931): 14–15. Wattis had been admitted to St. Francis Hospital in February, 1931. He was undergoing an experimental cancer treatment, developed by surgeons Walter Coffey and John Humber, that consisted of a series of sheep-cell extract injections.

2. The literature on the natural and human history of the Colorado River is vast. Two excellent general works are Frank Waters, *The Colorado,* and T. H. Watkins et al., *The Grand Colorado.*

3. Lieutenant J. C. Ives, *Report Upon the Colorado River of the West,* H. Exec. Doc. 90, 36th Cong., 1st sess., 1861.

4. John Wesley Powell, *Report on the Lands of the Arid Regions of the United States.*

5. The adventures of Dr. Wozencraft in the California desert are detailed in Otis B. Tout, *The First Thirty Years,* p. 26; David O. Woodbury, *The Colorado*

Conquest, pp. 31−33; Watkins et al., *The Grand Colorado*, pp. 206−207. A more complete account of Wozencraft's career is presented in Barbara Ann Metcalf, "Oliver M. Wozencraft in California, 1849−1887."

6. Woodbury, *Colorado Conquest*, p. 32.

7. *Report of the Explorations and Surveys to Ascertain the Most Practicable and Economical Route for a Railroad from the Mississippi River to the Pacific Ocean*, S. Exec. Doc. 78, 33d Cong. 2d sess., 1856.

8. For accounts of the birth of Imperial Valley, see Woodbury, *Colorado Conquest;* Tout, *The First Thirty Years;* Watkins et al., *The Grand Colorado*, pp. 205−221; Norris Hundley, *Water and the West*, pp. 17−36; Remi Nadeau, *The Water Seekers*, pp. 139−69; Robert J. Schonfeld, "The Early Development of California's Imperial Valley," *Southern California Quarterly* 50 (September, 1968): 279−307. For information about the lives and careers of Charles Rockwood and George Chaffey, see Charles R. Rockwood, *Born of the Desert;* Margaret D. Morrison, "Charles Robinson Rockwood: Developer of the Imperial Valley," *Southern California Quarterly* 44 (December, 1962): 307−330; J. A. Alexander *The Life of George Chaffey;* Frederick D. Kershner, Jr., "George Chaffey and the Irrigation Frontier," *Agricultural History* 27 (October, 1953): 115−22.

9. From a poem by Mr. Wesener of Yuma, delivered at the Conference on the Construction of the Boulder Dam, San Diego, California, December 12, 1921, in *Problems of Imperial Valley and Vicinity with Respect to Irrigation from Colorado together with Proceedings of the Conference on Construction of Boulder Canyon Dam*, S. Doc. 142, 67th Cong., 2d sess., 1922, p. 290.

10. For a detailed account of Swing's role in the legislative campaign for the All-American Canal, see Beverly B. Moeller, *Phil Swing and Boulder Dam*.

11. See Gene M. Gressley, "Arthur Powell Davis, Reclamation, and the West," *Agricultural History* 42 (July, 1968): 241−57; Charles A. Bissell and Frank Weymouth, "Memoirs of Deceased Members: Arthur Powell Davis, Past President, Am. Soc. C.E. Died August 7, 1933," *Transactions of the American Society of Civil Engineers* 100 (1935): 1582−91; "Arthur Powell Davis," *Dictionary of American Biography*, Vol. 11, pp. 224−25. See also Alfred R. Golze, *Reclamation in the United States;* William E. Warne, *The Bureau of Reclamation;* Michael C. Robinson, *Water for the West: The Bureau of Reclamation 1902−1977*.

12. See *Problems of Imperial Valley*.

13. For a detailed history of the Colorado River Compact, see Hundley, *Water and the West*.

14. E. C. LaRue, engineer with the U.S. Geological Survey, was the man chiefly responsible for identifying the eight dam sites during field surveys conducted in the late teens. In addition to Boulder and Black canyons, the potential sites were Glen Canyon, Diamond Creek, Bridge Canyon, Bullshead, Topock, and Parker. See E. C. LaRue, *Colorado River and Its Utilization*, U.S. Geological Survey Water Supply Paper No. 395; E. C. LaRue, *Water Power and Flood Control of Colorado River Below Green River, Utah*, U.S. Geological Survey Water Supply Paper No. 556. See also Ray Lyman Wilbur and Elwood Mead, *The Construction of Hoover Dam: Preliminary Investigations, Design of Dam, and Progress of Construction*, pp. 1−7; U.S. Department of the Interior, Bureau of Reclamation, *Boulder Canyon Project Final Reports, Part III—Preparatory Examinations, Bulletin 1, Geological Investigations*, pp. 2−3, 9−10; U.S. Department of the Interior, Bureau of Reclamation, *Boulder Canyon Project Final Reports, Part IV—Design and Con-*

struction, Bulletin 1, General Features, 33–44; Annual Project History, Boulder Canyon Project (unpublished project histories, prepared annually by Bureau of Reclamation Personnel; hererafter cited as APH), Vol. 1, From Beginning to December 31, 1931, pp. 14–20.

15. "The Dam," *Fortune* 8 (September, 1933): 74–88; "Walker R. Young," *Who's Who in America,* p. 2616.

16. APH, Vol. 1, pp. 20–27; Wesley R. Nelson, "Construction of the Boulder Dam: Government Engineers and Surveyors Have Made a Notable Record on Exacting, Perilous Work," *Compressed Air Magazine* 39 (November, 1934): 4585–88. *Compressed Air Magazine* was the monthly publication of the Ingersoll-Rand Company, which had a lucrative contract to provide all the jackhammers, rock drills, drill-steel sharpeners, and air compressors for the Hoover Dam project. The magazine published a series of detailed articles on various aspects of the dam construction. It is a particularly good source for technical information on the equipment and methods used by the builders. See also Bureau of Reclamation, *BCP Final Reports, Preparatory Examinations, Geological Investigations,* pp. 9–10.

17. Nelson, "Construction of the Boulder Dam: Government Engineers and Surveyors Have Made," p. 4586; APH, Vol. 1, 1931, Appendix, "Fatalities," p. 343.

18. Nelson, "Construction of the Boulder Dam: Government Engineers and Surveyors Have Made," p. 4586; "Surveying in Black Canyon," *Reclamation Era* 23 (October, 1932): 172–73. *New Reclamation Era,* which became *Reclamation Era* after 1931 (hereafter cited as *NRE* or *RE*), was the official monthly publication of the Bureau of Reclamation and is an excellent source for both technical and human-interest information on the Hoover Dam story.

19. For detailed accounts of the legislative battle over the Boulder Canyon Project Act, see Paul L. Kleinsorge, *The Boulder Canyon Project: Historical and Economic Aspects,* pp. 75–136; Moeller, *Phil Swing and Boulder Dam,* pp. 86–122.

20. Bureau of Reclamation, *BCP Final Reports, Design and Construction, General Features,* pp. 47–57, 72–76.

21. Ibid., pp. 219–34. For Kaufmann's description of his contribution to the Hoover Dam design, see Gordon B. Kaufmann, "The Architecture of Boulder Dam," *Architectural Concrete* 2 (No. 3, 1936): 2–5. For an interesting discussion of the modernist influences on Kaufmann's design, see Richard Guy Wilson "Machine-Age Iconography in the American West: The Design of Hoover Dam," *Pacific Historical Review* 54 (November, 1985): 463–93; Richard G. Wilson et al., *The Machine Age in America 1918–1941,* pp. 111–15; Richard Guy Wilson, "Massive Deco Monument," *Architecture* (December, 1983): 45–47. For a description of the design of the dam's interior spaces by Denver artist Allen True, see Allen True, "Color and Decoration at the Boulder Power Plant," *RE* 26 (January, 1936): 12–13; Allen True, "The Planned Use of Color at the Boulder Dam Power Plant," *RE* 26 (February, 1936): 48. For a description of the memorial sculptures and plaques by sculptor Oskar J. W. Hansen, see Oskar J. W. Hansen, *Sculptures at Hoover Dam.* For illustrations of the dam's design before Kaufmann's consultation, see Wilbur and Mead, *The Construction of Hoover Dam,* Appendix, "Specifications No. 519: Specifications, Schedules, and Drawings, Hoover Dam, Power Plant, and Appurtenant Works, Boulder Canyon Project Arizona-California-Nevada" Drawings 1–76.

22. See Ray Lyman Wilbur and Northcutt Ely, *The Hoover Dam Power and*

Water Contracts and Related Data; Kleinsorge, *The Boulder Canyon Project,* pp. 137–244; Edgar E. Robinson and Paul C. Edwards, eds., *The Memoirs of Ray Lyman Wilbur, 1875–1949,* pp. 441–65.

23. Public-power advocates objected to the final division of Boulder Canyon Project electricity because of the inclusion of a private utility, Southern California Edison, in the formula. The power allocation was actually a precedent-setting victory for them, however, given that more than 90 percent of the electricity generated went to city or state purchasers. For the first time, the federal government was squarely in the power business, in direct competition with private utilities. See Moeller, *Phil Swing and Boulder Dam,* pp. 125–31; Linda J. Lear, "Boulder Dam: A Crossroads in Natural Resource Policy," *Journal of the West* 24 (October, 1985): 82–94.

24. *Las Vegas Evening Review-Journal,* September 17, September 18, September 22, 1930; *Las Vegas Age,* September 18, 1930; Moeller, *Phil Swing and Boulder Dam,* p. 132. Newspaperman Elton Garrett covered Wilbur's speech. "It was totally a surprise," he recalled of the name-change announcement. "As the crowd broke up, different people had different comments about it depending on their politics. At that time, the Department of Interior began changing all its literature to read 'Hoover Dam.' But in the public voice and mind, it was still Boulder Dam." See Dennis McBride, "The Battle for Boulder," *Nevada* 45 (September–October, 1985): 78.

25. Wilbur and Mead, *The Construction of Hoover Dam,* Appendix, "Specifications No. 519," pp. 1–10, 19–20.

26. *Las Vegas Age,* March 4, 1931; Author interview with S. D. Bechtel, Sr., San Francisco, December 11, 1984.

27. See "The Earth Movers I," *Fortune,* p. 102; Wilson and Taylor, *The Earth Changers,* pp. 21–25; "Edmund O. Wattis," *Who's Who in America,* p. 2469; "Edmund Orson Wattis," *The National Cyclopaedia of American Biography,* Vol. 25, p. 294.

28. "Harry Winford Morrison," *The National Cyclopaedia of American Biography,* Volume G, p. 225; "The Earth Movers I," *Fortune,* p. 102.

29. "The Earth Movers I," *Fortune,* pp. 102–103.

30. *Los Angeles Examiner,* March 9, 1931.

31. S. O. Harper and Walker R. Young, "Memoirs of Deceased Members: Francis Trenholme Crowe, Hon. M. ASCE, Died February 26, 1946," *Transactions of the American Society of Civil Engineers* 113 (1948): 1397–1403; "The Dam," *Fortune,* pp. 74–88.

32. "Man Rules Supreme on Hoover Dam," *Salt Lake City Tribune* article reprinted in the *Las Vegas Evening Review-Journal,* January 12, 1932; Author interview with S. L. ("Red") Wixon, Boulder City, Nevada, July 27, 1984.

33. "The Earth Movers I," *Fortune,* p. 103.

34. Ibid., p. 106.

35. See Mark S. Foster, "Giant of the West: Henry J. Kaiser and Regional Industrialization, 1930–1950," *Business History Review* 59 (Spring, 1985): 1–23; John Gunther, "Life and Works of Henry Kaiser," in *Inside USA,* pp. 64–75; Susanne T. Gaskins and Warren Beck, "Henry J. Kaiser—Entrepreneur of the American West," *Journal of the West* 25 (January, 1986): 64–72; Leonard Lyons, "Unforgettable Henry J. Kaiser," *Reader's Digest* (April, 1968): 209–210; Wilson and Taylor, *The Earth Changers,* pp. 29–31.

36. Robert L. Ingram, *A Builder and His Family*, p. 31; "The River Wrestlers," in *The Kaiser Story*, p. 18.

37. "The Earth Movers I," *Fortune*, p. 104; Ingram, *A Builder and His Family*; Wilson and Taylor, *The Earth Changers*, pp. 25-29.

38. Foster, "Giant of the West," p. 3; Lyons, "Unforgettable Henry Kaiser," p. 210.

39. "The Earth Movers I," *Fortune*, p. 106.

40. Edgar Kaiser, "America's Builders," speech given at America's Builders Recognition Night, Pepperdine University, 1968, from Henry J. Kaiser Collection, Bancroft Library, University of California, Berkeley.

41. SCCR, Vol. 1, 1931, p. 15; "Minutes of Adjourned Meeting of Board of Directors, February 25, 1931," SCCR, Vol. 1, 1931, pp. 66-69.

42. "Minutes of Adjourned Meeting of Board of Directors, February 25, 1931"; "Minutes of Special Meeting of Board of Directors, March 4, 1931," Vol. 1, 1931, pp. 99-107; *Denver Post*, March 5, 1931; *Oakland Tribune*, March 5, 1931; "Hoover Dam Bids Opened," NRE 22 (April, 1931): 79.

43. "Hoover Dam Bids Opened," NRE 22 (April, 1931): 79; APH, Vol. 1, 1931, Appendix, "Abstract of Bids," p. 258.

44. "Damn Big Dam," *Time Magazine*, pp. 14-15.

CHAPTER 2. "A DEADLY DESERT PLACE"

1. APH, Vol. 1 1931, p. 264; *Las Vegas Age*, March 12, 1931; *New York Times*, March 13, 1931.

2. "Hoover Dam Bids Opened," NRE 22 (April, 1931): 79.

3. Engineering criticism of the Hoover Dam plans is summarized in Kleinsorge, *The Boulder Canyon Project*, pp. 190-91.

4. Ibid., p. 300; NRE 21 (December, 1930): 245; U.S. Department of the Interior, Bureau of Reclamation, *The Story of Hoover Dam*, pp. 18-19; "Man Rules Supreme on Hoover Dam," *Salt Lake City Tribune* article reprinted in the *Las Vegas Evening Review-Journal* January 12, 1932.

5. For descriptions of Las Vegas on the eve of the Hoover Dam project, see Waters, *The Colorado*, pp. 344-45; Duncan Aikman, "New Pioneers in Old West's Deserts," *New York Times Magazine*, October 26, 1930, pp. 7, 18; Phillip I. Earl, "The Legalization of Gambling in Nevada, 1931," *Nevada Historical Society Quarterly* 24 (Spring, 1981): 39-46. According to the *New York Times* reporter, "the only concession Las Vegas makes to prohibition is that saloon doors no longer swing."

6. See "Charles Pemberton 'Pop' Squires," *Nevada Historical Society Quarterly* 16 (Fall, 1973): 207-208; *Las Vegas Age*, January 7, 1931.

7. For descriptions of the mass migration to southern Nevada in 1930 and early 1931, see Aikman, "New Pioneers," pp. 7, 18; George Albert Pettitt, *So Boulder Dam Was Built*, pp. 22-23; Edmund Wilson, *American Earthquake*, pp. 368- 78; Victor Castle, "Well, I Quit My Job at the Dam," *The Nation* 133 (August 26, 1931): 207-208; "Job Mad Mobs at Boulder Dam," *Industrial Worker*, May 23, 1931.

8. *Las Vegas Age*, February 10, 1931; NRE 22 (March, 1931): 49.

9. Wilson, *American Earthquake*, p. 373; Castle, "Well, I Quit," p. 207; *Las Vegas Age*, March 26, 1931. In "Activities at Hoover Dam," NRE 22 (September,

1931): 186–87, C. H. Pease wrote: "Numerous [gambling] tables were crowded with players and things were running full blast. There was but one other place where we saw such crowds and upon investigation I found that it was the state employment agency. Crowds were around this place so thick that it was difficult to get by on the sidewalk." In May, 1932, Blood reported that a total of 36,000 letters and 20,000 personal applications for employment had been filed during the preceding 15 months. "Boulder Canyon Project Notes," RE 23 (May, 1932): 98; see also U.S. Labor Commission Folder, "Organization and Operation of the U.S. Department of Labor Employment Office," p. 3, Leonard Blood Papers, University of Nevada–Las Vegas Special Collections; John F. Cahlen, "Reminiscences of a Reno and Las Vegas, Nevada Newspaperman, University Regent, and Public-Spirited Citizen," transcript of an April, 1968, interview conducted by Mary Ellen Glass for the Oral History Project of the University of Nevada–Reno Library, p. 101.

10. Edmund Wilson, "Hoover Dam," *The New Republic* 68 (September 2, 1931): 66–69. The chapter on Hoover Dam in Wilson's *American Earthquake* is a revised version of this article.

11. Duncan Aikman, "A Wild West Town that is Born Tame," *New York Times Magazine*, July 26, 1931, pp. 6–7, 15.

12. Mrs. D. L. Carmody, "Impressions of an Engineer's Wife on her First Trip to the Site of Hoover Dam," *NRE* 22 (June, 1931): 136–37.

13. APH, Vol. 1, p. 159; Theo White, "Building the Big Dam," *Harper's Magazine* 171 (June, 1935): 115.

14. Dennis McBride, *In the Beginning: A History of Boulder City Nevada*, p. 9.

15. Erma O. Godbey, "Pioneering in Boulder City, Nevada," transcript of a March, 1966, interview conducted by Mary Ellen Glass for the Oral History Project of the University of Nevada–Reno Library, pp. 53–54; James Menard Green, "Life in Nelson Township, Eldorado Canyon, and Boulder City, Nevada 1860–1960," pp. 50–51, 77–79; Pettitt, *So Boulder Dam Was Built*, pp. 24–25; Mrs. D. L. Carmody, "A Visit to the Hoover Dam Site," *NRE* 22 (June, 1931): 173–74. "There before their eager eyes is work—work, yet like the feast of the Barmecide it is not for them; but they stay on and on, prisoners of hope," Mrs. Carmody wrote.

16. McBride, *In the Beginning*, p. 11; "Everyone paid the prices that they paid back home," Murl Emory recalled in 1974. "Therefore, a lady from Kansas and a lady from Wyoming paid a different price for a loaf of bread or the same jar of preserves or cut of meat. The prices people paid could have been 1930 or 1931 prices back home rather than the prevailing Las Vegas prices. If I had not had the small boat work contract with the Six Companies, Inc., I would have gone broke; no one went hungry and the honor system worked well enough to make life-long friends." Murl Emory interviewed by James M. Green, October 18, 1974, in Green, "Life in Nelson Township," p. 79.

17. Godbey, "Pioneering in Boulder City," p. 52. The intense heat required other adjustment. Storing and preparing food was very difficult: butter was sold in liquid form, in jars, because there was no way to keep it from melting when it was delivered to Ragtown. One woman found that beans she had put in water to soak overnight were spoiled by morning. Young children became lethargic by midday: Erma Godbey tried to keep her four youngsters alert and occupied by putting them in a lean-to made with a wet sheet and encouraging them to dig down into the dirt for coolness.

18. *New York Times,* March 13, 1931; Pettitt, *So Boulder Dam Was Built,* pp. 25–26.

19. *Las Vegas Evening Review-Journal,* February 28, 1946.

20. *Las Vegas Age,* April 14, 1931; C. H. Vivian, "Construction of the Hoover Dam: How the Contractors Handled the Huge and Costly Program of Preliminary Work," *Compressed Air Magazine* 37 (March, 1932): 3742–48; APH, Vol. 1, p. 218; Wilson, "Hoover Dam," pp. 66–69.

21. *Las Vegas Age,* May 8, 1931; "Careless Blast Kills Workers on Hoover Dam," *Industrial Worker,* May 23, 1931; "Big Six Threaten Workers' Lives," *Industrial Solidarity,* May 19, 1931.

22. APH, Vol. 1, Appendix, "Fatalities," p. 343; *Las Vegas Age,* May 19, 1931; "Life Held Cheap at Boulder Dam," *Industrial Worker,* May 30, 1931; "Three Men Killed on Boulder Dam Job," *Industrial Solidarity,* June 2, 1931.

23. *Las Vegas Age,* May 12, 1931; APH, Vol. 1, p. 264.

24. *Las Vegas Age,* April 25, 1931; Wilbur and Mead, *The Construction of Hoover Dam,* Appendix, "Specifications No. 519," pp. 19–20.

25. Castle, "Well, I Quit," p. 208.

26. *Las Vegas Evening Review-Journal,* August 14, 1931; John Gieck interviewed by James M. Green, January 8, 1975, in Green, "Life in Nelson Township," p. 80; APH, Vol. 1, Appendix, "Fatalities" p. 342.

27. John C. Page, office engineer for the Bureau of Reclamation in Boulder City wrote in 1931: "Many of the men starting work on the project during the summer were unused to living under desert conditions and the consequent number of fatalities due to heat prostration were to some extent alarming." APH, Vol. 1, p. 175; "Some heat prostrations have occurred, a few of them being fatal," wrote Six Companies director Charles Shea in a July 16, 1931, memo to the Six Companies Executive Committee. "But it seems evident that the fatal cases were due to fundamental lack of vitality or induced by overeating or some other ailment." Construction Department Memo from C. A. Shea to Executive Committee, July 16, 1931, SCCR, Vol. 1 , 1931, pp. 262–67; Imre Sutton, "Geographical Aspects of Construction Planning: Hoover Dam Revisited," *Journal of the West* 7 (July, 1968): 322; Pettitt, *So Boulder Dam Was Built,* p. 37.

28. *Las Vegas Evening Review-Journal,* June 11, June 16, August 11, 1932; "Boulder Canyon Project Notes," *RE* 23 (April, 1932): 81; *RE* 23 (September, 1932): 160; Pettitt, *So Boulder Dam Was Built,* p. 46; "Conversation with Dr. David Bruce Dill," from A. D. Hopkins, "The Legend Builders," *Nevada* 45 (May–June 1985): 63–64.

29. Erma Godbey, "Pioneering in Boulder City," pp. 48–54.

30. "The water supply at this [River] camp has been a difficult problem and the silt content in the river recently changed so as to make it almost impossible to clarify the [drinking] water." Construction Department Memo from C. A. Shea to Executive Committee, July 16, 1931, SCCR, Vol. 1, 1931, pp. 262–67; "Boulder Dam Mecca Proves to be Hell," *Industrial Worker,* June 27, 1931. See also Kleinsorge, *The Boulder Canyon Project,* p. 222; Sutton, "Geographical Aspects," p. 322.

31. Wilson, *American Earthquake,* p. 370. A worker described the River Camp environment as follows: "Have just come in from a 22 day hitch at the river camp of the wonderful Hoover Dam project. I wanted to stay longer but conditions were so rotten I had to make town to recuperate. . . . Sleeping quarters are the

worst seen in years. From 75 to 80 men in one bunkhouse. With men coming in and going out constantly sleep is almost impossible." From "Accident Talk Forbidden on Hoover Dam Project," *Industrial Solidarity*, June 16, 1931.

32. APH, Vol. 1, p. 175.

33. "The labor situation has been somewhat difficult in the past six weeks due to the summer heat which has made the turnover considerably greater than existed in May." Construction Department Memo from C. A. Shea to Executive Committee, July 16, 1931, SCCR Vol. 1, 1931, pp. 262–67; *Salt Lake City Tribune* reporter Owen Malmquist described Frank Crowe as follows: "The general superintendent is tall and lank, having sacrificed 27 pounds to heat and high tension work since the job was started. . . . One cannot talk to him without noting his long, tapering, nervous fingers, which seem entirely out of place on a rough construction job." In "Man Rules Supreme on Hoover Dam," article from *Salt Lake City Tribune* reprinted in *Las Vegas Evening Review-Journal*, January 15, 1932.

34. McBride, *In the Beginning*, p. 12.

35. Godbey, "Pioneering in Boulder City," pp. 53–54.

36. *San Francisco Examiner*, June 9, 1931; Wilson, "Hoover Dam," p. 66; "Boulder Dam Mecca Proves to be Hell," *Industrial Worker*, June 27, 1931.

37. "For several weeks . . . we had thought of suspending work because of the heat, but we continued in order that the men might not be deprived of employment," said Frank Crowe in the *New York Times*, August 10, 1931; Six Companies had negotiated a number of very large, exclusive contracts with such suppliers as the Crucible Drill Steel Company (drill steel), Ingersoll-Rand (pneumatic equipment), Union Oil Company (petroleum products) Westinghouse and General Electric (electrical equipment) and Hercules and Giant powder companies (blasting supplies). See "Minutes of Regular Meeting of Board of Directors, April 20, 1931," SCCR, Vol. 1 1931, pp. 164–79; author interview with S. D. Bechtel, Sr., San Francisco, December 11, 1984. Reclamation Commissioner Elwood Mead answered criticism of the injuries and deaths in Black Canyon as follows: "The summer heat in the canyon is terrific. . . . The deaths that have resulted are deplored by all, but they have not been excessive when the difficult and dangerous character of the work to be done and the desert location is taken into account. . . . Anyone who is familiar with what is going on knows that if it had not been for the very large expenditures to mitigate climatic conditions and protect the workers, the death toll would have been much heavier." "Letter to the Editor," *The New Republic* 68 (August 26, 1931): 48. See also Committee of Associated General Contractors of America and the American Engineering Council, "Report on the Hoover Dam Project and Present Status," Henry J. Kaiser Collection, Bancroft Library, University of California, Berkeley; Mead, "Letter," *The New Republic* 68 (August 26, 1931): 48.

38. For a description of Anderson, see Wilson, *American Earthquake*, p. 374; "IWW Delegates to Make Drive on Boulder Dam," *Industrial Solidarity*, September 30, 1930; "Boulder Dam Contract Let to Six Co.," *Industrial Worker*, March 14, 1931. The best account of IWW activities at Hoover Dam is Guy Louis Rocha, "The IWW and the Boulder Canyon Project: The Final Death Throes of American Syndicalism," *Nevada Historical Quarterly* 21 (Spring, 1978): 3–24.

39. The IWW's philosophy and goals were summed up in the "Preamble of the Industrial Workers of the World" as follows: "The working class and the employing class have nothing in common. There can be no peace as long as hunger

and want are found among millions of working people and the few, who make up the employing class, have all the good things in life. Between the two classes, a struggle must go on until the workers of the world organize as a class, take possession of the earth and the machinery of production, and abolish the wage system. . . . By organizing industrially we are forming the structure of the new society within the shell of the old." See Melvyn Dubofsky, *We Shall Be All: A History of the IWW*; John S. Gambs, *The Decline of the I.W.W.*

40. "Workers in Mood to Listen to Message of IWW," *Industrial Solidarity*, February 17, 1931.

41. "IWW Delegates to Make Drive on Boulder Dam," *Industrial Solidarity*, September 30, 1930.

42. See "Job Mad Mobs at Boulder Dam," *Industrial Worker*, May 23, 1931; "Life Held Cheap at Boulder Dam," *Industrial Worker*, May 30, 1931; "We'll Tell Our Kids the Story of Boulder Dam," *Industrial Worker*, June 6, 1931; "Big Six Threaten Workers' Lives," *Industrial Solidarity*, June 2, 1931; "Accident Talk Forbidden on Hoover Dam," *Industrial Solidarity*, June 16, 1931; *Las Vegas Evening Review-Journal*, August 24, 1932.

43. Wilson, *American Earthquake*, p. 376; author interview with S. L. ("Red") Wixon, Boulder City, Nevada, July 27, 1984.

44. *Las Vegas Age*, July 11, 1931.

45. "Boulder Dam Bosses Move Against IWW," *Industrial Solidarity*, July 21, 1931.

46. *Las Vegas Evening Review-Journal*, July 13, 1931.

47. *Las Vegas Evening Review-Journal*, July 11, 1931.

48. *Las Vegas Evening Review-Journal*, July 16, 1931.

49. *Las Vegas Age*, July 16, 1931; "All Cases Against IWW Members at Las Vegas Dismissed by Court," *Industrial Worker*, July 25, 1931; "Rebels at Boulder Dam Win Victory in Court," *Industrial Solidarity*, July 28, 1931.

50. *Industrial Solidarity*, July 28, 1931.

51. *Las Vegas Age*, July 19, 1931; "Big Six Installs Cooling Plant," *Industrial Worker*, August 1, 1931.

52. APH, Vol. 1, 1931, p. 175.

53. "The heat was such that Superintendent Frank T. Crowe said that most of the employees were able to perform only approximately a third of a normal day's work," reported Bureau of Reclamation Office Engineer John Page. "The men were in such a state of mind that they did not care whether they worked or not." From "Try These Out on Your Ukelele and See If You Can Harmonize Them," *Industrial Worker*, September 5, 1931. Wrote director Charles Shea: "Labor efficiency was very low, and the heat conditions under which the men had to work was a possible factor in the dissatisfaction producing the walkout of August 7." Memo from C. A. Shea, August 15, 1931, SCCR, Vol. 2, pp. 290–91.

54. APH, Vol. 1, pp. 51–52; "100 Percent Walk Out at Boulder Dam," *Industrial Worker*, August 15, 1931; "IWW Striking at Boulder Dam," *Industrial Solidarity*, August 18, 1931; *New York Times*, August 9, August 12, 1931.

55. *Las Vegas Evening Review-Journal*, August 8, 1931. "100 Percent Walk Out at Boulder Dam," *Industrial Worker*, August 15, 1931.

56. *Las Vegas Age*, August 8, 1931.

57. *Las Vegas Evening Review-Journal*, August 8, 1931; *New York Times*, August 10, 1931.

58. APH, Vol. 1, Appendix, "Fatalities" pp. 342–44.

59. *Las Vegas Evening Review-Journal,* August 9, 1931; *New York Times,* August 10, 1931.

60. *San Francisco Examiner,* August 9, 1931; *Los Angeles California Express,* August 10, 1931.

61. *Las Vegas Evening Review-Journal,* August 8, 1931; *New York Times,* August 9, 1931.

62. Wilson, *American Earthquake,* p. 373.

63. "Boulder Dam Strikers Carry On," *Industrial Worker,* August 22, 1931; "Big Six Efforts to Deport Are Nipped in Bud," *Industrial Solidarity,* August 18, 1931.

64. Wilson, *American Earthquake,* p. 374.

65. Ibid., p. 377; APH, Vol. 1, 1931, Appendix, "Memo from Walker R. Young to General Strike Committee, August 11, 1931," p. 294.

66. Wilson, *American Earthquake,* pp. 377–78, quotes Young as saying, "If you should refuse to go, we'll make you."

67. *New York Times,* August 19, 1931.

68. *Las Vegas Evening Review-Journal,* August 9, 1931; *New York Times,* August 12, August 22, 1931.

69. APH, Vol. 1, Appendix, "Memo from Walker R. Young to Six Companies, Inc.," p. 295; *Las Vegas Evening Review-Journal,* August 13, 1931; "Free and easy entrance to the reservation is over," said Young in the *Las Vegas Evening Review-Journal* on August 17, 1931.

70. *Las Vegas Evening Review-Journal,* August 14, 1931.

71. Within hours of the news that the job would reopen, "the highway was a maze of speeding cars and men afoot racing toward the employment office." From "Try These Out on Your Ukelele and See If You Can Harmonize Them," *Industrial Worker,* September 5, 1931.

72. *Las Vegas Evening Review-Journal,* August 15, 1931.

73. *Las Vegas Evening Review-Journal,* August 17, 1931.

74. "Boulder City Building Activities," NRE 22 (November, 1931):234–35; "Engineers-Contractors Committee Finds Hoover Dam Conditions Satisfactory," RE 23 (February, 1932):32.

75. *Las Vegas Evening Review-Journal,* August 24, 1931.

76. "Red Cards," *Industrial Solidarity,* August 25, 1931.

77. *Las Vegas Evening Review-Journal,* November 9, 1931. This widely circulated poem stung the IWW into penning a scathing parody titled "Us Old Scabs on Boulder Dam":

> There are thousands we know are dissatisfied
> And holler that we are cheap
> Such stuff won't penetrate our scabby hides
> Nor thru our thick domes creep;
> For we've been here since it started
> And unless we're made to lam
> We'll be sticking to the finish
> US OLD SCABS ON BOULDER DAM!
>
> We don't mind the falling rocks
> Nor the scorching rays of the sun
> For our heads are solid ivory blocks

And we think its lots of fun
We were ragged, buzzin' in the jungles;
It's better than Sally soup, you bet
So we're all mighty humble
With not a feeling of regret
And we're sticking to the finish—
There's Woody, Mac and me! Damn
But we're getting fat and scaley
US OLD SCABS ON BOULDER DAM!

Abe Lincoln freed the negroes,
Old Nero he burned Rome;
But the Big Six fills a graveyard
With the stiffs without a home.
In a bunkhouse barn we're sleeping
And toiling with might and main
But you'll never hear us crabbing
'Cause we hope to scab again—
So you'll find Shorty from Ossining
And Slim from Alabam'—
By golly, nearly all the crew are cussing
US OLD SCABS ON BOULDER DAM!

Oh that bacon and for breakfast
With gluey hotcakes and coffee cold
Over which a flag should wave at half mast
As the cause of belly pains untold;
Oh, that watery milk from kitchen sink
With soggy cake and leather steak
They may put our guts upon the blink
But we gotta work for Baldy's sake.
There's them that seems to think it tough
But just some beans and a hunk o' ham
By gosh, are plenty good enough
For US OLD SCABS ON BOULDER DAM!

Tho the rains may turn tornado
With whipping sleet and snow;
Tho the rest may damn the Colorado
And our dear super Baldy Crowe
We scabs are going to help him,
For we're solid and complete
While they pay us coolies' wages
With a place to flop and eat.
So in the slaughterhouse inferno
You'll find us every day
Where the muddy Colorado
Rushes madly on its way.
And if the boys don't ORGANIZE
And quickly make us scram
We're sticking to the finish—
US OLD SCABS ON BOULDER DAM!

"Us Old Scabs on Boulder Dam (With No Apologies to Claude Rader and his 'Us Old Boys on Boulder Dam,'" *Industrial Worker,* January 12, 1932.

CHAPTER 3.   TO TURN A RIVER

1. APH, Vol. 2, 1932, pp. 204–208; Pettitt, *So Boulder Dam Was Built,* pp. 53–57; Kleinsorge, *The Boulder Canyon Project,* p. 206; U.S. Department of the Interior, Bureau of Reclamation, *Boulder Canyon Project Final Reports, Part I—Introductory, Bulletin 1, General History and Description of the Project,* pp. 89, 91; Norman S. Gallison, "Construction of the Hoover Dam: Details of the Driving of the Four Huge Tunnels Which Will Divert the Colorado River Around the Dam Site," *Compressed Air Magazine* 37 (May, 1932): 3804–10.

2. *BCP Final Reports, General History,* p. 83.

3. Wilbur and Mead, *The Construction of Hoover Dam,* Appendix, "Specifications No. 519," pp. 19–20.

4. Gallison, "Construction of the Hoover Dam: Driving of the Four Huge Tunnels," p. 3806.

5. Vivian, "Construction of the Hoover Dam: How the Contractors Handled," pp. 3743–44; Kleinsorge, *The Boulder Canyon Project,* pp. 197–98; *BCP Final Reports, General History,* p. 85.

6. "The Earth Movers I," *Fortune,* p. 214; author interview with S. D. Bechtel, Sr., San Francisco, December 11, 1984.

7. Gallison, "Construction of the Hoover Dam: Driving of the Four Huge Tunnels," pp. 3806–3807; "Excavation of the Diversion Tunnels for Hoover Dam," *RE* 23 (September, 1932): 158–59.

8. Allen S. Park, "Mammoth Drill Carriages Speed Hoover Dam Tunnel Work," *Compressed Air Magazine* 37 (April, 1932): 3780–81.

9. Author interview with S. L. ("Red") Wixon, Boulder City, Nevada, July 27, 1984; Pettitt, *So Boulder Dam Was Built,* p. 71; *San Francisco News,* February 1, 1932; Park, "Mammoth Drill Carriages," p. 3780.

10. *New York Times,* March 13, 1931.

11. Park, "Mammoth Drill Carriages," pp. 3780–82; APH, Vol. 1, 1931, p. 256.

12. *BCP Final Reports, General History,* p. 85; Vivian, "Construction of the Hoover Dam: How the Contractors Handled," pp. 3743–44.

13. "The Earth Movers I," *Fortune,* p. 214; author interview with S. D. Bechtel, Sr., San Francisco, December 11, 1984.

14. APH, Vol. 2, 1932, pp. 204–208; Gallison, "Construction of the Hoover Dam: Driving of the Four Huge Tunnels," p. 3806.

15. *Las Vegas Evening Review-Journal,* September 14, 1931; SCCR, "Minutes of Regular Meeting of Board of Directors, September 21, 1931," Vol. 2, p. 306.

16. At times, fourteen headings were excavated simultaneously during the diversion tunnel phase of the Hoover Dam construction. See "Excavation of the Diversion Tunnels for Hoover Dam," pp. 158–59. For makeup of the tunneling crews, see Gallison, "Construction of the Hoover Dam: Driving of the Four Huge Tunnels," p. 3809; author interview with S. L. ("Red") Wixon, Boulder City, Nevada, July 27, 1984.

17. The story of Marion V. Allen is drawn from his privately published mem-

oir *Hoover Dam and Boulder City.* Copies can can be obtained by writing Marion V. Allen, P.O. Box 65, Shingletown, California.

18. Marion Allen provides an excellent first-hand description of the atmosphere in the diversion tunnels in *Hoover Dam,* pp. 47–52.

19. Ibid., p. 51. For definitions of mining terms, see P. W. Thrush, ed., *A Dictionary of Mining, Mineral, and Related Terms,* which was published by the Interior Department's Bureau of Mines in 1968.

20. Gallison, "Construction of the Hoover Dam: Driving of the Four Huge Tunnels," p. 3807.

21. Park, "Mammoth Drill Carriages," p. 3782.

22. Gallison, "Construction of the Hoover Dam: Driving of the Four Huge Tunnels," p. 3809; BCP *Final Reports, General History,* p. 89.

23. Gallison, p. 3808; author interview with S. L. ("Red") Wixon, Boulder City, Nevada, July 27, 1984. Approximately 1.5 million cubic yards of rock were removed during the excavation of the diversion tunnels (*BCP Final Reports, General History,* p. 89). The electric shovels with the 3.5-cubic-yard dippers removed 110 cubic yards, or 31.4 dipperfuls, per hour, on average (Gallison, "Construction of the Hoover Dam: Driving of the Four Huge Tunnels," p. 3808). Thus, approximately 13,636 hours (1.5 million cubic yards divided by 110 cubic yards per hour) were spent mucking in the tunnels. If the standard-issue 2.25-cubic-yard dippers had been used, only 70 cubic yards per hour would have been removed (31.4 dipperfuls times 2.25 cubic yards). Total hours mucking would have been 21,428 (1.5 million cubic yards divided by 70 cubic yards per hour). The difference, 7,792 hours, or 324.6 days divided by 8 shovels, came to 974 hours, or 40 days per shovel.

24. "Man Rules Supreme at Hoover Dam," *Las Vegas Evening Review-Journal,* January 15, 1932; Allen, *Hoover Dam,* p. 91; Letter to the author from John Meursinge, Six Companies engineer, February 12, 1985; author interview with S. L. ("Red") Wixon, Boulder City, Nevada, July 27, 1984.

25. "Correspondence showed that the government had been removing the I.W.W. and other agitators from the reservation. . . . Over 50 were separated from the payrolls." Ray L. Wilbur, Jr., "Boulder City: A Survey of Its Legal Background, Its City Plan and Its Administration," p. 152. Ray Wilbur, Jr., was the son of Interior Secretary Ray Wilbur. He interviewed the government officials running Boulder City and was given free access to correspondence and records of the Boulder City administration during his sojourn there in the summer of 1933. See also Judson King, "Open Shop at Boulder Dam," *The New Republic* 68 (June 24, 1931): 147–48; Duncan Aikman, "Amid Turmoil Boulder Dam Rises," *New York Times Magazine,* July 23, 1933, pp. 8–9, 13.

26. "Six Companies Ignore Nevada Pay-Day Laws," *Industrial Worker,* June 27, 1931. This article quoted a letter purportedly received from an Arizona state mine inspector as follows: "I got a promise from the [Six Companies] management that none of the Arizona mining laws would be violated on this side of the river. I forbid [*sic*] them using Hercules powder that I found in the magazine at the time, as there was no mark on the powder or the box that contained it concerning the date of manufacture. I informed Mr. Crowe . . . that the powder could not be used on the Arizona side." See also "Three More Men are Killed at Boulder Dam," *Industrial Worker,* July 4, 1931.

27. *San Diego Sun,* January 15, 1932.

28. APH, Vol. 2, 1932, Appendix, "Fatalities," p. 300, Vol. 3, 1933, Appen-

dix, "Fatalities," p. 310; Pettitt, *So Boulder Dam Was Built*, p. 72; Allen, *Hoover Dam*, p. 75.

29. Allen, *Hoover Dam*, pp. 152, 154.

30. Nevada Compiled Laws, Section 4229, Stats. Nev. 1931, c. 167, p. 274; *Six Companies, Inc. v. Stinson, State Inspector of Mines, et al.*, No. G-191, Dist. Ct., D. of Nev., April 18, 1932, 58 F.(2d), 649–53.

31. *Six Companies, Inc. v. Stinson*, 652.

32. *Six Companies, Inc. v. Stinson*, 2F. Supp., 689–92, February 15, 1933.

33. "Big Chief Crowe Weeps Muddy Tears," *Industrial Worker*, December 15, 1931.

34. Gallison, "Construction of the Hoover Dam: Driving of the Four Huge Tunnels," p. 3810.

35. John Meursinge, "What Price Dam?" p. 73.

36. "Six Days at Boulder Dam," *Industrial Worker*, April 12, 1932. The gas grievance was also the subject of a poem, titled "At Hoover Dam," published April 26, 1932, in *Industrial Worker:*

> What's that down there at Number One?
> The men are out en mass!
> They must be blasting! No, not that,
> It's carbon monoxide gas.
> They stagger out into the air
> And fall in gasping heaps
> While Huntington the walker is,
> So mad he almost weeps.

37. McBride, *In the Beginning*, p. 43; Sandy Klimek, "A History of Hospitals: Clark County, Nevada." The Bureau of Reclamation's fatality statistics show that forty-two deaths were attributed to pneumonia during the construction period, more than any other cause. In 1932, when tunneling activity was at its peak, sixteen pneumonia deaths were recorded. APH, Vol. 5, 1935, Appendix, "Fatalities," p. 305.

38. Meursinge, "What Price Dam?," p. 82.

39. Gallison, "Construction of the Hoover Dam: Driving of the Four Huge Tunnels," pp. 3808–3809; *Las Vegas Age*, January 22, 1932.

40. *Las Vegas Age*, January 29, 1932.

41. Norman S. Gallison, "Construction of the Hoover Dam: An Account of the Extensive Railroad System and of the Important Work It is Doing," *Compressed Air Magazine* 37 (September, 1932): 3908–12.

42. Pettitt, *So Boulder Dam Was Built*, p. 75.

43. Author interview with S. L. ("Red") Wixon, Boulder City, Nevada, July 27, 1984.

44. Pettitt, *So Boulder Dam Was Built*, p. 62; *Las Vegas Nevadan*, February 15, 1981.

45. Allen, *Hoover Dam*, p. 25.

46. *Las Vegas Age*, January 28, 1932.

47. "Boulder Canyon Project Notes," RE 24 (January, 1933): 12; Pettitt, *So Boulder Dam Was Built*, pp. 75–76.

48. J. B. Priestly, "Arizona Desert: Reflections of a Winter Visitor," *Harper's Magazine* 173 (March, 1937): 365.

49. APH, Vol. 2, 1932, p. 124; Pettitt, *So Boulder Dam Was Built*, p. 66.

50. APH, Vol. 2, 1932, pp. 11, 124; *Las Vegas Age,* February 10, 1932; Pettitt, *So Boulder Dam Was Built,* pp. 67–70; Author interview with S. L. ("Red") Wixon, Boulder City, Nevada, July 27, 1984.

51. "The Earth Movers I," *Fortune,* p. 210.

52. Pettitt, *So Boulder Dam Was Built,* p. 70; *Las Vegas Age,* February 12, 1932; APH, Vol. 2, 1932, p. 11.

53. Gallison, "Construction of the Hoover Dam: Driving of the Four Huge Tunnels," p. 3809.

54. Allen, *Hoover Dam,* p. 25.

55. *New York Times,* November 13, 1932; "Hoover Dam Has Distinguished Guests," *RE* 23 (December, 1932): 191; *Las Vegas Evening Review-Journal,* November 12, 1932.

56. *Las Vegas Evening Review-Journal,* November 13, 1932; "Colorado River Turned from Its Course," *RE* 23 (December, 1932): 198; Wesley R. Nelson, "Construction of the Hoover Dam: A Résumé of Current Activities and an Account of the Building of the Cofferdams," *Compressed Air Magazine,* 38 (March, 1933): 4069–73; Pettitt, *So Boulder Dam Was Built,* pp. 76–78.

57. Pettitt, *So Boulder Dam Was Built,* p. 77.

CHAPTER 4.   UNDER THE EAGLE'S WING

1. Aikman, "A Wild West Town that is Born Tame," pp. 6–7, 15.

2. *BCP Final Reports, General History,* p. 84; Wilbur and Ely, *The Hoover Dam Power and Water Contracts and Related Data,* p. 440; Kleinsorge, *The Boulder Canyon Project,* pp. 219–20.

3. Text of Mead address, "Hoover Dam, the World's Largest Irrigation Structure," at M.I.T., January 9, 1931, reprinted in *NRE* 22 (February, 1931): 22.

4. Aikman, "A Wild West Town that is Born Tame," p. 6

5. Wilbur and Mead, *The Construction of Hoover Dam,* Appendix, "Specifications No. 519," pp. 23–24; "Government Plans Model Town at Boulder City, Nev.," *NRE* 22 (February, 1931): 28.

6. Aikman, "A Wild West Town that is Born Tame," p. 6

7. John F. Cahlen, "Reminiscences," 104–105; "Reclamation Organization Activities and Project Visitors," *NRE* 20 (August, 1929): 22.

8. APH, Vol. 1, p. 49; Wilbur and Ely, *Hoover Dam Power and Water Contracts and Related Data,* pp. 449–53; Bureau of Reclamation Office of the Solicitor, District Counsel, Los Angeles, Files, Box 32; Wilbur, Jr., "Boulder City," pp. 4–19; Sutton, "Geographical Aspects," pp. 320–21.

9. SCCR, "Minutes of Special Meeting of Board of Directors, March 20, 1931"; Vivian, "Construction of the Hoover Dam: How the Contractors Handled," pp. 3745–46.

10. S. R. DeBoer, "Boulder City—The Proposed Model Town Near the Hoover Dam," *The American City* 44 (February, 1931): 146–49; Kleinsorge, *The Boulder Canyon Project,* p. 220; Sutton, "Geographical Aspects," pp. 323–26; McBride, *In the Beginning,* pp. 15–16.

11. Cahlen, "Reminiscences," pp. 111–12; Wilbur, Jr., "Boulder City," pp. 62–63; Sutton, "Geographical Aspects," pp. 325–26; McBride, *In the Beginning,* p. 16; Architect DeBoer was unhappy about the changes made in his city plan and about the way Boulder City ultimately grew. The rectangular residential blocks sprawling to the south and east, the monotonous architectural style of many of the

buildings, and the rise of an ugly commercial strip running west along Boulder Highway particularly displeased him. S. R. DeBoer, "Plan of Boulder City," *Architectural Record* 73 (1933): 154–58.

12. *Las Vegas Age,* April 14, 1931.

13. White, "Building the Big Dam," p. 116.

14. "Boulder City is Modeled Like Army Barracks," *Industrial Worker,* October 24, 1931.

15. "Boulder Canyon Project Notes," *NRE* 22 (July, 1931): 149; "The Building of Boulder City," *RE* 26 (May, 1936): 110–11; Vivian, "The Construction of Hoover Dam: How the Contractors Handled," p. 3745.

16. *Las Vegas Age,* July 14, 1931. "Some complaint has been made regarding [the food at] River Camp but we must concede that Anderson [Brothers] has worked under very trying conditions at that place," wrote Boulder City Company Manager V. G. Evans. "Several of their employees have died or had to leave during the past 30 days and this has naturally lowered the morale of the other employees to some extent." From SCCR, "Memo from V. G. Evans, Manager of Boulder City Co. to Executive Committee, July 17, 1931," Vol. 1, p. 261. See also "Boulder Dam Mecca Proves to be Hell," *Industrial Worker,* June 27, 1931.

17. Vivian, "The Construction of Hoover Dam: How the Contractors Handled," pp. 3746–47; Pettitt, *So Boulder Dam Was Built,* pp. 39–40; "Boulder City Building Activities," *NRE* 22 (November, 1931): 234; "The Building of Boulder City," *RE* 26 (May, 1936): 110–11; McBride, *In the Beginning,* pp. 38–39.

18. Vivian, "The Construction of Hoover Dam: How the Contractors Handled," pp. 3746–47; Pettitt, *So Boulder Dam Was Built,* pp. 37–38; "Boulder City Building Activities," *NRE* 22 (November, 1931): 234; Gene W. Segerblom, "Bugler Boy of Boulder," *Las Vegas Nevadan,* February 22, 1981. "We work[ed] it this way," recalled worker Buck Blaine. "I bring the sandwiches, another fella brings the fruit, another the cake. If we have anything left we give it to the guys who don't have anything to eat." *Las Vegas Nevadan,* August 26, 1973.

19. "Thanksgiving Day in Boulder City," *RE* 23 (January, 1932): 12.

20. Allen, *Hoover Dam,* p. 114. Frank Crowe had a sharp, dry wit. When humorist Will Rogers visited the dam site, he asked the superintendent how many men he had working. "Oh, about half of them," replied Crowe (Allen, p. 29). On another occasion, Crowe was showing Interior Secretary Ray Wilbur, who was reputed to be something of a prude, the diversion tunnels. How smooth did the concrete tunnel linings have to be to conform to specifications? the Secretary asked. Replied the superintendent: "About as smooth as a schoolmarm's leg, Mr. Wilbur, and if I remember my geography, that's pretty smooth" ("The Dam," *Fortune,* p. 86).

21. "The Building of Boulder City," *RE* 26 (May, 1936): 110–11; "Boulder City Building Activities," *NRE* 22 (November, 1931): 234.

It was Henry Kaiser who proposed that the guesthouse be built. He believed that high government officials, business leaders, and other VIPs would want to visit Boulder City and the dam site while the construction was going on. He thought it was important, in view of the government's control of the project's purse strings, and in the interest of winning future contracts, to receive the sightseeing politicians and their friends in style. The other directors resisted the proposal; a luxurious mansion at a construction site seemed an expensive frill to them. But Kaiser insisted and the house was built and staffed with several servants and an experienced chef. It was used extensively, both by visiting dignitaries and by the directors themselves, and it did much, as Kaiser knew it would, to enhance Six Companies' reputation as a first-class business organization. See Sidney Hyman, *Marriner S. Eccles*, p. 76.

"Frank Crowe often used to say that his most serious problem in the building of Hoover [Dam] was the guest house," recalled Edgar Kaiser in 1968. "You know, Las Vegas wasn't what it is today. It was just a few little buildings, and there was no place to stay. So, as Frank used to say, we [the Six Companies directors] spent an hour or two going over the serious problems on the job and the rest of the day discussing first whether to build the guest house, then who was going to stay in it, and finally whether you would have to pay to stay in it or whether you wouldn't." From Edgar Kaiser, "America's Builders," speech given at America's Builders Recognition Night, Pepperdine University, 1968, in the Henry J. Kaiser Collection, Bancroft Library, University of California, Berkeley.

22. "Boulder City Building Activities," *NRE* 22 (November, 1931): 234; Mrs. D. L. Carmody, "Boulder City—From a Woman's Viewpoint," *RE* 23 (March, 1932): 66–68; "The Building of Boulder City," *RE* 26 (May, 1936): 110–11; C. H. Vivian, "The Construction of Hoover Dam: Within a Year's Time the Government has Reared a Modern City in the Desert at a Cost of $1,600,000," *Compressed Air Magazine* 37 (April, 1932): 3774–79; McBride, *In the Beginning*, pp. 81–86; APH, Vol. 1, pp. 227–29.

23. According to the Bureau of Reclamation, 658 frame cottages were constructed at a total cost of $477,331, or approximately $725 per cottage. See "The Building of Boulder City," *RE* 26 (May, 1936): 110–11. This dollar figure, reported to the bureau by Six Companies, seems far too high, however. Carpenters told Marion Allen the cottages cost $140 each to build. See Allen, *Hoover Dam*, p. 9. *Industrial Worker* reported that the cost of each cottage did not exceed $250. See "Boulder Dam Workers Forced to Move into Cheap Built Shacks," *Industrial Worker*, March 29, 1932. A two-room cottage rented from January, 1932, to January, 1936, at $19 per month netted Six Companies $912. In February, 1936, Six Companies was paid $25 for each of the two-room cottages by the Wattis-Decker Company, which tore them down and sold the salvaged building materials. See Green, "Life in Nelson Township," p. 96. Thus, depending on which figure for the original construction cost of the cottages is accurate, Six Companies' profit was approximately $207 (29 percent), $682 (273 percent), or $897 (641 percent) for each two-room cottage.

24. McBride, *In the Beginning*, p. 82; "The Big Boss and the Strong Little People," *Las Vegas Nevadan*, February 15, 1981.

25. Allen, *Hoover Dam*, p. 9.

26. "Boulder City Water Supply," *RE* 23 (November, 1932): 182–83; Kleinsorge, *The Boulder Canyon Project*, pp. 221–22; Vivian, "Construction of the Hoover Dam: Within a Year's Time," pp. 3776–78.

27. Vivian, "Construction of the Hoover Dam: Within a Year's Time," p. 3774.

28. APH, Vol. 2, pp. 168–72; "Landscaping of Boulder City," *RE* 23 (April, 1932): 70–71; McBride, *In the Beginning*, pp. 22–25.

29. From Godbey, "Pioneering in Boulder City," pp. 54–60.

30. McBride, *In the Beginning*, p. 83.

31. Godbey, "Pioneering in Boulder City," pp. 66–70.

32. Pettitt, *So Boulder Dam Was Built,* p. 43.

33. Madeline Knighten, Oral History Interview, January, 1981, Boulder City Public Library.

34. Mary Ann Merrill, "Boulder Dam," transcript of a July 16, 1985, interview conducted by R. T. King for the Oral History Project of the University of Nevada–Reno Library, pp. 18–19; McBride, *In the Beginning,* pp. 23, 77; Allen, *Hoover Dam,* p. 103.

35. *Las Vegas Evening Review-Journal,* March 11, 1932.

36. Allen, *Hoover Dam,* p. 15.

37. "The Building of Boulder City," *RE* 26 (May, 1936): 110–11.

38. *Las Vegas Evening Review-Journal* March 11, 1932.

39. "Church Ahead of Satan in Boulder City," *Literary Digest* 114 (December 31, 1932): 18; Godbey, "Pioneering in Boulder City," pp. 70–71; Merrill, "Boulder Dam," pp. 20–21; "The Development of Boulder City as a Social Unit," *RE* 23 (December, 1932): 200–201; "Reverend Stevenson Arrives," "Stevenson Profiled," "First Service," "Weekly Service," "Stevenson Biography," "Stevenson Killed," *Las Vegas Evening Review-Journal,* October 2, October 6, October 17, 1931, April 4, 1932, February 25, 1933, December 27, 1937; McBride, *In the Beginning,* pp. 63–68.

40. "Boulder City Organizes American Legion Post," *NRE* 22 (November, 1931): 239; "Boulder City Legion Post, 'Baby' of State, Headed for Highest Enrollment," *Las Vegas Evening Review-Journal,* March 11, 1932.

41. Pettitt, *So Boulder Dam Was Built,* p. 36.

42. George Williams Lang, "A Study of the Boulder City, Nevada Schools"; Green, "Nelson Township," p. 103; Wilbur, Jr., "Boulder City," pp. 224–30; "Educational Facilities in Boulder City," *RE* 23 (November, 1932): 188–90; Godbey, "Pioneering in Boulder City," pp. 72–75; McBride, *In the Beginning,* pp. 57–62.

43. *Las Vegas Evening Review-Journal,* January 19, 1933; "Boulder Canyon Project Notes," *RE* 23 (March, 1933): 37; McBride, *In the Beginning,* p. 55.

44. White, "Building the Big Dam," p. 117.

45. Author interview with Elton Garrett, Boulder City, Nevada, July 26, 1984; Elton Garrett March 1, 1981 Oral History Interview by Allen Gurwitz, University of Nevada–Las Vegas Special Collections; "Boulder Canyon Project Notes," *RE* 22 (March, 1932): 56; "Aging Ex-scribe Matured with Planned City," *Las Vegas Nevadan,* February 15, 1981; McBride, *In the Beginning,* pp. 69–70.

46. Wilbur and Mead, *The Construction of Hoover Dam,* p. 28; Wilbur, Jr., "Boulder City," pp. 4–19, 116–38; McBride, *In the Beginning,* pp. 25–28; Kleinsorge, *The Boulder Canyon Project,* p. 21.

47. "Sims Ely, Pioneer in Arizona, Dies at 92," *New York Times,* November 12, 1954; "Sims Ely," *Who Was Who in America;* "Sims Ely Appointed City Manager Boulder City," *NRE* 22 (November, 1931): 246.

48. White, "Building the Big Dam," p. 117.

49. "Business Permits at Boulder City," *NRE* 22 (June, 1931): 118–20; "Filing System for Leases and Concessions at Boulder City," *NRE* 22 (July, 1931): 142; Wilbur, Jr., "Boulder City," pp. 116–38.

50. "Business Permits at Boulder City," *NRE* 22 (June, 1931): 118–20.

51. John Meursinge, "Engineering in the Wilderness: May 1931–October 1931," pp. 2–3 in "Memories: Boulder City, Nevada, and the Workers Who Built the Hoover Dam, May, 1931–October, 1935."

52. "Business Permits at Boulder City," *NRE* 22 (June, 1931): 118–20;

"Permits and Leases at Boulder City," *RE* 23 (August, 1932): 137–39; Wilbur, Jr., "Boulder City," pp. 116–38.

53. *New York Times,* May 18, 1931. See also "Free Competition in Business? Not by a Dam Site Says U.S.," *Business Week* (May 27, 1931): 14.

54. "Free Competition in Business?" p. 14.

55. Wilbur, Jr., "Boulder City," p. 133; McBride, *In the Beginning,* p. 32.

56. "Aging Ex-scribe Matured with Planned City," *Las Vegas Nevadan,* February 15, 1981.

57. "Permits and Leases at Boulder City," *RE* 23 (August, 1932): 137–39; Wilbur, Jr., "Boulder City," pp. 136–37.

58. SCCR, "Memo from V. G. Evans, Manager of Boulder City Company to Executive Committee, July 17, 1931," Vol. 1, pp. 257–61; McBride, *In the Beginning,* p. 32; "Lots of Six Companies workers wanted to go to Vegas to gamble, but all they had was scrip, and you couldn't gamble with that," remembered Boulder City pioneer Madeline Knighten. "So they'd sell it for cash for less than face value. I bought it all the time. That way, I could shop in the company store for things I couldn't get elsewhere, and the company workers could go off to Vegas and gamble." Madeline Knighten, Oral History Interview, January, 1981, Boulder City Public Library. For a more detailed discussion of the scrip controversy, see chapter 5.

59. *Las Vegas Age,* March 24, 1931.

60. McBride, *In the Beginning,* p. 11.

61. *New York Times,* May 19, 1931.

62. "Aging Ex-scribe Matured with Planned City," *Las Vegas Nevadan,* February 15, 1981.

63. Sutton, "Geographical Aspects," p. 337.

64. Wilbur, Jr., "Boulder City," pp. 246–47.

65. "Drinking Curbed in Boulder City," *RE* 24 (February, 1933): 28.

66. Wilbur, Jr., "Boulder City," pp. 157–59, 249; McBride, *In the Beginning,* p. 27.

67. Wilbur, Jr., "Boulder City," p. 251. Ely did not always get his way. Clarence Newlin, owner of the Green Hut restaurant in downtown Boulder City, hired a black cook. The city manager ordered him to fire the cook because of complaints from Boulder City residents. Newlin reportedly told Ely: "If you have $80,000, my Green Hut is yours tomorrow. But as long as I own the Green Hut, you're not telling me who to hire." Ely backed down, and the black cook stayed in Boulder City. See Erma Godbey, Leo Dunbar, Marion Allen, Carl Merrill, and Mary Ann Merrill, "Recollections of the Construction of Hoover Dam and Life in Boulder City, 1931–1935," transcript of September 17, 1985, panel discussion in Boulder City, moderated by T. R. King and Guy L. Rocha for the Oral History Project of the University of Nevada–Reno Library, p. 33.

68. *Las Vegas Evening Review-Journal,* August 24, 1931.

69. "Bud Bodell, Little Mussolini," *Industrial Worker,* December 8, 1931; *San Diego Union,* January 17, 1932.

70. See "Shootout," "Dam Strike Fracas," "Hunts Dillinger," "Crash," "Brawl," *Las Vegas Evening Review-Journal,* November 26, August 16, 1932, April 9, November 9, 1934, April 22, 1935.

71. "Boulder Canyon Project Notes," *RE* 23 (October, 1932): 174; See also Wilbur, Jr., "Boulder City," p. 152. *Industrial Worker* charged that the reservation rangers, acting as a self-appointed "purity squad," routinely peeped through the windows of Six Companies residences at night and illegally entered and searched

the dormitory rooms of suspected union sympathizers while they were away at work. "Anyone who is overheard protesting . . . is subject to arrest and deportation. The gunmen have on some occasions gone into the tunnel in the midst of a shift and taken a victim out for a kangaroo trial and final deportation." From "Iron Heel Used to Stifle All 'Squawks,'" *Industrial Worker*, January 26, 1932.

72. *San Diego Union*, January 17, 1932.

73. *Industrial Worker* reported that the recreation-hall gambling racket was finally shut down in late 1932 after the wife of a worker, who had lost his entire paycheck playing poker, wrote a letter of protest to authorities in Washington. *Industrial Worker*, January 17, 1933.

74. Mrs. D. L. Carmody, "Under the Eagle's Wing: Some Phases of Life in Boulder City," *RE* 23 (July, 1932): 129–30.

75. John Meursinge, letter to the author, March 14, 1985.

76. "Boulder Canyon Project Notes," *RE* 24 (February, 1933): 23. "In 1935 . . . I returned to Los Angeles [from Boulder City]. If I tried to say a few words about the [bad] conditions in Boulder City, the answer always was: 'John you were lucky. You had enough money to support your family.' I learned to keep my mouth shut." John Meursinge, letter to the author, March 14, 1985.

77. Of the sixteen heat-prostration deaths officially reported, all occurred in 1931, before Boulder city was finished. See APH, Vol. 1, Appendix, "Fatalities," pp. 342–43, and APH, Vol. 5, Appendix, "Summary of Fatalities," p. 325.

78. Aikman, "A Wild West Town That is Born Tame," p. 15.

CHAPTER 5.   "INCESSANT, MONSTROUS ACTIVITY"

1. There was a distinct Rube Goldberg flavor to the lighting of the Black Canyon dam site: banks of powerful arc lights were salvaged from a condemned minor-league baseball park in San Francisco and set up at strategic points on the cliffs and the river bottom. See "The Dam," *Fortune* 8 (March, 1933): 76. "These lamps are subjected to a variety of hard experiences and to keep them supplied with reflectors which are ordinarily made use of for this purpose would be a very expensive proposition by reason of the frequent breakage," reported *Reclamation Era*. "A bright new dishpan was experimented with for this purpose and the results were found to be so satisfactory that they were adopted all over the site. Ten thousand dishpans were ordered for this purpose." See "Boulder Canyon Project Notes," *RE* 24 (April, 1933): 47.

2. Diversion of the river exposed 2,624 feet of riverbed. The excavation for the dam foundation spanned approximately 650 feet. See Wilbur and Mead, *The Construction of Hoover Dam*, Appendix, "Specifications No. 519," Drawings No. 23 and No. 24.

3. "Construction of Diversion Cofferdams," *RE* 24 (April, 1933): 45–46; Pettitt, *So Boulder Dam Was Built*, p. 78.

4. APH, Vol. 2, p. 11.

5. "Construction of Diversion Cofferdams," *RE* 24 (April, 1933): 45–46; Wesley R. Nelson, "Construction of the Hoover Dam: A Résumé of Current Activities," p. 4072; APH, Vol. 2, pp. 223–27.

6. Allen, *Hoover Dam*, p. 85; A record minimum temperature for the Boulder Canyon Project—12 degrees—was recorded on February 12, 1933. See "Boulder Canyon Project Notes," *RE* 24 (April, 1933): 47.

7. Meursinge, "What Price Dam?" p. 51.

8. APH, Vol. 1, Appendix, "Fatalities," p. 344; "Death Toll at Dam Paid Again," *Industrial Worker*, October 3, 1931.

9. SCCR, "Minutes of Regular Meeting of Board of Directors," August 15, 1931, p. 370.

10. Author interview with S. L. ("Red") Wixon, Boulder City, Nevada, July 27, 1984; "The Big Boss and the Strong Little People," *Las Vegas Nevadan*, February 15, 1981.

11. Hopkins, "The Legend Builders," p. 61.

12. Meursinge, "What Price Dam?" pp. 50–51.

13. Nelson, "Construction of the Hoover Dam: A Résumé of Current Activities," p. 4073; Kleinsorge, *The Boulder Canyon Project*, p. 204.

14. A comparison of Six Companies bid figures with the government engineers' cost estimates for the project reveal the extent to which the winning bid was unbalanced. For the major construction items that were part of the diversion phase—stripping the canyon walls, excavating the construction highway and railroad rights-of-way, drilling and lining the diversion tunnels, pouring concrete for inlet and outlet structures, installing bulkhead and stoney gates, and building the cofferdams—the Bureau of Reclamation had estimated a cost of approximately $16.7 million. The Six Companies bid for the same items was approximately $21.7 million, a difference of $5 million, exactly the amount the partners had invested to secure the required performance bond. Six Companies made up the difference by underbidding on many of the major construction items for the latter part of the project, most notably the placing of concrete in the dam itself, which was the single biggest item. For this item Six Companies bid $2.70 per cubic yard; the government's estimate was $3.25. See APH, Vol. 1, Appendix, "Summary of Bids," p. 258.

15. SCCR, "Minutes of Regular Meeting of the Board of Directors, August 15, 1932," p. 372. Apparently not satisfied with these profits, the directors, at the same meeting, "resolved that a committee . . . study all the rates of wages at Hoover Dam with the view of reducing them if possible." See also "Minutes of Regular Meeting of the Board of Directors, February 13, 1933," p. 402.

16. "The Earth Movers I," *Fortune*, p. 212. In the same article, Crowe related the following anecdote about Felix Kahn's penchant for gambling: "One day I was driving down Montgomery Street in San Francisco. Kahn spotted 'Dad' Bechtel headed for the bank. 'Drive over,' he said; 'this is going to cost "Dad" some money!' I pulled alongside the curb and Felix shouted, ' "Dad," I'm matching you a double eagle.'

" 'Dad' didn't even say good morning. He just gave Felix a quick disapproving look, dug into his vest pocket for a coin, and slapped it on the car window. He took his hand off and said to Felix 'you lose' and walked off without another word."

17. Ibid.

18. Allen, *Hoover Dam*, p. 39.

19. "Contractors are all alike," Henry Kaiser once said. "They start out broke with a wheelbarrow and a piece of hose. Then, suddenly, they find themselves in the money. Everything's fine. Ten years later they are back where they started from—with one wheelbarrow, a piece of hose, and broke." Avoiding this fate was simple, according to Kaiser. "Before you work yourself out of the last job, line up a bigger one to pull yourself out." The members of the Six Companies consortium heeded this advice and moved to parlay their Boulder Canyon Project success into new jobs. Pacific Bridge and the J. F. Shea Company secured a $3.6 million contract in 1932 to build the piers for the Golden Gate Bridge. Contracts,

totaling $11.5 million, were awarded to two separate groups of Six Companies partners in 1933 to build segments of the San Francisco–Oakland Bay Bridge. In 1934, consortium members, led by Henry Kaiser, submitted the winning bid to build Bonneville Dam on the Columbia River. Six Companies was also awarded the contract to build Parker Dam on the Lower Colorado, following completion of Hoover Dam. See "The Earth Movers II," *Fortune* 28 (September, 1943): 119–22, 219–26.

20. *New York Times,* April 14, 1932; Kleinsorge, *The Boulder Canyon Project,* p. 212; "Boulder Canyon Project Notes," *RE* 23 (June, 1932): 112.

21. *New York Times,* April 14, 1932.

22. "A Paying Investment," *RE* 23 (June, 1932): 113.

23. *New York Times,* May 28, 1932; Kleinsorge, *The Boulder Canyon Project,* p. 212.

24. Herbert Hoover, *The Memoirs of Herbert Hoover, 1920–1933, The Cabinet and the Presidency,* pp. 228–29.

25. The government also aided Six Companies during the legal fight over applicability of Nevada mining safety laws on the project reservation. The filing of amicus curiae briefs in support of the Six Companies position by Bureau of Reclamation lawyers helped sway the district court's decision in favor of the contractors. See Bureau of Reclamation Office of the Solicitor, General Counsel, Los Angeles, Files, Box 24; *Six Companies, Inc. v. Stinson,* 649–53.

26. *Las Vegas Evening Review-Journal,* October 20, 1932.

27. *Las Vegas Evening Review-Journal,* November 1, 1932; John Meursinge, "Life in Boulder City, Nevada: 1932," p. 10, from "Memories: Boulder City, Nevada, and the Workers Who Built the Hoover Dam."

28. Meursinge, "What Price Dam?" p. 90.

29. *Las Vegas Evening Review-Journal,* November 9, 1932; Nevada Secretary of State, *Official Returns of the Election of November, 1932.*

30. Meursinge, "What Price Dam?" p. 93.

31. Ickes to Robert P. Crane, May 20, 1933; Harold Ickes, *The Secret Diary of Harold L. Ickes: The First Thousand Days, 1933–1936,* p. 37.

32. For example, in the 1932 film *I Loved You Wednesday,* the story of an engineer working on the dam, the structure was referred to as Boulder Dam. See Moeller, *Phil Swing and Boulder Dam,* p. 172. Best-selling author Zane Grey spent much of 1932 researching and writing a novel he titled *Boulder Dam* (the novel was published posthumously in 1963). See Frank Gruber, *Zane Grey,* pp. 231–32; Candace Kant, *Zane Grey's Arizona,* p. 80.

33. Herbert Hoover publicly maintained that Ickes' changing of the dam's name did not matter to him, but privately he viewed it with some bitterness, seeing it as part of a calculated campaign by the Roosevelt administration to smear him. On April 30, 1947, the name *Hoover Dam* was permanently restored by a joint resolution of Congress. Hoover wrote to one of the resolution's sponsors: "Confidentially, having had streets, parks, school houses, hills, and valleys named for me, as is done to all Presidents, I have not thought this item of great importance in the life of a nation. But when a President of the United States tears one's name down, that is a public defamation and an insult. Therefore, I am gratified to you for removing it." See Herbert Hoover, *The Memoirs of Herbert Hoover, 1929–1941, The Great Depression,* pp. 455–56. Public reaction to the controversy was summed up by an anonymous citizen who proposed, in a letter to the *Las Vegas Review-Journal,* that the name be changed a third time to "Hoogivza Dam." *Las Vegas*

Review-Journal, May 10, 1947. See also McBride, "The Battle for Boulder," p. 78.

34. *Las Vegas Evening Review-Journal,* May 23, August 19, August 27, December 8, 1932. See also Loren B. Chan, *Sagebrush Statesman: Tasker L. Oddie of Nevada,* pp. 149–50, 161–63.

35. *Las Vegas Evening Review-Journal,* August 19, 1932.

36. *Las Vegas Evening Review-Journal,* September 8, 1932.

37. *Las Vegas Evening Review-Journal,* December 5, 1932.

38. *Las Vegas Evening Review-Journal,* May 8, 1933; *New York Times,* May 10, 1933; *Las Vegas Evening Review-Journal,* May 22, 1933.

39. *Las Vegas Evening Review-Journal,* November 28, 1930.

40. Roosevelt Fitzgerald, "Blacks and the Boulder Dam Project," *Nevada Historical Society Quarterly* 24 (Fall, 1981): 256.

41. *Las Vegas Age,* January 7, 1932.

42. *Las Vegas Evening Review-Journal,* May 26, 1932.

43. *Las Vegas Evening Review-Journal,* January 1, 1931.

44. *Las Vegas Age,* June 18, 1932.

45. *Las Vegas Evening Review-Journal,* July 7, 1932.

46. *Las Vegas Age,* July 8, 1932.

47. *Las Vegas Evening Review-Journal,* September 1, 1932.

48. Mark W. Kruman, "Quotas for Blacks: The Public Works Administration and the Black Construction Worker," *Labor History* 16 (Winter, 1975): 40; Raymond Wolters, *Negroes and the Great Depression: The Problem of Economic Recovery,* pp. 199–200.

49. Walters, *Negroes and the Great Depression,* pp. 199–200; Graham White and John Maze, *Harold Ickes of the New Deal: His Private Life and Public Career,* pp. 117–18.

50. *Six Companies, Inc. v. DeVinney,* 693–99.

51. *Reno Gazette,* June 7, 1932.

52. Bureau of Reclamation Office of the Solicitor, General Counsel, Los Angeles, Files, Box 24.

53. *Reno Gazette,* June 7, 1932. See also Las Vegas Chamber of Commerce, *Taxation of Private Property Upon the Boulder Canyon Project Reservation.*

54. *Las Vegas Evening Review-Journal,* February 15, 1933.

55. APH, Vol. 3, pp. 113–14; *Las Vegas Evening Review-Journal,* August 7, 1933.

56. *New York Times,* August 29, 1933.

57. SCCR, "Minutes of Regular Meeting of Board of Directors, October 16, 1933," Vol. 2, p. 432.

58. Wilbur and Mead, *The Construction of Hoover Dam,* Appendix, "Specifications No. 519," pp. 55–57.

59. Norman S. Gallison, "Construction of the Hoover Dam: An Account of the Extensive Railroad System," pp. 3908–12.

60. "Concrete for Hoover Dam (Part I—Concrete Aggregates)," *RE* 23 (May, 1932): 94–96; Allen S. Park, "Construction of the Hoover Dam: A Description of the Methods of Obtaining and Preparing the Aggregates for the 4,400,000 Cubic Yards of Concrete to be Poured," *Compressed Air Magazine* 37 (October, 1932): 3937–42.

61. John Meursinge, "I Become a Track Shifter Operator: 1933," pp. 11–15, from "Memories: Boulder City, Nevada, and the Workers Who Built the Hoover Dam."

62. *Las Vegas Evening Review-Journal,* January 22, 1932.

63. APH, Vol. 3, pp. 164–69; Lawrence P. Sowles "Construction of the Hoover Dam: A Description of the Himix Concrete Plant and of the Cement Blending and Handling Equipment," *Compressed Air Magazine* 39 (September, 1934): 4533–37; "Concrete for Hoover Dam (Part II—Concrete Mixing)," *RE* 23 (July, 1932): 125–26.

64. APH, Vol. 3, pp. 164–69; Sowles, "Description of the Himix," pp. 4533–37; *RE* 23 (July, 1932): 125–26.

65. APH, Vol. 3, pp. 164–69; Sowles, "Description of the Himix," pp. 4533–37; *RE* 23 (July, 1932): 125–26.

66. "Report of the Colorado River Board," *RE* 24 (January, 1933): 4.

67. APH, Vol. 3, p. 204; *Las Vegas Evening Review-Journal,* December 30, 1932.

68. "Boulder Canyon Project Notes," *RE* 24 (March, 1933): 37.

69. Waters, *The Colorado,* pp. 347–48.

70. Letter to the author from Marion Allen, July 9, 1984.

71. Priestly, "Arizona Desert," pp. 365–67.

72. *Las Vegas Evening Review-Journal,* April 1, 1933.

73. *Las Vegas Evening Review-Journal,* July 3, 1933.

74. APH, Vol. 3, p. 205.

75. Ibid., p. 211; *Las Vegas Evening Review-Journal,* May 31, 1933; Lawrence P. Sowles, "The Construction of the Hoover Dam: Description of the Methods of Pouring the Concrete," *Compressed Air Magazine* 39 (April, 1934): 4394.

76. APH, Vol. 3, p. 211; *Las Vegas Evening Review-Journal,* June 6, 1933.

CHAPTER 6.   "A CALLOUS, CRUEL LUMP OF CONCRETE"

1. White, "Building the Big Dam," pp. 120–21.

2. Lawrence P. Sowles, "Construction of the Hoover Dam: How the Concrete Is Being Cooled as It Is Poured," *Compressed Air Magazine* 38 (November, 1933): 4265–71; APH, Vol. 3, pp. 216–18; Wilbur and Mead, *The Construction of Hoover Dam,* pp. 38–40; Kleinsorge, *The Boulder Canyon Project,* pp. 209–10.

3. "The Earth Movers I," *Fortune,* pp. 210, 212.

4. APH, Vol. 1, Appendix, "Summary of Bids," p. 258.

5. Sowles, "Description of the Methods of Pouring the Concrete," p. 4389.

6. *Las Vegas Evening Review-Journal,* January 4, 1935; Pettitt, *So Boulder Dam Was Built,* pp. 96–98.

7. *Las Vegas Evening Review-Journal,* February 19, February 20, 1935.

8. Allen, *Hoover Dam,* p. 92.

9. *Las Vegas Evening Review-Journal,* January 3, 1933; "Boulder Dam Slaves Put on the Spot," *Industrial Worker,* January 10, 1933.

10. Mildred Adams, "Taming the Untamable at Boulder Dam," *New York Times Magazine,* February 24, 1935, pp. 5, 19.

11. *Las Vegas Evening Review-Journal,* December 15, 1931, February 18, 1932.

12. Wilbur and Mead, *The Construction of Hoover Dam,* Appendix, "Specifications No. 519," p. 65.

13. "The Dam," *Fortune,* p. 85.

14. "Big Chief Crowe Weeps Muddy Tears," *Industrial Worker,* December 15, 1931.

15. *Las Vegas Evening Review-Journal,* August 24, 1932.

16. *Las Vegas Evening Review-Journal,* February 28, 1933.

17. *Las Vegas Evening Review-Journal,* August 16, August 17, 1933.

18. Ibid.; Allen, *Hoover Dam,* pp. 177–79; author interview with S. L. ("Red") Wixon, Boulder City, Nevada, July 27, 1984; Rocha, "The IWW and the Boulder Canyon Project," p. 20.

19. Fred Thompson, *The I.W.W. Its First 50 Years,* 1905–1955, p. 159.

20. For example, see *Las Vegas Evening Review-Journal,* January 6, April 13, June 7, July 6, July 28, August 1, 1933. See also *E. F. Kraus v. Six Companies, Inc., Frank Bryant, and John Tacke,* Complaint filed in Eighth Judicial District Court of Nevada, No. 4499, April 13, 1933; *Jack F. Norman v. Six Companies, Inc., Woody Williams, and Tom Regan,* Complaint filed in Eighth Judicial District Court of Nevada, No. 5256, May 24, 1934 (these records are on file at the Clark County Courthouse).

21. "Iron Heel Is Used to Stifle All Squawks," *Industrial Worker,* January 26, 1932.

22. SCCR, "Minutes of Regular Meeting of the Board of Directors, December 16, 1932," p. 396.

23. *Las Vegas Evening Review-Journal,* September 20, 1933.

24. *Las Vegas Evening Review-Journal,* September 22, 1933.

25. *E. F. Kraus v. Six Companies, Inc.; Las Vegas Evening Review-Journal,* October 17, 1933.

26. *Las Vegas Evening Review-Journal,* October 18, October 19, October 20, 1933.

27. *Las Vegas Evening Review-Journal,* October 25, 1933.

28. *Las Vegas Evening Review-Journal,* October 30, 1933.

29. *Las Vegas Evening Review-Journal,* November 2, November 13, November 21, November 29, 1933.

30. *Las Vegas Evening Review-Journal,* December 4, December 5, December 6, December 7, 1933.

31. *Jack Norman v. Six Companies, Inc.; Las Vegas Evening Review-Journal,* February 18, March 18, 1935.

32. Affidavit of Harry Austin, March 26, 1935 (transcript at Clark County Courthouse).

33. Affidavit of Clifton Hildebrand, March 23, 1935 (transcript at Clark County Courthouse).

34. *Las Vegas Evening Review-Journal,* March 27, 1935.

35. Affidavit of John L. Russell, March 25, 1935 (transcript at Clark County Courthouse).

36. *Las Vegas Evening Review-Journal,* May 22, 1935.

37. *Las Vegas Evening Review-Journal,* January 16, 1935.

38. "Boulder Dam Spillways," *RE* 25 (February, 1935): 31–32; Allen, *Hoover Dam,* pp. 87–88.

39. *Las Vegas Evening Review-Journal,* September 19, 1933; Pettitt, *So Boulder Dam Was Built,* pp. 101–102; Allen, *Hoover Dam,* pp. 88–89.

40. Wilbur and Mead, *The Construction of Hoover Dam,* pp. 67–68.

41. "Boulder Canyon Project Notes," *RE* 24 (January, 1933): 12; "Steel Plate Shipments Begin," *RE* 24 (May, 1933): 61; "Progress of Construction at Boulder Dam," *RE* 26 (January, 1935): 11.

42. Aikman, "Amid Turmoil Boulder Dam Rises," pp. 8–9, 13.

43. Pettitt, *So Boulder Dam Was Built,* pp. 89–90. The record one-day pour was 10,402 cubic yards on March 21, 1934. On that occasion Charlie Shea said of the workers: "They're a great bunch of boys, this crew of ours. I'll stack 'em up against the best anywhere and back 'em to the limit. They have to be good to pour concrete like that." *Las Vegas Evening Review-Journal,* March 21, 1934.

44. In reality it was impossible for a worker to be buried in the concrete; the pours never exceeded five feet in depth, and even if a bucket emptied directly on a puddler's head, there were at least half a dozen other puddlers in the same form to dig him out. This did not stop the men from having fun with the buried-alive myth, however. Concrete inspector Larry Morand remembered a practical joke played by fellow inspector Bob Skinner: "One day he [Skinner] cut the soles off a pair of rubber boots [and] put the soles against the inside of a concrete form for the lining of the diversion tunnel. After they filled the form and it dried, and the guys stripped the form off, they saw that pair of shoes stickin' out of the concrete, toes down. They were worried. An [assistant superintendent] ordered a guy with a chipping tool to chip away the concrete from around those soles—and Skinner got a good lacing down!" See Richard Meyer, "A Lot of Water Has Gone Over the Dam," *Los Angeles Times,* June 7, 1983, p. 3.

45. Ibid., pp. 92–94; *Las Vegas Evening Review-Journal,* November 9, 1933; APH, Vol. 3, Appendix, "Fatalities," p. 310.

46. Bruce Bliven, "The American Dnieperstroy," *The New Republic* 72 (December 11, 1935): 127.

47. White, "Building the Big Dam," pp. 118–19.

48. Bliven, "The American Dnieperstroy," p. 127.

49. Castle, "Well, I Quit," p. 207.

50. Ibid.

51. Allen, *Hoover Dam,* p. 26.

52. The *Las Vegas Evening Review-Journal,* reported on October 23, 1933, that there had been sixteen traffic fatalities on Boulder Highway since the start of dam construction in March, 1931.

53. *Las Vegas Evening Review-Journal,* July 20, 1933.

54. APH, Vol. 4, p. 132. See also litigation files for *U.S. v. Pansey,* Bureau of Reclamation Office of the Solicitor, General Counsel, Los Angeles, Files, Box 25.

55. White, "Building the Big Dam," p. 119.

56. Ibid., p. 120.

57. APH, Vol. 4, p. 14.

58. Aikman, "Amid Turmoil Boulder Dam Rises," p. 8.

59. "Boulder Canyon Project Transportation," *RE* 26 (May, 1935): 102.

60. McBride, *In the Beginning,* pp. 78–80; Boulder Dam Hotel Scrapbook.

61. *Las Vegas Evening Review-Journal,* December 5, 1934.

62. *Las Vegas Evening Review-Journal,* February 7, 1935.

63. APH, Vol. 5, p. 11; *Las Vegas Evening Review-Journal* February 1, 1935.

64. *Las Vegas Evening Review-Journal,* February 5, 1934; SCCR, "Minutes of the Regular Meeting of the Board of Directors," February 19, 1934.

65. *Las Vegas Evening Review-Journal,* February 26, 1935.

66. Wilbur and Mead, *The Construction of Hoover Dam,* Appendix, "Contract for Construction of Hoover Dam," p. 77.

67. See Linda Lear, "Boulder Dam: A Crossroads in Natural Resource Policy," p. 87.

68. *Las Vegas Evening Review-Journal,* March 6, 1935.

69. "The Earth Movers I," *Fortune,* p. 210; Mark Foster, "Giant of the West." pp. 4–5; author interview with S. D. Bechtel, Sr., San Francisco, December 11, 1984.

70. See Pettitt, *So Boulder Dam Was Built;* "The Question of Overtime on Boulder Dam," Pamphlet in Kaiser Collection, Bancroft Library.

71. Foster, "Giant of the West." p. 4.

72. SCCR, "Minutes of Regular Meeting of the Board of Directors, December 16, 1935; "The Earth Movers I," *Fortune,* p. 214; *Las Vegas Evening Review-Journal,* February 17, 1936.

73. *Las Vegas Evening Review-Journal,* March 30, June 21, 1934.

74. Ragnald Fyhen, "Labor Notes by Ragnald Fyhen, Secretary-Treasurer of the Central Labor Council of Clark County from 1933–1947, Las Vegas, Nevada, October 1968," p. 4.

75. Ibid., p. 6.

76. "Labor's Memorial to Its Dead at Boulder Dam," *RE* 25 (July, 1935): 143; *Las Vegas Evening Review-Journal,* May 30, 1935.

77. *Las Vegas Evening Review-Journal,* June 4, 1935; copy of the coroner's inquest on file at the Clark County clerk's office.

78. Allen, *Hoover Dam,* p. 169.

79. *Las Vegas Evening Review-Journal,* July 12, 1935.

80. Allen, *Hoover Dam,* p. 177.

81. *Las Vegas Evening Review-Journal,* July 13, 1935; Fyhen, "Labor Notes," p. 6.

82. *Las Vegas Evening Review-Journal,* July 15, July 17, 1935; Allen, *Hoover Dam,* p. 179.

83. *Las Vegas Evening Review-Journal,* July 15, 1935.

84. *Las Vegas Evening Review-Journal,* July 19, 1935.

85. Allen, *Hoover Dam,* pp. 180–81.

CHAPTER 7.   "TWENTIETH-CENTURY MARVEL"

1. *Las Vegas Evening Review-Journal,* September 30, 1935; *New York Times,* October 1, October 6, 1935.

2. David McCullough, *The Great Bridge,* pp. 533–36; Waters, *The Colorado,* p. 337.

3. *Las Vegas Evening Review-Journal,* September 30, 1935.

4. "Honorable Harold L. Ickes, Secretary of the Interior, Delivers Address at Dedication of Boulder Dam," *RE* 25 (November, 1935): 209–10.

5. *Las Vegas Evening Review-Journal,* September 20, 1935.

6. "President Roosevelt Dedicates Boulder Dam, September 30, 1935, Text of Dedicatory Address," *RE* 25 (October, 1935): 193–94, 196.

7. "The Earth Movers I," *Fortune,* p. 214.

8. APH, Vol. 1, Appendix, "Fatalities," p. 343; APH, Vol. 5, Appendix, "Fatalities," p. 291; *Las Vegas Evening Review-Journal,* December 20, 1935.

9. *Annual Report of the Secretary of the Interior for the Fiscal Year Ended June 30, 1935,* p. 53; "Six Companies' Boulder Dam Contract Completed," *RE* 26 (March, 1936): 69.

10. "The Earth Movers I," *Fortune,* p. 214.

11. *New York Times,* January 27, 1936.

12. "Six Companies Boulder Dam Contract Completed," *RE* 26 (March, 1936): 69; Kleinsorge, *The Boulder Canyon Project,* p. 213.

13. *Las Vegas Evening Review-Journal,* February 29, 1936.

14. "The Earth Movers I," *Fortune,* p. 214; James R. Kluger, "Elwood Mead: Irrigation Engineer and Social Planner," pp. 208f.; author interview with S. D. Bechtel, Sr., San Francisco, December 11, 1984.

15. "The Dam," *Fortune,* p. 85.

16. *Las Vegas Evening Review-Journal,* March 3, 1936.

17. Kleinsorge, *The Boulder Canyon Project,* pp. 242–44; Harper and Young, "Memoirs of Deceased Members: Francis Trenholme Crowe," p. 1401; Pacific Constructors, Inc., *Shasta Dam and Its Builders,* pp. 23–46, 67–107; "The Earth Movers II," *Fortune,* p. 121.

18. *Las Vegas Evening Review-Journal,* February 28, 1946.

19. "Walker R. Young," *Who's Who in America,* p. 2694.

20. Walker Young to Virginia ("Teddy") Fenton, May 18, 1972.

21. McBride, *In the Beginning,* p. 28; *New York Times,* November 12, 1954.

22. "The Earth Movers II," *Fortune,* p. 219; Gaskins, "Henry J. Kaiser," p. 66.

23. Gaskins, "Henry J. Kaiser," pp. 66–68; "The Earth Movers II," *Fortune,* pp. 222–25.

24. "The Earth Movers III," *Fortune* 28 (October, 1943): 139–44.

25. Gaskins, "Henry J. Kaiser," pp. 68–69; Foster, "Giant of the West," p. 5; "The Earth Movers III," *Fortune,* p. 193.

26. "The Earth Movers III," *Fortune,* p. 193.

27. *New York Times,* January 26, 1942; *San Francisco Chronicle,* June 6, 1958.

28. Gaskins, "Henry J. Kaiser,", pp. 70–71; Foster, "Giant of the West," pp. 1–23.

29. See M. Kolbenschlag, "Bechtel's Biggest Job—Constructing Its Own Future," *Forbes* 128 (December 7, 1981): 138–42; "G.E.'s Giant Deal," *Time* 106 (December 29, 1975); "Bigger May Not Be Better for Broken Hill," *Business Week,* March 19, 1984, p. 38.

30. Author interview with S. D. Bechtel, Sr., San Francisco, December 11, 1984.

31. "The Earth Movers I," *Fortune,* p. 214.

32. Kleinsorge, *The Boulder Canyon Project,* pp. 214–19; U.S. Department of the Interior, Bureau of Reclamation, *Hoover Dam: 50 Years,* pp. 36–40.

33. See Gerald D. Nash, *The American West in the Twentieth Century.*

34. Lear, "Boulder Dam," pp. 88–92.

35. *Los Angeles Times,* September 22, 1985.

36. Eugene P. Moehring, "Public Works and the New Deal in Las Vegas 1933–1940," *Nevada Historical Society Quarterly* 24 (Summer, 1981): 107–29.

37. Eugene P. Moehring, "Las Vegas and the Second World War," *Nevada Historical Society Quarterly* 29 (Spring 1986): 1–4.

38. *Hoover Dam: 50 Years,* p. 32.

39. U.S. Bureau of the Census, *Sixteenth Census of the United States: 1940. Population. Unincorporated Communities, United States by States. Total Population of Unincorporated Communities Having 500 or More Inhabitants for Which*

Separate Figures Could be Compiled (Washington: Government Printing Office, 1943), p. 15.

40. Green, "Nelson Township," p. 96.

41. Moehring, "Las Vegas and the Second World War," pp. 1–4; Godbey, "Pioneering in Boulder City," pp. 97–105.

42. *Hoover Dam: 50 Years*, p. 27; McBride, *In the Beginning*, pp. 28–29.

43. Angela Brooker and Dennis McBride, *Boulder City: Passages in Time*, p. 8.

44. *Hoover Dam: 50 Years*, pp. 36–40; Author interview with Julian Rhinehart, Regional Public Affairs Officer, Bureau of Reclamation Lower Colorado Region, Boulder City, July, 1986.

45. *Hoover Dam: 50 Years*, p. 30.

46. Author interview with Julian Rhinehart, Boulder City, July 1986.

47. May Sarton, *The Lion and the Rose*, p. 22.

Bibliography

Collections of Primary Source Materials

Bancroft Library, University of California, Berkeley, California. Henry J. Kaiser Collection. Six Companies Corporate Records, documents and papers by or concerning Henry Kaiser, scrapbooks of Hoover Dam project newspaper clippings, many Hoover Dam construction photographs.

Boulder City Library, Boulder City, Nevada. Hoover Dam and Boulder City Collection. Newspaper clippings, pamphlets, articles, oral history audiotapes, and photographs pertaining to the construction of Hoover Dam and the history of Boulder City.

Bureau of Reclamation Lower Colorado Region Library, Photo Laboratory, and Map Collection, Boulder City, Nevada. Documents, photographs, and maps pertaining to the construction of Hoover Dam and the development of the Colorado River Basin.

Federal Records Center and National Archives, Denver, Colorado. Annual Project Histories, Boulder Canyon Project. Hoover Dam construction photographs, Boulder Canyon Project administrative and engineering correspondence, personnel records.

Federal Records Center and National Archives, Laguna Niguel, California. Bureau of Reclamation Office of the Solicitor, District Counsel, Los Angeles, Files.

Documents pertaining to the establishment of the Boulder Canyon Project Federal Reservation, lease and rental contracts for residential and commercial property in Boulder City, and condemnation and right-of-way proceedings. Litigation files for *Six Companies, Inc. v. Stinson State, Inspector of Mines, et al, Six Companies, Inc. v. Devinney, County Assessor, U.S. v. Pansey.*

Fenton, Virginia and William Harbour, Boulder City, Nevada. Teddy and Bill Collection. Letters, memoranda, and photographs pertaining to the construction of Hoover Dam and the history of Boulder City.

James R. Dickenson Library, University of Nevada–Las Vegas, Las Vegas, Nevada. Special Collections. Leonard Blood Papers, including letters, memoranda, and records pertaining to administration of the U.S. Department of Labor Employment Office, which managed hiring of Hoover Dam construction workers for Six Companies. Ragnald Fyhen, "Labor Notes by Ragnald Fyhen, Secretary-Treasurer of the Central Labor Council of Clark County from 1933–1947, Las Vegas, Nevada, October, 1968." Books, articles, pamphlets, manuscripts, numerous oral history audiotapes, and photographs pertaining to the construction of Hoover Dam and early Las Vegas.

McBride, Dennis. Boulder City, Nevada. Boulder City Oral History Project Collection. Videotaped interviews with Hoover Dam construction workers, photographs, and film footage of Hoover Dam construction.

Unpublished Manuscripts

Andrews, Rena M. "A Federal City," 1975.

Director of Power, Boulder Canyon Project. "A Critique of Boulder City Administration," July 15, 1949.

Green, James Menard. "Life in Nelson Township, Eldorado Canyon, and Boulder City, Nevada 1860–1960." M. A. thesis, University of Nevada–Las Vegas, 1975.

Johnson, Arlin Rex. "Certain Economic Aspects of the Boulder Dam Project," Ph. D. thesis, George Washington University, 1931.

Klimek, Sandy. "A History of Hospitals: Clark County, Nevada," M. A. thesis, University of Nevada–Las Vegas, 1985.

Kluger, James R. "Elwood Mead: Irrigation Engineer and Social Planner." Ph. D. thesis, University of Arizona, 1970.

Lang, George Williams. "A Study of the Boulder City, Nevada Schools." M. A. thesis, Leland Stanford Junior University, 1933.

Metcalf, Barbara Ann. "Oliver M. Wozencraft in California, 1849–1887." M. A. thesis, University of Southern California, 1963.

Meursinge, John. "Memories: Boulder City, Nevada, and the Workers Who Built the Hoover Dam, May 1931–October 1935."

———. "What Price Dam?" Undated.

Wilbur, Ray Lyman, Jr. "Boulder City: A Survey of Its Legal Background, Its City Plan and Its Administration." M. A. thesis, Syracuse University, 1935.

Oral History Materials

Boulder City Public Library, Audiotape interviews:
Baker, A. F., April 4, 1975.
Baker, Nan, April 31, 1975.
Burt, Lucille, January, 1981.
Dill, David B., March 13, 1975.

Dunbar, Leo, February 24, 1974.
Eaton, Bruce and Mary, January 30, 1975.
Ferguson, Vera, April, 1975.
Francis, Richard C. ("Curley"), March 4, 1975.
French, Edna, April 31, 1975.
French, William, November 18, 1974.
Garrett, Elton, January, 1981.
Garrett, Perle, June 23, 1975.
Garrett, Theodore, February 12, 1975.
Godbey, Erma, January, 1981.
Holland, Mrs. Fred, April 6, 1975.
Holmes, Helen, September 17, 1974.
Holmes, Neil, February 12, 1975.
Kine, Mildred, April 7, 1975.
Knighten, Madeline, September 17, 1974; January, 1981.
Lawson, Rose, July 26, 1974.
McKenzie, Dan, February 12, 1975.
Members of '31ers Club, April, 1982.
Parker, Bob, January, 1981.
Stice, Sherl, January, 1981.
Voss, Nadean, March 3, 1975.
Voss, Wilfred T., March 2, 1975.
Whalen, Lillian, September 20, 1974.
Young, Lila, January, 1981.
Young, Walker, June 23, 1975.
Boulder City Oral History Project, Boulder City, Nevada, videotape interviews by
    Dennis McBride:
Allen, Marion, April 14, 1986.
Burt, Leroy, Joe Kine, and Tommy Nelson, July 3, 1986.
Dunbar, Leo, Undated.
Dunbar, Leo, June 16, 1986.
Eaton, Bruce, Undated.
Hall, Harry, June 20, 1986.
Merrill, Carl, June 24, 1986.
Merrill, Carl and Mary Ann, June 24, 1986.
Nunley, Altus E. ("Tex"), June 9, 1986.
Nunley, Altus E. ("Tex") and Dorothy, June 9, 1986.
Parker, Robert, June 2, 1986.
Reining, Henry, Jr., Undated.
James Dickenson Library, University of Nevada–Las Vegas, Special Collections,
    audiotape interview with Garrett, Elton M., March 1, 1981.

Interview Transcripts

Cahlen, John F. "Reminiscences of a Reno and Las Vegas, Nevada Newspaperman,
    University Regent, and Public-Spirited Citizen." Transcript of an April, 1968,
    interview conducted by Mary Ellen Glass for the Oral History Project of the
    University of Nevada–Reno Library.
Godby, Erma O. "Pioneering in Boulder City, Nevada." Transcript of a March,
    1966, interview conducted by Mary Ellen Glass for the Oral History Project of
    the University of Nevada–Reno Library.

Godbey, Erma, Leo Dunbar, Marion Allen, Carl Merrill, and Mary Ann Merrill. "Recollections of the Construction of Hoover Dam and Life in Boulder City, 1931–1935." Transcript of September 17, 1985, panel discussion in Boulder City, moderated by T. R. King and Guy L. Rocha for the Oral History Project of the University of Nevada–Reno Library.

Merrill, Mary Ann. "Boulder Dam." Transcript of a July 16, 1985, interview conducted by R. T. King for the Oral History Project of the University of Nevada–Reno Library.

Rockwell, Leon H. "Recollections of Life in Las Vegas, Nevada 1906–1908." Transcript of an April, 1968, interview conducted by Mary Ellen Glass for the Oral History Project of the University of Nevada–Reno Library.

Government Publications

Burkhardt, R. W., and E. R. Schultz. *Examination of Hoover Dam and Appurtenant Works and Hoover Power Plant, Boulder Canyon Project, Arizona-Nevada* (Appendix 1 of U.S. Bureau of Reclamation Condition of Major Structures and Facilities, Region 3). Denver: U.S. Bureau of Reclamation, 1957.

Hansen, Oskar J. W. *Sculptures at Hoover Dam.* Washington: Government Printing Office, 1942.

Hewett, Donnel Foster, et al. *Mineral Resources of the Region Around Boulder Dam.* Washington: Government Printing Office, 1936.

LaRue, E. C. *Colorado River and Its Utilization.* U.S. Geological Survey Water Supply Paper No. 395. Washington: Government Printing Office, 1916.

———. *Water Power and Flood Control of Colorado River Below Green River, Utah.* U.S. Geological Survey Water Supply Paper No. 556. Washington: Government Printing Office, 1925.

Las Vegas Chamber of Commerce. *Taxation of Private Property Upon the Boulder Canyon Project Reservation.* Las Vegas, Nev.: Chamber of Commerce, 1932.

Neilson, C. J. *Boulder Dam: Boulder Canyon Project Arizona-California-Nevada.* Washington: Government Printing Office, 1948.

Nevada Compiled Laws, 1931.

Nevada Secretary of State. *Official Returns of the Election of November, 1932.*

Powell, John Wesley. *Report on the Lands of the Arid Regions of the United States.* Washington: Government Printing Office, 1879.

Richardson, Joe T. *The Structural Behavior of Hoover Dam.* Denver: U.S. Bureau of Reclamation Foundations and Structural Behavior Section, Dams Branch, Commissioner's Office, 1957.

U.S. Congress. House. Lieutenant J. C. Ives, *Report Upon the Colorado River of the West,* 36th Cong., 1st sess., H. Exec. Doc. 90. Washington, 1861.

———. *Division and Apportionment, Waters of the Colorado.* 67th Cong., 1st sess., H. Rept. 191. Washington, 1921.

U.S. Congress. Senate. *Report of the Explorations and Surveys to Ascertain the Most Practical and Economical Route for a Railroad from the Mississippi River to the Pacific Ocean,* 33d Cong., 2d sess., S. Exec. Doc. 78. Washington, 1856.

———. *Problems of Imperial Valley and Vicinity with Respect to Irrigation from the Colorado together with Proceedings of the Conference on Construction of Boulder Canyon Dam,* 67th Cong., 2d sess., S. Exec. Doc. 142. Washington, 1922.

————. 72d Cong., 1st sess., Committee on Irrigation and Reclamation, *Hearings on S. 2885, A Bill Providing for the Application of State Laws within the Boulder Canyon Project Reservation.* Washington, 1932.

U.S. Department of the Interior. *Annual Reports, 1922–36.* Washington.

————. Bureau of Mines. P. W. Thrush, ed. *A Dictionary of Mining, Mineral, and Related Terms.* Washington: Government Printing Office, 1968.

————. Bureau of Reclamation. *Boulder Canyon Project Final Reports, Part I— Introductory, Bulletin 1, General History and Description of the Project.* Boulder City, 1948.

————. *Boulder Canyon Project Final Reports, Part I—Introductory, Bulletin 2, Hoover Dam Power and Water Contracts and Related Data.* Washington, 1950.

————. *Boulder Canyon Project Final Reports, Part III—Preparatory Examinations, Bulletin 1, Geological Investigations.* Denver, 1950.

————. *Boulder Canyon Project Final Reports, Part IV—Design and Construction, Bulletin 1, General Features.* Denver, 1941.

————. *Boulder Canyon Project Final Reports, Part IV—Design and Construction, Bulletin 2, Boulder Dam.* Denver, 1941.

————. *Boulder Canyon Project Final Reports, Part IV—Design and Construction, Bulletin 3, Diversion, Outlet, and Spillway Structures.* Denver, 1949.

————. *Boulder Canyon Project Final Reports, Part IV—Design and Construction, Bulletin 4, Concrete Manufacturing, Handling, and Control.* Denver, 1949.

————. *Boulder Canyon Project Final Reports, Part IV—Design and Construction, Bulletin 5, Penstocks and Outlet Pipes.* Denver, 1949.

————. *Boulder Canyon Project Final Reports, Part IV—Design and Construction, Bulletin 6, Imperial Dam and Desilting Works.* Denver, 1949.

————. *Boulder Canyon Project Final Reports, Part V—Technical Investigations, Bulletin 1, Trial Load Method of Analyzing Arch Dams.* Denver, 1938.

————. *Boulder Canyon Project Final Reports, Part V—Technical Investigations, Bulletin 2, Slab Analogy Experiments.* Denver, 1938.

————. *Boulder Canyon Project Final Reports, Part V—Technical Investigations, Bulletin 3, Model Tests of Boulder Dam.* Denver, 1939.

————. *Boulder Canyon Project Final Reports, Part V—Technical Investigations, Bulletin 4, Stress Studies for Boulder Dam.* Denver, 1939.

————. *Boulder Canyon Project Final Reports, Part V—Technical Investigations, Bulletin 5, Penstock Analysis and Stiffener Design.* Denver, 1940.

————. *Boulder Canyon Project Final Reports, Part V—Technical Investigations, Bulletin 6, Model Tests of Arch and Cantilever Elements.* Denver, 1940.

————. *Boulder Canyon Project Final Reports, Part VI—Hydraulic Investigations, Bulletin 1, Model Studies of Spillways.* Denver, 1938.

————. *Boulder Canyon Project Final Reports, Part VI—Hydraulic Investigations, Bulletin 2, Model Studies of Penstocks and Outlet Works.* Denver, 1938.

————. *Boulder Canyon Project Final Reports, Part VI—Hydraulic Investigations, Bulletin 3, Studies of Crests for Overfall Dams.* Denver, 1948.

————. *Boulder Canyon Project Final Reports, Part VI—Hydraulic Investigations, Bulletin 4, Model Studies of Imperial Dam, Desilting Works, All-American Canal Structures.* Denver, 1949.

————. *Boulder Canyon Project Final Reports, Part VII—Cement and Concrete Investigations, Bulletin 1, Thermal Properties of Concrete.* Denver, 1940.

———. *Boulder Canyon Project Final Reports, Part VII—Cement and Concrete Investigations, Bulletin 2, Investigations of Portland Cements.* Denver, 1949.
———. *Boulder Canyon Project Final Reports, Part VII—Cement and Concrete Investigations, Bulletin 3, Cooling of Concrete Dams.* Denver, 1949.
———. *Boulder Canyon Project Final Reports, Part VII—Cement and Concrete Investigations, Bulletin 4, Mass Concrete Investigations.* Denver, 1949.
———. *Boulder Canyon Project—Questions and Answers.* Washington: Government Printing Office, 1933.
———. *Dams and Control Works.* Washington: Government Printing Office, 1938.
———. *Hoover Dam: Fifty Years.* Washington: Government Printing Office, 1985.
———. *Mineral Resources and Possible Industrial Development in the Region Surrounding Boulder Dam.* Washington: Government Printing Office, 1934.
———. *President Franklin Delano Roosevelt Visits Boulder Dam.* Washington: Government Printing Office, 1935.
———. *The Story of Boulder Dam.* Washington: Government Printing Office, 1941.
———. *The Story of Hoover Dam.* Washington: Government Printing Office, 1976.
Wilbur, Ray Lyman, and Elwood Mead. *The Construction of Hoover Dam: Preliminary Investigations, Design of Dam, and Progress of Construction.* Washington: Government Printing Office, 1933.
Wilbur, Ray Lyman, and Northcutt Ely. *The Hoover Dam Power and Water Contracts and Related Data.* Washington: Government Printing Office, 1933.

Court Decisions

Six Companies, Inc. v. Stinson, State Inspector of Mines, et al., No. G-191, Dist. Ct., D. of Nev., April 18, 1932, 58 F.(2d), 649–53.
Six Companies, Inc. v. Stinson, State Inspector of Mines, et al., 2 F. Supp., 689–92, February 15, 1933.
Six Companies, Inc. v. DeVinney, County Assessor, No. G-195, Dist. Ct., D. of Nev., February 15, 1933, 2F. Supp., 693–99.

Court Records

E. F. Kraus v. Six Companies, Inc., Frank Bryant, and John Tacke. Complaint filed in Eighth Judicial District Court of Nevada, No. 4499, April 13, 1933.
Jack F. Norman v. Six Companies, Inc., Woody Williams, and Tom Regan. Complaint filed in Eighth Judicial District Court of Nevada, No. 5256, May 24, 1934.

Newspapers and Periodicals

Boulder City (Nev.) *News*
Chicago (Ill.) *Industrial Solidarity*
Compressed Air Magazine, 1931–1935
Denver (Colo.) *Post*
Henderson (Nev.) *Home News*
Las Vegas (Nev.) *Age*
Las Vegas (Nev.) *Evening Review-Journal*
Las Vegas (Nev.) *Review-Journal* (Sunday Nevadan)
Los Angeles (Calif.) *Examiner*

Los Angeles (Calif.) *Evening Herald*
Los Angeles (Calif.) *Express*
Los Angeles (Calif.) *Times*
New Reclamation Era, 1929–1931
New York (N.Y.) *Times*
Oakland (Calif.) *Tribune*
Reclamation Era, 1931–1936
Reno (Nev.) *Gazette*
Salt Lake City (Utah) *Tribune*
San Diego (Calif.) *Sun*
San Diego (Calif.) *Union*
San Francisco (Calif.) *Chronicle*
San Francisco (Calif.) *Examiner*
San Francisco (Calif.) *News*
Seattle (Wash.) *Industrial Worker*

Articles

Adams, Mildred. "Taming the Untamable at Boulder Dam," *New York Times Magazine*, February 24, 1935, pp. 5, 19.
Aikman, Duncan. "New Pioneers in Old West's Deserts," *New York Times Magazine*, October 26, 1930, pp. 7, 18.
———. "A Wild West Town That is Born Tame," *New York Times Magazine*, July 26, 1931, pp. 6–7, 15.
———. "Amid Turmoil Boulder Dam Rises," *New York Times Magazine*, July 23, 1933, pp. 8–9, 13.
Anderson, P. Y. "Boulder Dam Dynamite," *The Nation* 130 (February 12, 1930): 173–74.
"Another View of Boulder Dam," *The Nation* 130 (February 26, 1930): 130, 245.
"Big Business at Boulder Dam," *Literary Digest* 106 (September 6, 1930): 45–46.
Bird, F. L. "Who Will Benefit by Boulder Dam?" *The New Republic* 63 (July 30, 1930): 310–13.
Bissell, Charles A., and Frank Weymouth. "Memoirs of Deceased Members: Arthur Powell Davis, Past President, Am. Soc. C. E. Died August 7, 1933," *Transactions of the American Society of Civil Engineers* 100 (1935): 1582–91.
Bliven, Bruce. "The American Dnieperstroy," *The New Republic* 72 (December 11, 1935): 125–27.
Castle, Victor. "Well, I Quit My Job at the Dam," *The Nation* 133 (August 26, 1931): 207–208.
"Charles Pemberton 'Pop' Squires," *Nevada Historical Society Quarterly* 16 (Fall, 1973): 207–208.
"Church Ahead of Satan in Boulder City," *Literary Digest* 114 (December 31, 1931): 18.
"The Construction of Boulder Dam," *Literary Digest* 116 (November 3, 1933): 10–14.
"Costly Economy at the Hoover Dam," *Literary Digest* 113 (June 4, 1932): 19.
"The Dam," *Fortune* 8 (September, 1933): 74–88.
"Damn Big Dam," *Time Magazine* 21 (March 23, 1931): 14–15.
Davis, Arthur P. "Development of the Colorado River," *Atlantic Monthly* 143 (February, 1929): 254–63.

DeBoer, S. R. "Boulder City—The Proposed Model Town Near the Hoover Dam," *The American City* 44 (February, 1931): 146–49.

———. "Plan of Boulder City," *Architectural Record* 73 (1933): 154–58.

Didion, Joan. "A Piece of Work for Now and Doomsday," *Life* 63 (March 13, 1970): 20B.

Earl, Phillip I. "The Legalization of Gambling in Nevada, 1931," *Nevada Historical Society Quarterly* 24 (Spring, 1981): 39–46.

"The Earth Movers I," *Fortune* 28 (August, 1943): 99–107, 210–14.

"The Earth Movers II," *Fortune* 28 (September, 1943): 119–22, 219–26.

"The Earth Movers III," *Fortune* 28 (October, 1943): 139–44, 193–99.

Fitzgerald, Roosevelt. "Blacks and the Boulder Dam Project," *Nevada Historical Society Quarterly* 24 (Fall, 1981): 255–60.

Fleming, Russell C. "Construction of the Hoover Dam: A Description of the Tunnels for the Penstock Headers, the Penstocks, and the Canyon-Wall Outlets," *Compressed Air Magazine* 38 (July, 1933): 4171–76.

Foster, Mark S. "Giant of the West: Henry J. Kaiser and Regional Industrialization, 1930–1950," *Business History Review* 59 (Spring, 1985): 1–23.

"Free Competition in Business? Not by a Dam Site Says U.S.," *Business Week* (May 27, 1931): 14.

Gallison, Norman S. "Construction of the Hoover Dam: Details of the Driving of the Four Huge Tunnels which will Divert the Colorado River Around the Dam Site," *Compressed Air Magazine* 37 (May, 1932): 3804–10.

———. "Construction of the Hoover Dam: An Account of the Extensive Railroad System and of the Important Work It is Doing," *Compressed Air Magazine* 37 (September, 1932): 3908–12.

Gaskins, Susanne T., and Warren Beck. "Henry J. Kaiser—Entrepreneur of the American West," *Journal of the West* 25 (January, 1986): 64–72.

Gressley, Gene M. "Arthur Powell Davis, Reclamation, and the West," *Agricultural History* 42 (July, 1968): 241–57.

Harper, S. O., and Walker R. Young, "Memoirs of Deceased Members: Francis Trenholme Crowe, Hon. M. ASCE, Died February 26, 1946," *Transactions of the American Society of Civil Engineers* 113 (1948): 1397–1403.

"Henry J. Kaiser," *Fortune* 28 (October, 1943): 147–49, 249–61.

Hopkins, A. D. "The Legend Builders," *Nevada* 45 (May–June 1985): 63–64.

Hundley, Norris. "Clio Nods: Arizona v. California and the Boulder Canyon Act—A Reassessment," *Western Historical Quarterly* 3 (January, 1972): 17–51.

———. "The Politics of Reclamation: California, the Federal Government, and the Origins of the Boulder Canyon Act—A Second Look," *California Historical Quarterly* 52 (Winter, 1974): 292–325.

Kaufmann, Gordon B. "The Architecture of Boulder Dam," *Architectural Concrete* 2 (1936) no. 3: 2–5.

Kershner, Frederick D., Jr. "George Chaffey and the Irrigation Frontier," *Agricultural History* 27 (October, 1953): 115–22.

King Judson. "Open Shop at Boulder Dam," *The New Republic* 68 (June 24, 1931): 147–48.

Kruman, Marc W. "Quotas for Blacks: The Public Works Administration and the Black Construction Worker," *Labor History* 16 (Winter, 1975): 37–51.

Lake, Copeland. "Construction of the Hoover Dam: Compressed Air Plays a Part of Vital Importance in this Huge Undertaking," *Compressed Air Magazine* 37 (June, 1932): 3834–39.

―――. "Huge Trailer Hauls Penstock Pipe at Boulder Dam," *Compressed Air Magazine* 39 (October, 1934): 4549–50.

Lear, Linda J. "Boulder Dam: A Crossroads in Natural Resource Policy," *Journal of the West* 24 (October, 1985): 82–94.

―――. "The Boulder Canyon Project: A Re-examination of Federal Resource Management," *Materials and Society* 7 (1983), nos. 3–4: 329–37.

Lyons, Leonard. "Unforgettable Henry J. Kaiser," *Reader's Digest* 92 (April, 1968): 209–210.

Martin, Leo J. "A Gigantic Battle to Subdue a River," *New York Times Magazine,* July 31, 1932, p. 4.

McBride, Dennis. "The Battle for Boulder," *Nevada* 45 (September–October, 1985): 78.

Mead, Elwood. "Letter to the Editor re 'Boulder Dam,'" *New Republic* 68 (August 26, 1931): 48.

―――. "Construction of Boulder Dam," *Literary Digest* 114 (September 17, 1932): 14.

Moehring, Eugene. "Las Vegas and the Second World War," *Nevada Historical Society Quarterly* 29 (Spring, 1986): 1–30.

―――. "Public Works and the New Deal in Las Vegas 1933–1940," *Nevada Historical Society Quarterly* 24 (Summer, 1981): 107–29.

Morrison, Margaret D. "Charles Robinson Rockwood: Developer of the Imperial Valley," *Southern California Quarterly* 44 (December, 1962): 307–330.

Nelson, Wesley R. "Construction of the Hoover Dam: A Résumé of Current Activities and an Account of the Building of the Cofferdams," *Compressed Air Magazine* 38 (March, 1933): 4069–73.

―――. "Construction of the Hoover Dam: Description of the Aerial Cableways for the Transportation of Men, Materials, and Machinery," *Compressed Air Magazine* 38 (April, 1933): 4099–4104.

―――. "Construction of the Boulder Dam: Government Engineers and Surveyors Have Made a Notable Record on Exacting, Perilous Work," *Compressed Air Magazine* 39 (November, 1934): 4583–88.

―――. "Construction of the Boulder Dam: How the $35,000,000 Power Plant Will Appear When Completely Equipped," *Compressed Air Magazine* 40 (November, 1935): 4881–88.

Park, Allen S. "Mammoth Drill Carriages Speed Hoover Dam Work," *Compressed Air Magazine* 37 (April, 1932): 3780–82.

―――. "Construction of the Hoover Dam: A Description of the Methods of Obtaining and Preparing the Aggregates for the 4,400,000 Cubic Yards of Concrete to be Poured," *Compressed Air Magazine* 37 (October, 1932): 3937–42.

Priestly, J. B. "Arizona Desert: Reflections of a Winter Visitor," *Harper's Magazine* 173 (March, 1937): 358–67.

"Remaking the World," *Colliers* 95 (March 16, 1935): 66.

Richardson, Elmo. "Western Politics and the New Deal Policies," *Pacific Northwest Quarterly* 54 (January, 1963): 9–18.

Rocha, Guy Louis. "The IWW and the Boulder Canyon Project: The Final Death Throes of American Syndicalism," *Nevada Historical Society Quarterly* 21 (Spring, 1978): 3–24.

Schonfeld, Robert J. "The Early Development of California's Imperial Valley," *Southern California Quarterly* 50 (September, 1968): 279–307.

Segerblom, Gene W. "Boulder City," *Arizona Highways* 40 (May, 1964): 7–11.

Skerrett, R. G. "America's Wonder River—The Colorado: How Nature Formed This Amazing Watercourse and How Man Has Had to Face Many Hazards in Discovering It and in Completing Its Exploration," *Compressed Air Magazine* 36 (November, 1931): 3628–33.

———. "America's Wonder River—The Colorado: Something About Its Continually Changing Delta and the Adjacent Arid Region Which It Has Created and Which Man Is Now Making Abundantly Fruitful," *Compressed Air Magazine* 36 (December, 1931): 3664–69.

———. "America's Wonder River—The Colorado: Work on the Hoover Dam Now Underway After the Disposal of Many Puzzling Questions," *Compressed Air Magazine* 37 (January, 1932): 3694–3700.

Sowles, Lawrence P. "Construction of the Hoover Dam: How the Concrete Is Being Cooled as It Is Poured," *Compressed Air Magazine* 38 (November, 1933): 4265–71.

———. "Construction of the Hoover Dam: Description of the Methods of Pouring the Concrete," *Compressed Air Magazine* 39 (April, 1934): 4385–96.

———. "Construction of the Hoover Dam: A Description of the Himix Concrete Plant and of the Cement Blending and Handling Equipment," *Compressed Air Magazine* 39 (September, 1934): 4533–37.

Stevens, Joseph E. "Building a Dream: Hoover Dam," *American West* 21 (July–August, 1984): 16–27.

Sutton, Imre. "Geographical Aspects of Construction Planning: Hoover Dam Revisited," *Journal of the West* 7 (July, 1968): 301–44.

Swain, Donald C. "The Bureau of Reclamation and the New Deal, 1933–1940," *Pacific Northwest Quarterly* 61 (July, 1970): 137–46.

Trani, Eugene. "Conflict or Compromise: Harold L. Ickes and Franklin D. Roosevelt," *North Dakota Quarterly* 36 (Winter, 1968): 20–29.

Vivian, C. H. "Construction of the Hoover Dam: Some General Facts Regarding the Undertaking and the Men Who are Directing It," *Compressed Air Magazine* 37 (February, 1932): 3720–24.

———. "Construction of the Hoover Dam: How the Contractors Handled the Huge and Costly Program of Preliminary Work," *Compressed Air Magazine* 37 (March, 1932): 3742–48.

———. "Construction of the Hoover Dam: Within a Year's Time the Government has Reared a Modern City in the Desert at a Cost of $1,600,000," *Compressed Air Magazine* 37 (April, 1932): 3774–79.

———. "Construction of the Hoover Dam: The Concrete Mixing Plant Surpasses in Capacity and Refinements any Structure of Its Kind," *Compressed Air Magazine* 37 (November, 1932): 3970–74.

———. "Construction of the Hoover Dam: Lining of the Diversion Tunnels with Concrete," *Compressed Air Magazine* 37 (December, 1932): 4010–17.

Warshall, Peter. "The Great Colorado River War," *American West* 23 (September–October, 1986): 42–48.

White, Theo. "Building the Big Dam," *Harper's Magazine* 171 (June, 1935): 113–20.

Wilson, Edmund. "Hoover Dam." *The New Republic* 68 (September 2, 1931): 66–69.

Wilson, Richard Guy. "Massive Deco Monument," *Architecture* (December, 1983): 45–47.

———. "Machine-Age Iconography in the American West: The Design of Hoover Dam," *Pacific Historical Review* 54 (November, 1985): 463–93.

Books

Alexander, J. A. *The Life of George Chaffey*. Melbourne, Australia, 1928.

Allen, Marion V. *Hoover Dam and Boulder City*. Shingletown, Calif.: Privately published, 1983.

Anderson, Ellis L., ed. *History of Public Works in the United States 1776–1976*. Chicago: American Public Works Association, 1976.

Bartlett, Richard A. *Great Surveys of the American West*. Norman: University of Oklahoma Press, 1962.

Berkman, Richard L., and W. K. Viscusi. *Damming the West*. New York: Grossman Publishers, 1973.

Bolton, Herbert E. *Coronado: Knight of Pueblos and Plain*. Albuquerque: University of New Mexico Press, 1949.

Brooker, Angela, and Dennis McBride. *Boulder City: Passages in Time*. Boulder City: The Boulder City Library, 1981.

Burnside, Wesley M. *Maynard Dixon: Artist of the West*. Provo, Utah: Brigham Young University Press, 1974.

Chan, Loren B. *Sagebrush Statesman: Tasker L. Oddie of Nevada*. Reno: University of Nevada Press, 1973.

Darrah, William C. *Powell of the Colorado*. Princeton, N.J.: Princeton University Press, 1951.

Dictionary of American Biography. 22 vols. New York: Charles Scribner's Sons, 1958.

Didion, Joan. *The White Album*. New York: Simon & Schuster, 1979.

Dubofsky, Melvyn. *We Shall Be All: A History of the IWW*. Chicago: Quadrangle Books, 1969.

Flint, Timothy. *The Personal Narrative of James O. Pattie*. Philadelphia: J. B. Lippincott Company, 1962.

Freeman, Lewis R. *The Colorado River Yesterday, Today and Tomorrow*. New York: Dodd, Mead and Co., 1923.

Fradkin, Philip L. *A River No More: The Colorado River and the West*. New York: Alfred A. Knopf, 1984.

Gambs, John S. *The Decline of the I.W.W.* New York: Columbia University Press, 1932.

Gates, William H. *Hoover Dam, Including the Story of the Turbulent Colorado River*. Los Angeles: Wetzel Publishing Company, 1932.

Golze, Alfred R. *Reclamation in the United States*. New York: McGraw-Hill, 1952.

Grace Community Church. *Grace Community Church, Boulder City, Nevada, 1933–1983*. Boulder City, Nev.: Privately published, 1983.

Gruber, Frank. *Zane Grey*. New York: World Publishing Company, 1970.

Gunther, John. *Inside USA*. New York: Harper & Brothers, 1947.

Hoover, Herbert. *The Memoirs of Herbert Hoover, 1920–1933, The Cabinet and the Presidency*. New York: Macmillan, 1952.

———. *The Memoirs of Herbert Hoover, 1929–1941, The Great Depression*. New York: Macmillan, 1952.

Hundley, Norris. *Dividing the Waters*. Berkeley: University of California Press, 1966.

———. *Water and the West: The Colorado River Compact and the Politics of Water in the American West*. Berkeley: University of California Press, 1975.

Hyman, Sidney. *Marriner S. Eccles: Private Entrepreneur and Public Servant*. Stanford, Calif.: Stanford University Graduate School of Business, 1976.

Ickes, Harold L. *The Secret Diary of Harold L. Ickes: The First Thousand Days, 1933–1936*. New York: Simon & Schuster, 1953.

Ingram, Robert L. *A Builder and His Family*. San Francisco: Privately published, 1961.

The Kaiser Story. Oakland, Calif.: Kaiser Industrial Corporation, 1968.

Kant, Candace. *Zane Grey's Arizona*. Flagstaff, Ariz.: Northland Press, 1984.

Kleinsorge, Paul L. *The Boulder Canyon Project: Historical and Economic Aspects*. Stanford, Calif.; Stanford University Press, 1940.

Lingenfelter, Richard E. *Steamboats on the Colorado River, 1852–1916*. Tucson: University of Arizona Press, 1978.

Lowitt, Richard. *The New Deal and the West*. Bloomington: Indiana University Press, 1985.

McBride, Dennis. *In the Beginning: A History of Boulder City, Nevada*. Boulder City: Boulder City Chamber of Commerce, 1981.

McCullough, David. *The Great Bridge*. New York: Simon & Schuster, 1982.

Moeller, Beverly B. *Phil Swing and Boulder Dam*. Berkeley: University of California Press, 1971.

Nadeau, Remi. *The Water Seekers*. Garden City, N.Y.: Doubleday, 1950.

Nash, Gerald D. *The American West Transformed: The Impact of the Second World War*. Bloomington: Indiana University Press, 1985.

———. *The American West in the Twentieth Century*. Englewood Cliffs, N.J.: Prentice-Hall, 1973.

The National Cyclopaedia of American Biography, Vol. 25. New York: James T. White Company, 1936. Vol. G, 1946.

Pacific Constructors, Inc. *Shasta Dam and Its Builders*. San Francisco: Privately published, 1945.

Pettitt, George Albert. *So Boulder Dam Was Built*. Berkeley, Calif.: Lederer, Street & Zeuss, 1935.

Reisner, Marc P. *Cadillac Desert: The American West and Its Disappearing Water*. New York: Viking, 1986.

Robinson, Edgar E., and Paul C. Edwards, eds. *The Memoirs of Ray Lyman Wilbur, 1875–1949*. Stanford, Calif.: Stanford University Press, 1960.

Robinson, Michael C. *Water for the West: The Bureau of Reclamation 1902–1977*. Chicago: Public Works Historical Society, 1979.

Rockwood, Charles R. *Born of the Desert*. Calexico, Calif., 1930.

Sarton, May. *The Lion and the Rose*. New York: Rinehart & Company, 1948.

Simmons, Ralph B. *Boulder Dam and the Great Southwest*. Los Angeles: Pacific Publishers, 1936.

Stanley, Mildred. *The Salton Sea Yesterday and Today*. Los Angeles: Triumph Press, 1966.

Stegner, Wallace. *Beyond the Hundredth Meridian*. Boston: Houghton Mifflin Co., 1953.

———. *The Sound of Mountain Water*. Lincoln: University of Nebraska Press, 1985.

Stein, Mimi. *A Special Difference: A History of Kaiser Aluminum & Chemical Corporation.* Oakland, Calif.: Kaiser Aluminum & Chemical Corporation, 1980.

Thompson, Fred. *The I.W.W. Its First 50 Years, 1905–1955: The History of an Effort to Organize the Working Class.* Chicago: Industrial Workers of the World, 1955.

Tout, Otis B. *The First Thirty Years.* San Diego: Arts and Crafts Press, 1931.

Warne, William E. *The Bureau of Reclamation.* New York: Praeger, 1973.

Waters, Frank. *The Colorado.* New York: Rinehart & Company, 1946.

Watkins, T. H., et al. *The Grand Colorado: The Story of a River and Its Canyons.* Palo Alto, Calif.: American West Publishing Company, 1969.

White, Graham, and John Maze. *Harold Ickes of the New Deal: His Private Life and Public Career.* Cambridge, Mass.: Harvard University Press, 1985.

Who Was Who in America. Chicago: A. N. Marquis Company, 1950–61.

Who's Who in America. Chicago: A. N. Marquis Company, 1934–35.

Wiley, Peter, and Robert Gottlieb. *Empires in the Sun: The Rise of the American West.* New York: Putnam, 1982.

William, Albert N. *The Water and the Power.* New York: Duell, Sloan and Pearce, 1951.

Wilson, Edmund. *American Earthquake.* Garden City, N.Y.: Doubleday, 1958.

Wilson, Neill C., and Frank J. Taylor. *The Earth Changers.* Garden City, N.Y.: Doubleday, 1957.

Wilson, Richard G., et al. *The Machine Age in America 1918–1941.* New York: Harry N. Abrams, 1986.

Wolters, Raymond. *Negroes and the Great Depression: The Problem of Economic Recovery.* Westport, Conn.: Greenwood Publishing Company, 1970.

Woodbury, David O. *The Colorado Conquest.* New York: Dodd, Mead & Co., 1941.

Wollet, William. *Hoover Dam: Drawings, Etchings and Lithographs.* Santa Monica, Calif.: Hennessey & Ingalls, 1986.

Yates, Richard and Mary Marshall. *The Lower Colorado River: A Bibliography.* Yuma: Arizona Western College Press, 1974.

Index

Imperial Press: 12
Imperial Valley: 11–16, 231, 250,
  252; birth and blossoming of, 11–
  12; crops of, 12; shantytowns of,
  12; flooded by Colorado R., 14–15
  (*see also* cutback); irrigation of,
  258; *see also* All-American Canal
Indio, Calif.: 8
Industrial Solidarity: 66 ff., 79
Industrial Worker: 64 ff., 154, 206
Industrial Workers of the World: *see*
  IWW
Ingersoll-Rand Company: 271, 276
Injuries: *see* Boulder Canyon Project,
  accidents/injuries/fatalities of
Intake towers: 228–29, 241, 251
Invert: 110–11
Irrigation: Colorado River and, 9–16;
  Hoover Dam and, 258; *see also* All-
  American Canal; Colorado River
  Aqueduct; Imperial Valley Isthmian
  Canal Commission: 16
Ives, Joseph C.: 6–7
IWW (Industrial Workers of the
  World): 65–79, 154–55, 204–
  207, 234, 235, 278–79; back-
  ground of, 65; disowned by strik-
  ers, 71; and August 1933 strike
  attempt, 205–206; failure of, 206;
  philosophy of, 276–77; *see also*
  Anderson, Frank

J. F. Shea Company: 39, 44, 256, 257;
  and Golden Gate Bridge, 289; *see
  also* Shea, Charlie
Jameson, W. A.: 221, 247
Joe McGees: 200, 222
John Bernard Simon Company: 46
Johnson, Eddie: 67, 68
Johnson, Hiram: 18, 26, 27, 173
Johnson, Ike: 199
Jumbos: 88–90 ff., 110
June (mystery woman of Kraus case):
  210

Kahn, Felix: 40, 44, 89, 122, 167–68,
  227, 245, 248; death of, 257; as
  gambler, 289; *see also* MacDonald
  & Kahn; Six Companies, Inc.
Kaibab plateau: 5, 6

Kaiparowits plateau: 5
Kaiser, Edgar: 43, 285
Kaiser, Henry: 40–44, 89, 168, 170,
  244, 250; and Warren Bechtel,
  41–42; as Six Companies' Wash-
  ington-based lobbyist, 233–34;
  after Hoover Dam, 255–57; and
  design of Boulder City, 284–85; on
  contractors, 289; and Bonneville
  Dam, 290; *see also* Six Companies,
  Inc.
Kaiser Paving Company: 40
Kaufmann, Gordon B.: 30
Keane, Rev. Arthur: 140
Kirman, Richard: 239, 245
Kittitas Valley: 32
Knudsen, Morris: 36, 42; *see also*
  Morrison-Knudsen Company
Koebig, Dr. Walter: 209
Kraus, Ed F.: 207–11, 213
Kraus, Mrs. Ed F.: 209

Lake Havasu: 258
Lake Mead: 250–51, 261 ff.
Lake Powell: 263
Lamey, Jack: 91
Lane, Andrew: 57–58
Lange, Harry: 57–58
LaRue, E. C.: 270
Las Vegas, Nev.: 19, 33, 47–52; dam
  workers commuting from, 57; as
  IWW headquarters, 66; vs. IWW,
  67–68; as place for after-hours entertainment, 118, 120; as possible
  project base, 122; carbon monoxide
  poisoning trials in, 208–13; as
  Hoover Dam boomtown, 222–26;
  bars and casinos of, 223–24; pros-
  titutes of, 224–25 (*see also* Kraus,
  Ed F.); electrified by Hoover Dam,
  259; as gambling mecca, 260–61;
  heavy industry in, 260–61; Strip of,
  261
Las Vegas Age: 49, 50, 58, 61, 67, 71,
  125, 142, 177
Las Vegas Central Labor Council: 75
Las Vegas Evening Review-Journal:
  60, 68, 72, 124, 137, 138, 142,
  171, 176, 208, 210, 213, 227, 252
Las Vegas Wash: 7, 52, 159